Book of Abstracts for the

10th World Conference on Animal Production

Sponsors

CEVA Sante Animale

Ceva Animal Health Southern Africa (Pty) Ltd, a division of the French multinational company, CEVA Sante Animale, specialises in the supply of veterinary products to the livestock industry. The product range includes feed additives, vaccines, water-soluble medications, vitamins/electrolytes, anthelmintics, anti-parasitic agents, anti-babesials and injectable antibiotics.

CEVA is totally devoted to animal health, and ranks among the world's top twenty companies in this market sector with strong and growing international business activities in the Western and Eastern Europe, United States, Latin America, Australia, Southeast Asia, the Middle East and the Indian subcontinent. In Africa, it has a strong position in the northern Mahoreb countries of Algeria, Tunisia and Morocco, as well as French-speaking West Africa, with additional business growth in Eastern Africa.

CEVA's product portfolio includes poultry and ovine vaccines, injectables, oral and injectable antibiotics, products for ruminants, central nervous system and animal behaviour, as well as a growing range of products in the companion animal and aquine markets. The feed additive range comprises growth promoters, anti-bacterials, anti-coccidials and a mould inhibitor.

Chemoforma

Chemoforma Ltd. produces feed additives based on balanced formulation of purified RNA and nucleotides to meet the increase demand for these building blocks of life upon development, growth, stress and health challenges in livestock industry. Increasing feed prices, demand for higher numbers of animals and increased health concerns in animal production evoke the need for feed additives or feed supplements with proven physiological benefits for the animals and simultaneously provide protection against diseases in the target species.

VETPAC Animal Health cc

VETPAC Animal Health

We pride ourselves in our involvement with the emerging agricultural sector as well as the large commercial farmers, such as poultry, pig and cattle producers. We currently represent 3 Swiss based companies on an exclusive basis in the Sub Saharan region.

Book of abstracts for the

10TH WORLD CONFERENCE ON ANIMAL PRODUCTION

This event is held under the auspices of the World Association for Animal Production, the South African Society for Animal Science and the All-Africa Association of Animal Production.

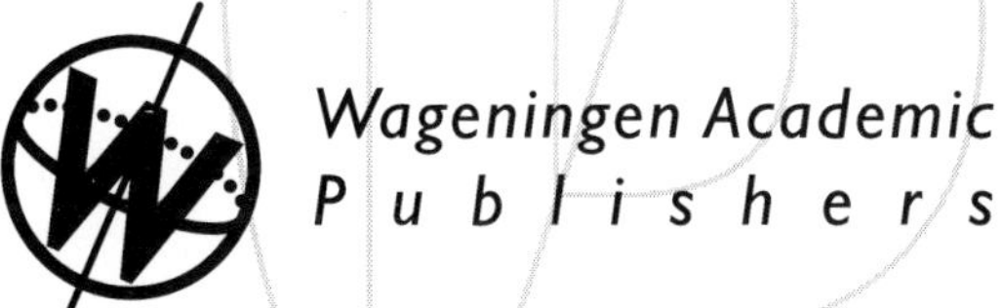

ISBN 978-90-8686-100-2

First published, 2008

Wageningen Academic Publishers
The Netherlands, 2008

Foreword

A World Conference on Animal Production is held under the auspices of the World Association for Animal Production at five-yearly intervals. This, the 10th WCAP is being hosted by the South African Society for Animal Science, a founder member of the WAAP, and the All-Africa Association for Animal Production. The Conference includes the 42nd Conference of SASAS and the 5th All-Africa Conference on Animal Production. The WCAP is a trend-setting conference that examines the scientific progress in livestock production and the effects that these have on the environment, economies and the quality of life of people.

The theme of the 10th WCAP, *New World; Future World*, takes in the philosophy of the WAAP, conceptualising the place of animals in our daily existence. The theme focuses the sessions on constructing a future scenario for the various systems within animal production. The sessions alternate between and address the important issues that advance animal production and the quality of animal products through the responsible application of current and new technologies in animal physiology, animal genetics, production-enhancing molecules and resource and environmental management. Four special sessions address feed and food safety, dairy production and the role of livestock in developing communities.

The World is at a critical stage where the rapid growth of populations and the general increase in prosperity are placing greater demands on natural resources, especially feeds and foods. While a significant proportion of the human population is experiencing a surge in welfare, the paradox remains that people around the world are struggling to cope with the basic needs for human dignity – enough food and the correct types of food, health management systems and social securities. Animal production is a non-negotiable factor in this complex equation. Animal production, though, relies on a range of input factors that collectively drive the systems to new levels of achievement. How knowledge is respected and applied in an ethical, responsible manner is fundamental to this process.

It is a pleasure to welcome you to this integrated international conference on animal production. I hope that you will find it interesting and that you will be returning to your respective positions enthusiastic and brimming with new ideas. It will be particularly important that the principles discussed are taken forward and applied in every aspect of the livestock industry.

Norman Casey

President: 10th WCAP
President: WAAP
Honorary President: SASAS

Executive committee

President: N Casey	President of WCAP 2008 Conference
P Bevan	President South African Society for Animal Science
J Meyer	Scientific Programme
M Neuhoff	Marketing
J van Ryssen	CEO South African Society for Animal Science

Secretariat	K Herring	WCAP 2008 Conference Secretariat
	D Ströh	WCAP 2008 Conference Secretariat
	D Cloete	WCAP 2008 Conference Secretariat

Scientific advisory committee

J Boyazoglu
L Bull
C Cruywagen
C Devendra
J du Plessis
J Flanagan
J Furquay
W Gertenbach
R Gous
J Greyling
M Hlatswayo
H Köster
A Magadlela
N Maiwashe
J Mc Dermott
J Meyer
B Mostert
A Oelofse
E Pieterse
A Rosati
M Scholtz
K Sejrsen
A Stroebel
F Swanepoel
A Tewolde
H Theron
A van der Zijpp
E van Marlé-Köster
J B J van Ryssen
E Webb
M Wulster-Radcliffe

PROGRAMME 10th WCAP

Cape Town International Convention Centre, South Africa (CTICC)

Viewing posters and exhibitions, and networking will take place in the WCAP exhibition and restaurant venue daily at 11:00, and 15:00, and lunch at 13:00.

SATURDAY 22 NOVEMBER 2008

10:00 - 17:00	SASAS Board meeting	**Meeting Suite 1.53**
15:00 - 17:00	WAAP Board meeting	**Meeting Suite 1.54**

SUNDAY 23 NOVEMBER 2008

11:00 - 17:00	Registration at *CTICC*	**CTICC Foyer**
18:30 - 20:30	Welcoming function. Dress casual.	**Auditorium 2 and Restaurants**

MONDAY 24 NOVEMBER 2008

09:00	Call to order by the president of the 10th WCAP	**Auditorium 2**
09:15	Keynote address	**Auditorium 2**

PLENARY 1

10:15 - 11:00	**Livestock systems and land tenure**	**Auditorium 2**
10:15	• Land tenure: Policies for rural governance and economic development	S Sibanda

SESSION 1

12:00 - 16:30	**Investing in genetics**	**Auditorium 2**
12:00	Chairperson	A Rosati
	Co-chairperson	E van Marlé-Köster
12:00	• Genetic engineering in livestock production	E Rege
12:30	• Investing in genetic evaluations in DECs	N Maiwashe
14:00	Chairperson	M Macneil
	Co-chairperson	E van Marlé-Köster
14:00	• Investing in international genetic evaluations	F Fikse
14:30	• Returns on investment in goat and sheep breeding in South Africa	S Schoeman
15:30	Chairperson	K Dzama
	Co-chairperson	E van Marlé-Köster
15:30	• Animal breeding and genomics in future animal production	J Pollack
16:00	Concluding discussion	E van Marlé-Köster

SESSION 2

12:00 – 17:00	**Human nutrition and livestock products**	**Meeting Room 1.4**
12:00	Chairperson	A Dhansay
	Co-chairperson	A Oelofse
12:00	• Animal products in nutrition, health and well-being – a global perspective	M Bloem
12:30	• Nutrition of vulnerable communities of affluent societies	N Steyn
14:00	Chairperson	H Schönfeldt
	Co-chairperson	A Oelofse
14:00	• Nutrition of vulnerable communities in economically marginal societies	M Faber
14:30	• Strategic use of naturally Se-rich milling co-products to eliminate Se deficiency and create Se-enriched foods	J Taylor
15:30	Chairperson	A Oelofse
	Co-chairperson	H Schönfeldt
15:30	• Validity of using a constant value for heme iron	H Schönfeldt
15:45	• Traceability and the eating-out-paradox	C Rogge
16:00	• Determination of German consumer attitudes towards meat traceability by means-end chain theory	L Lichtenberg
16:15	Forum discussion on Animal Products in Human Nutrition	A Oelofse
16:30	**WAAP General Assembly**	Meeting Suite 1.53
	Free evening for socialising or workshops	

TUESDAY 25 NOVEMBER 2008

PLENARY 1

08:30 – 13:00	**Physiological limits to production efficiency**	**Auditorium 2**
08:30	Chairperson	K Sejrsen
	Co-chairperson	E Webb
08:30	• Review of reproduction technologies and future perspectives	D Kessler
09:00	• Nutritional limitations on growth and development in poultry	R Gous
09:30	• Physiological limitations to growth and the related effects on meat quality	E Webb
10:00	• Going beyond the limits?	J Hodges
10:30	• Optimum growth rate of Belgian Blue double-muscled heifers	L Fiems
10:45	• Feeding n-3 fatty acid source improves semen quality by increasing n3/n6 fatty acids ration in sheep	A Towhidi
12:00	Chairperson	K Sejrsen
	Co-chairperson	E Webb
12:00	• Electrolyte losses in sweat, urine, blood and faeces of horses exposed to moderate and hot environments	M Etchichury
12:15	• Assessing feed efficiency in beef steers through infrared thermography, feeding behaviour and glucocorticoids	Y Montanholi
12:30	• Effect of cold shock on frozen-thawed spermatozoa of Nili-Ravi buffalo (*Bubalus bubalis*) and Sahiwal bulls	A Ijaz
12:45	Concluding discussion	E Webb

SESSION 3

14:00 – 18:00	**Technological surges and livestock production: Scenarios for developed and developing economies**	**Meeting Room 1.6 A**
14:00	Chairperson	H Köster
	Co-chairperson	C Coetzer
14:00	• Potential nutrigenomics in livestock production	F Dunshea
14:30	• The relationship between a satellite derived vegetation index and wool fibre diameter profiles The applications of biotechnology in livestock production	M Whelan
14:45	• Integrating farm animal recording with research and development: the South African model	J van der Westhuizen
15:30	Chairperson	H Köster
	Co-chairperson	C Coetzer
15:30	• Science and technology policies for nanotechnology in livestock production	J Kuzma
16:00	• Horizons in animal production – towards being clean, green and ethical	G Martin
16:30	• The applications of biotechnology in livestock production	B Glenn
17:00	• Single tube guiding in conventional milking parlours	R Brunsch
17:15	• Wattle tannins have the potential to control gastro-intestinal nematodes in sheep	I Nsahlai
17:30	Concluding discussion	H Köster

SESSION 4

14:00 – 17:45	**Intellectual capital for livestock production**	**Meeting Room 1.4**
14:00	Chairperson	I Mpofu
	Co-chairperson	J Greyling
14:00	• Indigenous, learned and lifelong learning for livestock production	M Wurzinger
14:45	• A model for achieving sustainable improvement and innovation in agricultural research and development interventions for maximizing socio-economic service delivery	R Clark
15:30	Chairperson	I Mpofu
	Co-chairperson	J Greyling
15:30	• Intellectual capital for livestock field services	E Nesamvuni
16:00	• The contributions of NGOs to intellectual capital for livestock production	N Hegde
16:30	• Integrated higher learning	N Casey
17:00	• A comparative analysis of alternative methods for capturing indigenous selection criteria	M Wurzinger
17:15	Concluding discussion	J Greyling

SESSION 5

14:00 – 15:00	**Open communication: genetics**	**Auditorium 2**
14:00	Chairperson	A Rosati
	Co-chairperson	E Rege
14:00	• Genetic analysis of carcass characteristics and indicator traits recorded using ultrasound	M Macneil
14:15	• Improving the quality of wool through the use of genetic markers	T Itenge
14:30	• Economic values for dairy production traits under different milk payment systems in South Africa	C Banga
14:45	• Expression of trypanotolerant quantitative trait loci (QTL) in cattle population under natural tsetse and trypanosomosis challenge	C Orenge

WORKSHOP

15:30 – 18:00	**Conservation, management and use of marginal breeds**	**Auditorium 2**
15:30	Chairperson	M Scholtz
	Co-chairperson	K Ramsey
15:30	Introduction by facilitator	M Scholtz
15:40	• Roles and services of livestock in marginal areas	I Hoffman
16:00	• Description of production environments to support management of farm animal breeds in marginal areas	B Scherf
16:20	• Using GIS (geo-referenced information systems) as a tool to describe the production environment of farm animal breeds in marginal areas	M Hererro
16:35	• Is it necessary to conserve marginal farm animal breeds?	J van der Westhuizen
16:50	Debate and discussion	
17:50	Summary of the outcome of the workshop	M Scholtz
18:00	Closure	K Ramsey

SESSION 6

14:00 – 17:30	**Open communication: livestock production**	**Meeting Room 1.6B**
14:00	Chairperson	J van Ryssen
	Co-chairperson	I Nsahlai
14:00	• The application of DNA technologies to combat stock theft and improve food security in South Africa	C Pilane
14:15	• Latest development in cryopreservation of Zebrafish (Danio rerio) oocytes	T Zhang
14:30	• Evaluation of supplementation with ration based on poultry litter, corn flour, sugar cane molasses and meat meal in dual purpose calf at grazing	G Nouel Borges
14:45	• Does restricted feeding affect development oof intramuscular fat across the musculature in Hanwoo carcasses?	J Thompson
15:30	Chairperson	J van Ryssen
	Co-chairperson	I Nsahlai
15:30	• Livestock systems and livelihoods: Responding to competing claims for land at the tourism/livestock interface	P Chaminuka
15:45	• Smallholder pig production and consumption dynamics in relation to porcine cysticercosis challenges in Tanzania	E Kimbi

16:00	• Development of beef production in Brazil during the last 10 years (1997-2006)	P Meyer
16:15	• Fat sources in dairy cows rations: Intake, milk yield and protein fraction composition	F Rennó
16:30	• Livestock losses and risk management in Zimbabwe's dryland areas	N Mpofu
16:45	• Successful Pro-poor Goat Development Under Heifer Uganda	W Ssendagire
17:00	Concluding discussion	I Nsahlai

17:30	**WAAP General Assembly**	**Meeting Suite 1.53**

WEDNESDAY 26 NOVEMBER 2008

SYMPOSIA: SEE SYMPOSIA PROGRAMME

08:30 – 13:00	AFMA: Ensuring feed and food safety towards 2010 and beyond	**Auditorium 2**
14:00 – 17:30	WPSA: Poultry production towards 2010 and beyond	**Auditorium 2**
14:00 – 17:30	EAAP, AAAP, ADSA and SASAS: Dairy production: Dairy production in difficult environments	**Meeting Room 1.4**
09:00 – 17:30	SASAS, US, WCDA: Developments in livestock production in southern Africa: Research to practice	**Meeting Room 2.6B**
09:00 – 17:30	I LRI and UFS: Role of livestock in developing communities: Enhancing multifunctionality	**Meeting Room 1.6A, to break away to 1.6B and 2.6A**
17:30 – 18:30	SA Society for Animal Science AGM	**Auditorium 2**

Free evening for socialising or workshops

THURSDAY 27 NOVEMBER 2008

SESSION 7

07:00 – 08:00	**Animal Biotechnology**	**Auditorium 2**
07:00	• Information discussion on Animal Biotechnology	B Glenn

PLENARY 3

08:30 – 13:00	**Managing resources for sustainable livestock production, human dignity and social security**	**Auditorium 2**
08:30	Chairperson	J Flanagan
	Co-chairperson	J Meyer
08:30	• Resource and waste management in livestock systems	J Radcliffe
09:00	• Considerations for sustainable production: Technologies and social aspects	J Mc Dermott
09:30	• Challenges of new and re-emergent infectious diseases in livestock , wildlife and humans	B Gummo
10:00	• A risk-based approach to water resource management	J Meyer
10:30	• Animal welfare regulations: a structural model of pig production	C Liljenstolpe

PLENARY 3

12:00	**Open communication on Managing resources for sustainable livestock production, human dignity and social security**	**Auditorium 2**
12:00	Chairperson	J Flanagan
	Co-chairperson	J Meyer
12:00	• Livestock emergency guidelines and standards	S Mack
12:15	• Development of livestock based sustainable family farms: Heifer International's approach	D Bhandari
12:30	• Effect of TDS plus Br on the accumulation of As and Pb from drinking water in broilers	M Mamabolo
12:45	Concluding discussion	J Flanagan

SESSION 8

14:00 – 17:15	**Concerning climate change and livestock production**	**Meeting Room 1.6A**
14:00	Chairperson	L Ndlovu
	Co-chairperson	J van Ryssen
14:00	• Livestock genetic diversity and adaptations to climate change	I Hoffman
14:30	• Livestock and climate change: is it possible to combine adaptation, mitigation and sustainability?	K Van'thooft
14:45	• Role of livestock production in climate change: A review	C Antwi
15:30	Chairperson	L Ndlovu
	Co-chairperson	J van Ryssen
15:30	• Options for the abatement of methane and nitrous oxide from ruminant production: A review	R Eckard
16:15	• Effects of climate change on the sustainability of livestock production systems	A Nardone
17:00	Concluding discussion	J van Ryssen

SESSION 9

14:00 – 17:30	**Molecules, production and quality: The role of production and product modifying substances**	**Meeting Room 1.6B**
14:00	Chairperson	M Neuhoff
	Co-chairperson	M Moroe-Rulashe
14:00	• A case for protocol reviews for registering production enhancing molecules	J Edwards
14:30	• A case of species specificity	J Swan
		Meeting Room 1.6B
15:30	Chairperson	M Neuhoff
	Co-chairperson	L Hoffman
15:30	• Feed additives: From design to practice	S Traylor
16:00	• Walking a new road – The case of introducing pharmaceuticals to developing communities	T de Bruyn
16:30	• Herbal technology in commercial animal production	S Chatterjee
17:00	• Ethnoveterinary medicine – an alternative to managing gastrointestinal parasites in goats	V Maphosa
17:15	Concluding discussion	M Moroe-Rulashe

SESSION 10

14:00 – 17:30	**Open communication on animal nutrition**	**Meeting Room 1.4**
14:00	Chairperson	F Siebrits
	Co-chairperson	M Lehmann-Maritz
14:00	• Effect of limestone particle size on egg production and eggshell quality during late production	F de Witt
14:15	• Evaluating rations for high producing dairy cows using three metabolic models	N Swanepoel
14:30	• Response of Danish Holstein, Red and Jersey cows to supplementation with saturated or unsaturated fat	M Weisbjerg
14:45	• Capability of yeast derivates to bind pathogenic bacteria	A Ganner
15:30	Chairperson	F Siebrits
	Co-chairperson	M Lehmann-Maritz
15:30	• Robustness of cutin, chromic oxide and acid detergent lignin as a markers to determining apparent digestibility of diets in horses	A Gobesso
15:45	• Accuracy and precision of cutin as an internal marker for determining apparent digestibility of diets in horses	A Gobesso
16:00	• Efficacy of phytogenic feed additives in farm animals	M Rouault
16:15	• Effect of full fat flaxseed and antioxidant supplementation on production performance and egg quality of layers	T Pasha
16:45	• Factors influencing energy demand in dairy farming	R Brunsch
17:00	• The effect of polyethylene glycol and polyvinylpyrrolidone on fermentation dynamics of indigenous varieties of sorghum grain	L Ndlovu
17:15	Concluding discussion	F Siebrits

SESSION 11

14:00 – 17:30	**Open communication on livestock production**	**Meeting Room 2.6**
14:00	Chairperson	K Dzama
	Co-chairperson	W Gertenbach
14:00	• A systems approach to the South African dairy industry	M Scholtz
14:30	• Small-scale milk production in South Africa: why is it not working?	S Grobler
14:45	• The impact of dairy crossbreds on herd productivity in smallholder farms in western Kenya	W Muhuyi
15:30	Chairperson	K Dzama
	Co-chairperson	W Gertenbach
15:30	• Working to fatten calves in Sweden: working time, working environment and managerial style	E Bostad
15:45	• Sheep-for-meat farming systems in French semi-upland area: adapting to increased cereal prices and new European agricultural policy	M Benoit
16:00	• Sheep systems to increase profit, manage climate risk and improve environmental outcomes in high-rainfall zones of southern Australia	M Friend
16:15	• Reproduction and production potential of communal sheep in the Eastern Cape	C Nowers

16:30	• Non-food-production functions of livestock and livestock systems	S Oosting
16:45	• Local knowledge on relevant traits in sheep and goats under pastoral management in Kenya: a contribution to characterization of animal genetic resources in their production system context	H Warui
17:00	• Live animal and carcass characteristics of South African indigenous goats	L Simela
17:15	Concluding discussion	W Gertenbach

15:30 **WAAP General Assembly**

18:30	Transport to Gala Dinner departs from the CTICC by luxury coaches	
19:30	Gala Dinner. Dress Smart Casual.	

FRIDAY 28 NOVEMBER 2008

07:00	WAAP Board	**Meeting Suite 1.53**

PLENARY 4

09:00	**A forum discussion and formulation of the "Cape Town Declaration on Principles for Animal Production"**	**Auditorium 2**
	Chairperson	J Boyazoglu
	Co-chairperson	L Ndlovu

PLENARY 4

11:00	**A forum discussion and formulation of the "Cape Town Declaration on Principles for Animal Production"**	**Auditorium 2**
12:30	Closing session	N Casey

Symposia programme

Themes, topics and speakers of symposia

WEDNESDAY 26 NOVEMBER 2008

08:30 - 13:00	AFMA: Ensuring feed and food safety towards 2010 and beyond	**Auditorium 2**
14:00 - 17:30	WPSA: Poultry production towards 2010 and beyond	**Auditorium 2**
14:00 - 17:30	EAAP, AAAP, ADSA and SASAS: Dairy production: Dairy production in difficult environments	**Meeting Room 1.4**
09:00 - 17:30	SASAS, US, WCDA: Developments in livestock production in southern Africa: Research to practice	**Meeting Room 2.6B**
09:00 - 17:30	I LRI and UFS: Role of livestock in developing communities: Enhancing multifunctionality	**Meeting Room 1.6A, to break away to 1.6B and 2.6A**
17:30 - 18:30	SA Society for Animal Science AGM	**Auditorium 2**

08:30 – 13:00

SYMPOSIA: Feed and food safety towards 2010 and beyond	**Auditorium 2**
Hosted by the *Animal Feed Manufacturers' Association (AFMA)*	
Chairpersons	J du Plessis, P Fisher
Opening	E Briedenhann
The role the SA Feed Manufacturers' Association in enhancing the industry	De Wet Boshoff
Global challenges for feed companies beyond 2010	S Ferrari
Raw material challenges in the feed industry	E Briedenhann
Progress in quality control in the SA feed industry	H Köster
New developments in NIR technology and the application in the feed industry	N Dominy
Custom formulation of supplementary feeding for extensive rangeland production	W van Niekerk
Concluding discussion lead by	L Dunn

14:00 – 18:00

SYMPOSIA: Poultry production towards 2010 and beyond	**Auditorium 2**
Hosted by *World Poultry Science Association of South Africa (WPSA)*	
Chairpersons	R Gous, D Barnard
Opening	R Gous
Local and international poultry industries: Prospects and challenges beyond 2010	K Lovell
Technical developments in poultry nutrition and management	A Lemme
Poultry breeds and genetics: What to expect beyond 2010	I Pevzner
Poultry disease and health management beyond 2010	S Bisschop
Poultry research challenges beyond 2010	R Gous
Concluding discussion lead by	D Barnard

Time	Programme	Speaker / Room
14:00 – 18:00	**SYMPOSIA: Dairy production in difficult environments**	**Meeting Room 1.4**
	Hosted by the *EAAP, AAAP, ADSA* and *SASAS*	
	Chairperson	C Cruywagen
	Product quality	F Miglior
	Nutrition	L Erasmus
	Genetics	A Rosati
	Environmental consequences	J Meyer
	Capitalising on small and pastoral production systems	S Taal
	Concluding discussion lead by	C Cruywagen

Time	Programme	Speaker / Room
09:00 – 18:00	**SYMPOSIA: Role of livestock in developing communities: enhancing multifunctionality**	**Meeting Room 1.6**
	Hosted by *ILRI* and the *University of the Free State*	
	Chairperson	N Nduli
	Setting the scene	A Pell, A Stroebel
	Promoting gender equality and empowering women	A Waters-Baker
	Livestock and the environment	M Herrero
	Food, nutrition and health systems	A Oelofse
	Review discussion lead by	A Nesamvuni, A Freedman
	Topics of parallel working groups	
	Vulnerability and risk	M D'Haese, B Kruger
	Sustainable intensification	A van der Zijpp
	Value chains and innovation	H Burrows, J Mc Dermott
	Concluding discussion: Synthesis, integration and the way forward, lead by	C Devendra, J van Rooyen, F Swanepoel

Time	Programme	Speaker / Room
09:00 – 18:00	**SYMPOSIA: Developments in livestock production in southern Africa: Research to practice**	**Meeting Room 2.6B**
	Hosted by the *Western Cape Branch of SASAS, University of Stellenbosch* and *Department of Agriculture of the Western Cape*	
08:00	Registration	
08:30	Opening	P Bevan

SESSION 1

08:40 – 10:40	**Farming for the future**	
08:40	Chairperson	K Dzama
08:40	• Getting through the chasm: adoption and institutionalization mechanisms for improving and maximising the benefits from agricultural research and development interventions	N Nengovhela
09:00	• The South African beef profit partnerships project: the estimated aggregate socio-economic results and impacts to date	N Nengovhela
09:20	• Validating indigenous veterinary practice in Cameroon	H Taboh
09:40	• A farmer-centred approach to the development of smallholder livestock systems in South Africa	M Motiang
10:00	• Opportunities for improving Nguni cattle production in the smallholder farming systems of South Africa	C Mapiye
10:20	Forum discussion on Farming for the Future	K Dzama

SESSION 2

11:00 – 12:40	**Nutrition and production performance**	
11:00	Chairperson	C Cruywagen
11:00	• Effect of different levels of rapeseed meal with enzymes in diets formulated based on digestible amino acid on performance and carcass characteristics of broiler chicks	M Toghyani
11:20	• Ruminal and post-ruminal digestibility of two different dietary fats by dairy bulls	G Ghorbani
11:40	• The effect of tannin-rich feeds on gas production and microbial enzyme activity	I Nsahlai
12:00	• Influence of dietary energy level and production in breeding ostriches	T Olivier
12:20	• Modelling the long-term consequences of undernutrition of cows grazing semi-arid range for the growth of their progeny	F Richardson

SESSION 3

12:40 – 14:50	**Reproductive biotechnologies for commercial efficiency**	
12:40	Chairperson	H Lambrechts
12:40	• Piglets born from frozen-thawed epididymal sperm obtained post mortem: a case study	L Schwalbach
13:30	• Effect of different extenders and preservation periods on the survival rate of South African indigenous ram semen stored at 5 °C	T Nedembale
13:50	• Brilliant cresyl blue staining improve the quality of selected indigenous cattle oocytes and their developmental capacity to blastoderm in	M Raito
14:10	• Effect of addition of BSA to extenders on survival time and motility rate of South African Landrace boar semen	T Nedembale
14:30	• The effect of season on aspects of in vitro embryo production (IVEP) in sub-fertile beef cows	J Rust
14:50	Concluding discussion	

SESSION 4

15:30-17:30	**Genetics for optimal breeding**	
15:30	Chairperson	S Cloete
15:30	• Can repeated super ovulation and embryo recovery in Boer goats limit donor participation in a MOET programme?	K Lehloenya
15:50	• Genetic parameter estimates for certain production traits of ostriches	M Fair
16:10	• Evaluation of the genetic performance of goats in an on-field research environment	C Otoikhian
16:30	• Parentage verification of South African Angora goats using micro-satellite markers	C Visser
16:50	• Quantifying the relationship between birth coat score and wool traits in Merino sheep	W Olivier
17:10	• Reproduction parameters for Holstein cows towards creating a fertility index	C Muller
17:30	• Body weight growth trends in Döhne Merino lambs from birth to 16 month age	T Rust
17:50	Closing session	F Siebrits

Livestock systems and land tenure

Investing in genetics

Human nutrition and livestock products

Physiological limits to production efficiency

Technological surges and livestock production: scenarios for developed and developing economies

no. Page

Intellectual capital for livestock production

Managing resources for sustainable livestock production, human dignity and social security

Concerning climate change and livestock production

Molecules, production and quality: the role of production and product modifying substances

International aspects of livestock and livestock production: are fences and boundaries an enigma?

Animal nutrition

Livestock productions systems

no. Page

Miscellaneous topics

no. Page

Ensuring feed and food safety towards 2010 and beyond (AFMA)

Poultry production towards 2010 and beyond (WPSA)

Developments in livestock production in Southern Africa

no. Page

The role of livestock in developing communities: enhancing multifunctionality

no. Page

Dairy production in difficult environments

no. Page

Evaluation of different levels of *Mimosa arenosa* leaves and *Acacia macracantha* pods in rations for rabbits

G.E. Nouel Borges, J. Salas Araujo, J.B. Rojas Castellanos, M.A. Espejo Díaz and R.J. Sánchez Blanco, Universidad Centroccidental Lisandro Alvarado, Unidad de Investigación en Producción Animal, Tarabana, Cabudare, Estado Lara, 3023, Venezuela

Were evaluated the incorporation of different levels of dehydrated *Mimosa arenosa* leaves and *Acacia macracantha* pods in rations for growing rabbits. In all random design, experimental period of 77 days, were evaluated the effect five rations on live weight gain and food intake; with tree repetitions for treatment and two experimental units by repetition (rabbits 30 to 45 days old and 691±18 g). the rations were: T_0: 100% balanced commercial food; T_1:12.5% Ma, 50% Am; T_2: 25% Ma, 37.5% Am T_3: 37.5% Ma, 25% Am T_4: 50% Ma, 12.5% Am and 36,5% *Manihot esculenta* root flour:sugar cane molasses (2:1) and 1% mineral premix to 100% in T_1 to T_4. Food intake was different, 191 ± 18^a, 95 ± 12^b, 56 ± 3^c, 16 ± 10^d and 9 ± 7^d g/rabbit/day for T_0, T_1, T_2, T_3 y T_4. The live weight gain was different, 32.2 ± 4.3^a, 9.9 ± 13.0^b, -23.6 ± 17.8^c, -47.3 ± 4.4^d y -51.3 ± 11.5^d g/rabbit/day for T_0, T_1, T_2, T_3 and T_4 respectively. The secondary compounds in the rations, $T_1$to T_4, were not permitted adequate live weigh gains, but the rations less toxic was with mayor quantity of Am, the rations with more Ma were less intake

---

no. 2

Evaluation of supplementation with ration based on poultry litter, corn flour, sugar cane molasses and meat meal in dual purpose calf at grazing

G.E. Nouel Borges, V.R. Hernández, R.J. Sánchez Blanco, F.F. Villasmil and J.B. Rojas Castellanos, Universidad Centroccidental Lisandro Alvarado, Unidad de Investigación en Producción Animal, Tarabana, Cabudare, Estado Lara 3023, Venezuela

Was evaluated the supplementation of dual purpose post weaning calf with different mixes of agro industrial byproducts, in assay at Silva Municipality of Falcon State. Were used 45 post weaning calf. 30 females and 15 males, initial weight was 121.3±7.8 kg and age of 4 to 5 months. Treatments were T_0: commercial balanced food 2 kg/animal/day for 15 females; T_1: 40% poultry litter(PL), 35% maize meal (MM), 6% meat meal (ME), 18% sugar cane molasses (CM), 0.8% salt and 0.2% sulfur flower, for 15 males; T_2: 40% PL, 35% MM, 8% ME, 16% CM, 0.8% salt and 0.2% sulfur flower, for 15 females; animals of T1 and T2 receive 1 kg/animal/day first 78 days and 2 kg/animal/day the last 28 days of assay, in all random design. All animals range on African star grass. Live weight gain were different with 223^a; 175^b; 188^b gr/animal/day for T_0, T_1 and T_2 respectively. The cost for weigh gain was different 937^a; 483^b y 512^b Bs/kg LW for T_0, T_1 and T_2 respectively. The results indicate were possible use agro industrial byproducts as supplement in bovine post weaning calf under experimental conditions.

Synchronization after pre-implantation with growth hormone to determine the fertility of heifers

M.J. Chipa, Agricultural Research Council, Animal Nutrition, Private Bag X2, Irene, 0062, South Africa

Nineteen Nguni heifers where implanted with a growth hormone (Revalor®-H) and raised in a feedlot for 110days. The growth implant used to suppose to retard and even damage the development of ovaries according to manufacturer. After achieving about 70% of their mature weight in the feedlot averaging 375kg, animal were moved to the veld. Animals where treated and implanted, after three months in the veld, with a commercial synchronization product to come into oestrus. The heifers where divided into three groups of 6, 7 and 6 animal per group. The three groups were synchronized on different dates. The synchronization involved intramuscular injection and ear implant which was removed after nine days (Crester®, Intervet). The heifers were placed with 2 Nguni bulls for natural mating. Observations showed that most heifers come into heat and were receptive to bull. After six weeks from the last day of removal of implants from the last group, the animals will be tested for pregnancy by rectal examination. The result will indicate if Nguni heifers can be made to grow faster with hormonal implants without affecting their further reproductive potential. The conception rate was found 32%. More investigation still needs to be done.

---

no. 4

The effect of Bonsilage mais Flussig and Lalsil Fresh LB on the fermentation and aerobic stability of whole crop maize silage

T. Langa[1], B.D. Nkosi[1], R. Meeske[2], I.B. Groenewald[3] and D. Palic[1], [1]ARC-LBD: Animal Production Institute, Animal Nutrition, P/Bag x2, Irene, 0270062, South Africa, [2]Outeniqua Experimental Farm, Agriculture, P.O. Box 249, George, 0276530, South Africa, [3]University of the Free State, Centre for Sustainable Agriculture, P.O. Box 339, Bloemfontein, 0279300, South Africa

The effects of heterolactic inoculants Bonsilage mais Flussig (BM) and Lalsil Fresh LB (LFLB) as additives for whole crop maize (300 g/kgDM) were studied. The silage was produced with BM, LFLB and untreated maize in 1.5l jars under laboratory conditions. Samples were collected on day 0, 4, 10, 21 and 60 of ensiling and analysed for pH, water-soluble carbohydrates (WSC), volatile fatty acid (VFA), lactic acid, ammonia-N, crude protein (CP), crude fibre (CF), dry matter (DM), energy, fat and ash. The aerobic stability of silage was determined on 90 of ensiling. The BM and LFLB caused lower ($P<0.05$) pH, ammonia-N, butyric acid and DM, while increasing WSC, lactic acid, CP and acetic acid than the control. Treatments did not impair ($P>0.05$) CF, fat, ME, ash and propionic acid contents of the silages. The BM and LFLB treated silage were more ($P<0.05$) aerobically stable than the control silages as indicated by lower CO_2 production. It was concluded that BM and LFLB improved the aerobic stability of maize silage due to their higher acetic acid productions.

Kernel-based classification of crop-livestock systems

A. Gonzalez[1], G. Russell[2], A. Marquez[1], C. Dominguez[1] and O. Colmenares[1], [1]Romulo Gallegos University, IDESSA, Guarico, 2301, Venezuela, [2]The University of Edinburgh, GeoSciences, Scotland, EH8 9NG, United Kingdom

The main objective of this research is to asses the supervissed classification accuracy of crop-livestock systems after linear and kernel (non linear) feature extraction preprocessing. This conparison is based on percentage of individuals correctly allocated; and by five statistics: Wilks Lambda, Pillai's trace, Hotelling's test, Roy's minimum root, and squared average canonical correlation. The assesment included atributive information as recorded in census data of 168 holdings located in Aragua and Guarico states in Venezuela. We have found that classification accuracy after kernel feature extraction resulted significantly greater than linear preprocesing. There were also significant differences for the amount of variance explained by the classes achieved under the non linear approach. It is concluded that non linear preprocesing using kernel functions may empower extracted features to yield better quality information for eventual clustering.

---

no. 6

Comparison of growth rates in the primal cuts of Canadian composites

L.A. Goonewardene[1], Z. Wang[2], R. Seneviratne[2], J.A. Basarab[1], E.K. Okine[2], M.A. Price[2] and J. Aalhus[3], [1]Alberta Agriculture and Rural Development, Agriculture Research Division, # 204-7000-113 street, Edmonton, AB Canada T6H 5T6, Canada, [2]University of Alberta, Agriculture Food and Nutritional Science, University of Alberta, Edmonton, AB Canada T6G 2P5, Canada, [3]Agriculture and Agri-Food Canada, C & E trail, Lacombe, Canada

Growth rates (GR) of muscle (M) and fat (F) in primals of Beefbooster® composite (C) types (SM=C of small breeds, AH=C of Angus and Hereford, GLC=C with Gelbvieh, Limousin or Charolais) from 274-456 days (d) were compared to determine harvest times maximizing %M and minimize %F. Covariance determined GR/d for tissue, M and F within cut and C type. The GR of M+F+bone was highest in the chuck (SM=106.05, AH=137.85, GLC=118.01 g/d) followed by the round, rib, plate or flank. The GR of M was 50.63 – 70.31, 32.57 – 48.01and 12.89 –17.98 g/d in the chuck, round and rib respectively. The decrease %M and increase in %F was highest in the plate –0.095 to -0.074%/d and 0.093–0.123%/d respectively in all C's. The round lost less %M and gained less %F. The GR of M was higher ($P<0.10$) in AH than SM in the round, loin, chuck, rib, plate and brisket. Over all cuts SM, AH and GLC lost 9.63, 9.24 and 7.76% of M respectively. The gain in F in SM, AH and GLC was 13.22, 13.67 and 11.33% respectively. Based on GR of M and F, SM and AH should be harvested earlier than GLC to maximize %M and minimize %F.

Performances and prospects of landless small scale dairy production system in Tunisia

M. Ben Salem[1] and R. Bouraoui[2], [1]INRA, Rue Hédi Karray, 2049 Ariana, Tunisia, [2]ESA Mateur, Mateur, 7030, Tunisia

Tunisia has always given high priority to the development of its local dairy production. This resulted in the emergence of a new landless small scale cattle production system in parts of the country not traditionally known for dairying. Such system showed a strong development over the last 2 decades despite the fact that almost all the feed is external to the farm. However, with recent increases in feed prices the sustainability of this system became the a current debate raising the need for its close investigation in the absence of research studies. The objective of this work is to analyze the production potential of the landless dairy production system and to review the prospects of its durability and contribution to the sustainable rural development in connection with current issues resulting from increases in livestock feed prices on the international market. Comprehensive surveys covering as many areas of a dairy operation which would be of utmost use in describing and analyzing the production systems were conducted. A total of 220 producers in 3 different zones were studied. Results showed production performances well below the animal's genetic potential and a low product quality due to inadequate feeding management. In the future, the durability of the production system as well as its capacity to contribute to the sustainable rural depend both on the ability of the farmers to improve their production efficiency and the local policy of milk pricing.

---

no. 8

Variation in carcass traits relative to diet & gender and analysis by principal components

L.A. Goonewardene[1], C.K. Gunawardena[2], E. Beltranena[1], R. Zijlstra[2], Z. Wang[2] and W. Robertson[3], [1]Alberta Agriculture and Rural Development, Research Division, # 204-7000 113 street, Edmonton, AB T6H 5T6, Canada, [2]University of Alberta, Department of AFNS, Edmonton, AB, Canada, [3]Agriculture and Agri-Food Canada, 6000 C & E trail, Lacombe, AB, Canada

The trial determined the efficacy of four diets (n=24/diet) (Soy, Fababean, 50% Soy+50% Faba & Pea) fed to market pigs. The objective was to compare the variances in carcass weight, quality traits and muscle (M), fat (F) and bone (B) distribution of primal cuts (butt, picnic, loin and ham) relative to diet and gender and use principal component (PC) analysis to identify linear trait combinations. Carcass traits associated with F were the most variable followed by M and B traits. Except for random minor differences, the variances in carcass quality and tissue traits were similar for diets. Whenever, the gender variances were different, it was higher ($P<0.05$) in barrows. Certain cuts from barrows were less uniform than from gilts. Across all traits and within primals, %F and %M were negatively correlated and F grew at the expense of M. The first PC containing the F traits and the second PC containing the M traits accounted for 47.8% of the variation in all traits. Increased leanness could be achieved by limiting fat growth. Harvesting pigs at lighter weights will increase percent M in the carcass especially in gilts. Sorting of animals on weight between and within gender improves the uniformity of carcass quality and tissue traits.

Effect of selenium and vitamin E on chronic mastitis and reproduction activities in Iranian Baloochi fat tailed ewe

M. Forghani, F. Amir Saifadini and A. Esmaili, Uni.Sh.Bahonar, Animal sci., Kerman, 133, Iran

The study on 12 heads of second pregnancy and 12heads of third pregnancy in a CRD design with factorial arrangement of 2 × 2 were used. This experimental include 4 treatments, and each were 6replicate (3 heads of second pregnancy and 3 heads of third pregnancy). Treatment No.1 used twice vitamin E injection (2 * 500 IU=1000 IU/head) on days 145 and 150 of ewe pregnant. Treatment No.2 used one injection of selenium 15 mg/head) on day 145 and treatment No.3 used twice vitamin E injection (2 * 500 IU=1000 IU/head) and selenium (15 mg/head) on days145 and 150 of ewe pregnant, and control treatment: No.4 used without any injection of selenium and vitamin E. The level of Beta-globulin in colostrum and lambs blood serum of all treatments also has increased. The correlation coefficient of Beta-globulin between lambs and ewes on delivery day is positive ($P<0.05$). Gamma-globulin concentration in different treatments, on delivery day shows increasing in comparison with control group ($P<0.05$). The level of Gamma-globulin in colostrum of treatments No.2 &3 and lambs blood serum in different treatments has increased. The correlation coefficient of Gamma-globulin concentration between serum of ewes blood on delivery day with that of in colostrum and ewes blood serum shows positive relation ($P<0.05$). The level of milk chlorine in different treatments on delivery day and one week after that was lower than control ($P<0.05$).

---

no. 10

Monensin in dairy cows rations: blood parameters and ruminal fermentation

J.R. Gandra, J.E. Freitas Jr, M. Maturana Filho, A.P.C. Araújo, B.C. Venturelli, L.F.P. Silva and F.P. Rennó, University of Sao Paulo, School of Veterinary Medicine and Animal Science, Av Duque de Caxias Norte, 225, 13635900, Brazil

The objective was to evaluate the effect of monensin in ration for dairy cows on blood parameters and ruminal fermentation. Twelve Holstein cows were allocated in four balanced 3x3 latin squares, and fedwith the rations: 1) Control, basal diet without monensin addition, 2) Monensin 24, addition of 24 mg/kg DM of sodic monensin in the ration, and 3) Monensin 48, addition of 48 mg/kg DM of monensin in the ration, Blood samples were collected in vacuolizade tubes with venipuncture of coccygeal artery. Blood concentration of glucose, total triglycerides, total cholesterol,cholesterol-HDL, cholesterol-LDL, cholesterol-VLDL, total protein, albumin, urea and serum ureicnitrogen were determined. Samples of ruminal fluid were collected using esophageal probe two times, before and three hours after morning feeding. The level of pH, ammonia nitrogen and volatile fatty acids were determined in the ruminal fluid. The experimental rations had no effect ($P>0.05$) on blood parameters. The pH, concentrations of ruminal ammonia nitrogen and volatile fatty acids were not different ($P>0.05$) for different levels of monensin used in rations, in both times of collection. The obtained resultssuggest that the monensin levels used in this trial had no effect on bloodparameters or ruminal fermentation.

Fat sources in dairy cows rations: intake, milk yield and protein fraction composition

J.E. Freitas Jr, J.R. Gandra, M. Maturana Filho, L.S. D'Angelo, A.P.C. Araújo, M.V. Santos and F.P. Rennó, University of Sao Paulo, School of Veterinary Medicine and Animal Science, Av Duque de Caxias Norte, 225, 13635900, Brazil

This study was carried out to evaluate the effect of different fat sources in dairy cows rations on dry matter intake, milk yield and milk protein fraction composition. Twelve Holstein cows were allocated in three balanced 4x4 latin square, and fed with the rations: 1) Control, with 2.5% of ether extract (EE) in dry matter; 2) Refined soybean oil, with 5.5% of EE; 3) Whole soybeans, with 5.5% of EE; and 4), Calcium salts of fatty acids (Megalac-E) (CSFA), with 5.5% of EE. The milk yield and intake were measured daily during the trial period. The samples used for milk composition analysis were collected at two alternated days, and at the two daily milkings. The concentration of milk total nitrogen, non protein nitrogen and non casein nitrogen were evaluated. Cows receiving CSFA had lower ($P<0.05$) intake in relation to control diet. Lower milk yield ($P<0.05$) were observed in cows that received whole soybeans in the diets. Fat sources added to the rations had no effect ($P>0.05$) on milk crude protein, non protein and non casein nitrogen, true protein, casein, ratio casein/true protein and whey protein. Similarly, the experimental rations did not influence all these fractions when expressed as percentage of milk crude protein. The use of fat sources in dairy cows rations influenced the dry matter intake, milk yield and composition, depending on the fat sources characteristics added to the rations.

---

no. 12

Fat sources in dairy cows rations: blood parameters and ruminal fermentation

J.E. Freitas Jr, J.R. Gandra, M. Maturana Filho, B.C. Venturelli, C. Foditsch, L.F.P. Silva and F.P. Rennó, University of Sao Paulo, School of Veterinary Medicine and Animal Science, Av Duque de Caxias Norte, 225, 13635900, Brazil

This study was carried out to evaluate the effect of different fat sources in dairy cows rations on blood parameters and ruminal fermentation. Twelve Holstein cows were allocated in three balanced 4x4 latin square, and fed with the rations: 1) Control, with 2.5% of eter extract (EE) in dry matter; 2) Refined soybean oil, with 5.5% of EE; 3) Whole soybean, with 5.5% of EE; and 4), Calcium salts of fatty acids (Megalac-E) (CSFA), with 5.5% of EE. Samples of ruminal fluid were collected two times using an esophageal probe, before and three hours after morning feeding. No differences ($P>0.05$) were observed for experimental rations on ruminal pH and volatile fatty acids in the evaluated times. Ruminal ammonia nitrogen was greater ($P<0.05$) for cows receiving whole soybeans on time zero. However, after three hours of feeding no difference ($P>0.05$) had observed in the concentration of ruminal ammonia nitrogen among rations. The concentrations of total cholesterol, cholesterol LDL and HDL were higher ($P<0.05$) for cows that received rations with fat sources. The concentrations of urea and blood urea nitrogen were similar among rations, except for the diet containing CSFA, in which the concentrations were lower ($P<0.05$). The use of fat sources in dairy cows rations influenced the blood parameters, especially for parameters related to lipidogram, however it did not influence the parameters of ruminal fermentation.

Monensin in dairy cows rations: intake, milk yield and protein fraction composition

J.R. Gandra, J.E. Freitas Jr, M. Maturana Filho, C. Foditsch, L.S. D'Angelo, M.V. Santos and F.P. Rennó, University of Sao Paulo, School of Veterinary Medicine and Animal Science, Av Duque de Caxias Norte, 225, 13635900, Brazil

This study evaluated the use of monensin in ration for dairy cows on intake, milk yield and protein fraction composition. Twelve Holstein cows were allocated in four balanced 3x3 latin squares, and fed with the rations: (1) Control, basal diet without monensin addition, (2) Monensin 24, addition of 24 mg/kg DM of monensin in the ration, and 3) Monensin 48, addition of 48 mg/kg DM of monensin in the ration. The milk yield and intake were measured daily during the experimental period. The samples used for analysis of milk composition and milk protein faction composition were collected at two alternated days, and at thetwo daily milkings. Addition of monensin in the diet reduced linearly ($P<0.05$) the intake (kg/day and %PV), increased quadraticaly milk yield ($P<0.05$), and increased linearly production efficiency ($P<0.05$). No differences ($P>0.05$) were observed among rations for milk fat, lactose, total dry and non fat extract composition, body weight and body condition score ($P>0.05$). There were no effects ($P>0.05$) on percentage of milk crude protein, non protein and non casein nitrogen, true protein, casein, ratio casein/true protein and whey protein. Similarly, the experimental rations did not influence all these fractions expressed as percentage of milk crude protein. The use of monensin at 24 mg/kg DM increased the productive performance of dairy cows.

---

no. 14

Grassland in organic dairy farming of Latvia

A. Jemeljanovs, B. Osmane, J. Miculis and J. Zutis, Research Institute of Biotechnology and Veterinary Medicine, 1 Instituta Street, LV-2150 Sigulda, Latvia

Efficiency of grassland is very significant in milk production: if it ensures cheap and full value feed in summer and winter then high yield of milk with low cost price can be obtained. Qualitative and cheap grass forage can be obtained only from good botanic composition grasses, where the main place are taken by papilionaceae (red clover, white clover, lucerne, galega etc.) and at least 3 – 5 grasses species (timothy, fescue, rye-grass etc.). Pasture grass, hay, silage, straw are the main energy and nutritive substances sources for herbivore feeding in the organic farms. Grass feed in organic farms contained higher total protein level by 0.3%, phosphorus by 0.15%, sugar by 1.0% and digestible protein by 1.6% per feed unit, but less dry mater by 1.0%, fiber by 4.0%, total fat by 0.4%, nitrogen free extracts by 5.0% and calcium by 0.1%, better calcium and phosphorus ratio – 1.86:1 in comparison with conventional farms. In the grass of organic farms was observed a tendency of higher amino acids – aspartic acid, serine, glutamic acid, proline, glycine, valine, methionine, isoleucine, hystidine and lysine levels but less cystine, thyrozine, phenilalanine and arginine levels in comparison with conventional grass feed. Mentioned above differences of amino acids levels were not statistically credible ($P>0.05$). Organic farms milk had higher essential amino acids level and contained significantly less amount of cholesterol and it can be considered as more healthy than milk obtained in conventional farms.

no. 15

Intensification drivers of dual-purpose livestock systems

A. Marquez[1], G. Russell[2], N. Dendoncker[2] and A. Gonzalez[1], [1]Romulo Gallegos University, IDESSA, Guarico, 2301, Venezuela, [2]The University of Edinburgh, GeoSciences, Scotland, EH9 3JF, United Kingdom

To investigate the main socio economic factors influencing dual-purpose livestock intensification in the Monagas and Guaribe municipalities in the Guarico state of Venezuela, a logistic regression procedure was applied to a sample of 103 farms from a population of 939. The farm intensification level classified as extensive and semi intensive was used as the dependant binary variable. A backward stepwise exploratory analysis was conducted to help select the covariates to be included in the final model based on their ability to explain livestock intensification. A total of eight variables were initially evaluated but only farm size, land tenure rights and farmers' educational background were significant ($P<0.05$) at explaining livestock intensification. The percentage of cases that could be correctly classified by the model rose to 70.9% after the inclusion of the above referred variables. The predicted probabilities of livestock intensification were found to dramatically increase with farmer level of education and private land rights, while farm size had a slightly negative impact on intensification.

---

no. 16

The effect of optimal management strategies on the production and reproduction of communal cattle on sourveld in the Eastern Cape

C.B. Nowers and J. Welgemoed, Dohne Agricultural Development Institute, Agriculture, Private Bag X15, Stutterheim 4930, South Africa

The production potential of communal livestock is largely unknown as they have never been challenged to prove their true potential under favourable conditions. The objective of this trial was to determine the true production and reproduction potential of cattle in these communal sourveld areas when they are managed under favourable commercial conditions. Cattle from the communal sourveld farming areas at Wartburg community were randomly divided into one of two treatment groups. One group (Communal treatment) continued to be managed under the normal communal farming practices and remained in the communal area whilst the other group was transferred to the Dohne A.D.I. where they was managed under sound practises for commercial beef farming (Commercial treatment). The average calving percentage of communal cows and their progeny managed under optimal conditions at Dohne (80.4%) was significantly higher than the calving rates (41.7%) of communal cows at the Wartburg communal area. Data suggests that the production potential of the progeny of communal cows managed at Dohne could be favourably compared to that of commercial beef breeds (Dohne and Bonsmara) managed at Dohne A.D.I. Calves reared from cows under communal conditions weighed only 58.5% (at approximately 205 days) of the live mass of calves reared under commercial conditions. Weaning data of the calves at Dohne indicates pre-weaning growth rates of 0.745kg/day and average cow efficiency rates of 49.8%.

Carcass and haematological characteristics of Mubende goats fed graded levels of *Moringa oleifera*

M.L. Wasswa[1], F.B. Bareeba[2], J.D. Kabasa[1] and F. Kabi[2], [1]Makerere University, Physiological Sciences, Faculty of veterinary Medicine, P.O. Box 7062 Kampala, Uganda, [2]Makerere University, Animal Science, Faculty of Agriculture, P.O. Box 7062 Kampala, Uganda

The effect of substituting *Moringa oleifera* leaf meal (MOLM) for cottonseed cake (CSC) on carcass and haematological characteristics of Mubende goats was determined using thirty two intact male goats (18 – 20 weeks old and 13 kg+2.2 kg BWt). The experimental diets fed for 18 weeks comprised of a basal diet of *Brachiaria* hay offered *ad lib.* and four isonitrogenous (CP 16%) and isocaloric (10 MJ ME/kg DM) diets, (247g/d DM) consisting of 0 (control), 25, 50, and 100% MOLM mixed with complementary proportions of CSC. Carcass yield, pH, meat to bone ratio, kidney fat, carcass CP, ash and weights of internal organs (pluck, kidney) were not different (P>0.05) among the diets. However, the weight of the liver, spleen, thickness of *Longissimus dorsi*, omental and carcass fat differed (P<0.05). There were no differences (P>0.05) in the hematological parameters (red blood cells, packed cell volume, haemoglobin, white blood cells, eosinophils, fibrinogen and total plasma proteins) of different treatment groups. The results showed that the inclusion of Moringa as a protein supplement to low quality diets did not affect carcass characteristics and health of Mubende goats.

---

no. 18

Influence of management systems on growth and survival of young indigenous chickens in Uganda

D. Nampanzira and C.C. Kyarisiima, Makerere University, Department of animal Science, P.O. Box 7062, Kampala, Uganda

Two experiments were conducted to investigate the effect of management systems on performance of indigenous chickens. In each experiment, 135 chicks were allotted to three treatments. In Experiment I, chicks were either confined to 12 weeks (CONF121); or confined to six weeks (CONF6), or allowed to access the run with broody hens (RUN1). In Experiment II, chicks were either confined to 12 weeks (CONF122), or creep fed (CREEP) or allowed into the run with broody hens (RUN2). In both experiments, chicks which were confined were brooded for two weeks. At the end of the confinement period, all birds had access to the run. All birds were allowed free access to feed, except hens whose chicks were creep fed. In Experiment I, chicks under RUN1 quickly adapted to the run conditions and grew faster than those that were first confined. Growth rate of birds under CONF6 and CONF121 treatments declined when they were released into the run. In Experiment II, birds under CREEP treatment were significantly smaller than those under RUN2 and CONF122 treatments. Birds that were confined to 12 weeks were smaller than those under RUN2 and their growth was depressed when they were released into the run. Hens which moved and fed with their chicks lost broodiness within four weeks post hatching, but hens whose chicks were creep fed remained broody. Mortality of chicks was 18.6% and 22% under RUN1 and RUN2, respectively. Confining birds to 12 weeks increased feed consumption by over 14%, but resulted into over 95% survival rate.

Lysine and zinc chelate for brown hens: effects in the egg composition

M.A. Trindade Neto[1], B.H.C. Pacheco[1], R. Albuquerque[1] and E.A. Schammass[2], [1]Universidade de Sao Paulo, Faculdade de Medicina Veterinaria e Zootecnia, Nutrição e Produção Animal, Av. Duque de Caxias Norte, 225, Jd., 13635-900 Elite, Pirassununga, SP, Brazil, [2]Instituto de Zootecnia, Analises Quantitativas, Heitor Penteado 56, 13635002 Nova Odessa, Sao Paulo, Brazil

It was evaluate eggs quality of Isa Brown hens under different lysine and zinc chelate levels. Were used 720 hens at 48th to 60th week age allotted in randomized design and the treatments were disposed in factorial scheme (5x3). The analyzed lysine levels were: 0.560; 0.612; 0.677; 0.749; 0.851% and zinc: 20, 40 e 80 ppm in chelate form. There was interaction of lysine and zinc in fat and mineral yolk. The zinc increase in diet coincided with linear decrease on yolk and egg mineral matter, however the inverse was observed in albumen. The mineral increase limited the amino acid use on egg composition. There was effect of lysine in egg, yolk and albumen weight and egg dry matter. There was effect of lysine in albumen composition (lipid, protein, mineral) in total egg (lipid and protein) in water accretion on yolk, albumen and total egg. The estimated digestible lysine in these variables based in amino acid analysis was 0.644±0.07%. The water percentage in total egg decrease linearly when the lysine was increased in diet this result was inverse for egg lipid concentration. In concentration studied the increase of dietetic zinc prejudices the egg composition and 20ppm attends the egg quality characteristic.

---

no. 20

Digestible lysine: metabolizable energy to barrows and gilts diets at 20 to 50 kg – Aparent digetibility and blood parameters

E. Gandra[1], D. Berto[1], J. Gandra[1], F. Budiño[2], E. Schammass[2] and M. Trindade Neto[1], [1]Universidade de Sao Paulo, Nutrição e Produção Animal, Av. Duque de Caxias Norte, 225, Campus USP, Pirassununga, SP, 13635-900, Brazil, [2]Instituto de Zootecnia, Produção Animal, Av. Heitor Penteado, 56, Nova Odessa, SP, 13460-000, Brazil

The effects of digestible lysine were studied in apparent digestibility and blood parameters assays with specific line of barrows and gilts at 20 to 50 kg body weight allotted in randomized bloc design. In digetibility assay were used 20 pigs, 4 replications and one animal for experimental unit. In blood profile assay were used 20 barrows and 20 gilts, 4 replications for sex and one animal for experimental unit. The treatments were 0.80, 0.90, 1.00, 1.10 and 1.20% digestible lysine in diet with 16.18 CP and 3425 kcal ME. In digestibility assay the feed intake based in dry matter for metabolic unit ($kg^{0.75}$). The control blood parameters occurred at first study (farm condition) using 5 gilts and 5 barrows and the final blood parameters occurred at finish study (segregate condition) using 20 gilts and 20 barrows. There was quadratic effect of lysine in N retention and neutrophil/linphocit suggesting 1.04% and 1.00% of digestible amino acid, respectively, independent of sex. The Dunnet test indicates significant differences between farm and segregate conditions.

Lamb production at confinement, feeding with rations using ammoniated sugar cane bagasse, corn byproducts and poultry litter

G.E. Nouel Borges, P. Hevia, R.J. Sanchez Blanco, M.A. Espejo Díaz and J.B. Rojas Castellanos, Universidad Centroccidental Lisandro Alvarado, UIPA, Agronomía, Tarabana, Cabudare, Estado Lara 3023, Venezuela

The effect of four levels of poultry litter (C) and corn hominy (H), with ammoniated sugar cane bagasse (B, 40% of total diet), in diets for growing lambs, feeding in total confinement. Experiment was carried out to estimate the productive response in commercial exploitation conditions. Diets were: C (11,4; 20,4; 29,4 and 38,4%) and H (48, 39, 30 and 21% respectively) during a period of 112 days in all random design, with three repetitions and four lambs by repetition. Results permitted to conclude that the C (20,4 to 29,4%), H (39 to 30%) and B (40%) can be utilized to raise and fattening of lambs, with weight gains of 100 and 83 g/lamb/day, cost of ration 709 to 683 Bs/kg of LW and without compromising the health of the animals, with normal liver (393 and 364 g) and kidneys (63 and 68 g). Weight of hot carcass of 11,67 and 10,27 kg, carcass yield (49,1 and 45,2%), with weight of thighs (3,62 and 3,11 kg) and backs (2,53 and 2,51 kg) of size and adequate proportion al standard degree of carcass lambs. Being able to recommend these rations on intensive lambs production in confinement on the influence area of corn, sugar cane and poultry agro industry, with costs of rations that would guarantee their profit value

---

no. 22

Phosphorus apparent efficiency in dairy cattle herds

D. Biagini and C. Lazzaroni, University of Torino, Dept. Animal Science, Via L. da Vinci 44, 10095 Grugliasco, Italy

Identification of P as one of the main polluting nutrients has focused attention on P excretions from animals rose by feeds ingredients and mineral supplementation. So a study was carried on in several commercial dairy farms in North Italy to evaluate P intake and estimated P excretion (by the P balance method) and P apparent efficiency. In 15 herds (3,331 heads and 1,702 t of live weight) feeds consumption, mineral supplementation, P requirements, live weight and productions (weight gains, born calves and milk production) were recorded for lactating cows (LC), dry cows (DC), heifers (H, 12-24 months) and young heifers (YH, 6-12 months). Feeds samples were collected and analysed to determine P intake. The diets mean P content of dry matter intake and the standard deviation were 0.43±0.03, 0.42±0.09, 0.37±0.10 and 0.38±0.08% respectively for LC, DC, H and YH. The results are due to an overfeeding common in several herds. The P intake was 30.6±3.4, 11.6±3.1 and 8.7±1.7 kg/year respectively for LC plus DC, H and YH. The estimated P excretion was 22.1±3.0, 9.4±2.9 and 7.69±2.3 kg/year respectively for LC plus DC, H and YH, corresponding to 30.5±4.7 kg/year/t of live weight or 28.9±4.0 kg/year/remount animal group. P apparent efficiency was 28.0, 19.0 and 11.7% respectively for LC plus DC, H and YH. P efficiency in the examined dairy cattle herds could be increased, particularly with a reduction of the P intake and giving major attention to the replacing animals' diet formulating feeds as closely as possible to their requirements.

no. 23

Pevalence of ticks and their effects on growth rate of calves

S.K. Kibiru, Kenya Agricultural Research Institute, Animal Health and Production, P.O. Box 57811, Nairobi, 254, Kenya

Tick counts and their effects on calf growth rate were monitored for twelve months in Maasai pastoral herds in Kenya. The study which involved two-stage sampling of 23 household herds in two contrasting sites representing former communal group ranches (now subdivided) used 1624 calves less than 12 months old. The mean total tick counts varied between sites and herds. Mean total tick counts for *R. appendiculatus, R. evertsi A. gemma, A. variegatum, B. decoloratus* and *R. pulchellus* were significantly different between the sites. An increase in total number of individual tick species including *R. appendiculatus, R. pulchelus* and *B. decoloratus* caused significant decrease in average daily weight gain. The study suggests that differences in total tick counts and their effects on average daily weight gains were affected by the environment and type of management at household level.

---

no. 24

Smallholder pig production and consumption dynamics in relation to porcine cysticercosis challenges in Tanzania

E. Kimbi, F. Lekule, E. Komba, H. Ngowi, G. Ashimogo, A. Willium and S. Thamsborg, Sokoine University of Agriculture, Box, 3004, Morogoro, +255, Tanzania

Cysticercosis caused by the zoonotic tapeworm Taenia solium has being an endemic problem affecting pigs and human in Tanzania, hence causing significant constraints to nutrition and economical wellbeing of farming communities. The study using participatory rural appraisal and questionnaire interviews was conducted to evaluate the smallholder pig production and utilization dynamics in relation to porcine cysticercosis. A random sample of 30 villages in two districts of Mbeya region in Tanzania was used. Pig keeping ranked the first to third among important livestock enterprises in crop-livestock mixed farming systems. Four types of pig production systems were practiced, namely; total confinement, semi confinement, free range, herding, and tethering. Each system has specific management practices such as feeding and housing which are influenced by seasons of the year and feed availability. About 87% of rural families consumed pork from their respective villages, indicating high local utilization of pork. The amount consumed per household per month varied between 50 and 2500g with mean frequency of use of 4.8 times per month. Cysticercosis was mentioned as one of the important challenges facing smallholder pig keepers. About 93% of pig keepers are aware of the disease, though at different levels of knowledge and perceptions. For ensuring sustainable pig production and control of cysticercosis holistic approaches are recommended

Socioeconomic aspects influencing productivity and utilisation of camels in selected arid and semi-arid (ASALs) districts in Kenya

S.O. Nyamwaro, J.K. Chemuliti, M.O.K. Mochabo, K.B. Wanjala, A. Kagunyu and F.M. Murithi, KARI, Socioeconomics, P.O. Box 12, Makindu, 90138, Kenya

Research on socio-economic aspects influencing productivity and utilisation of camels was undertaken in three selected ASAL districts of Kenya. The study was designed with an aim to improving productivity and utilisation of camels through improved knowledge synthesis and utilisation. Informal socio-economic surveys were conducted through PRAs, which mostly relied on the instruments of FGDs. A maximum of 21 FGDs were surveyed whose organisation was based on gender as dictated by culture. Data and information collected were of qualitative nature, which were subjected to qualitative data analysis SPSS. Highlights of the research indicate that the most preferred livestock species to keep was ranked as camels with 90% of FGDs in consensus. Major camel production constraints were cited as diseases. Agrovet shops (71%) were ranked as the most important source in supplying camel drugs. First order of preference for milk production was bestowed to camels (81% of the FGDs). Major role of camels was given as milk production and draft power. Camel marketing was most affected by poor pricing (50%). Main channels of marketing of live camels were through middlemen (55%) and neighbours (20%), while camel milk was marketed through middlemen (60%). The most critical roles of camels in the ASALs study areas were observed to be milk production and draught power; hence camels could be critical in safeguarding food security in the Kenyan ASALs.

---

no. 26

Land tenure and indigenous pastoral system in Cameroon: the Wum case study

N. Justice Sama, Fedev, P.O Box 593, Bamenda, NWP, Cameroon

Land, pasture and water are the prime resources for pastoralist worldwide. Access to these resources is thus an invaluable asset in the sustainability and wellbeing of indigenous pastoralist communities. Pastoralist communities in Cameroon, regard land as a common property or common resource. They value land more in terms of use than permanent ownership. The itinerant or nomadic lifestyle of most pastoralists, accord more credence to their perception of land use. Conversely, the post colonial statutory land tenure introduce private property regime of land ownership. This implies that land either belongs to the state (national land) or is private property evidenced by a land certificate. Practically, this private property regime has negatively impacted on pastoral lifestyle, access to resources and indigenous livestock development. Wum is in the Donga-Mantung Division of North West Province of Cameroon. It is composed of many villages made up of pastoralist communities who graze predominantly cattle and goats. On the other hand are equally crop farmers. Unfortunately Wum has witnessed persistent and at times violent farmer-grazer conflicts from the late 80s till date. This case elucidates the loss of range land, resource inaccessibility, dwindling livestock and increase poverty caused by tenural injustice. It postulates that where statutory land tenure polices practically exclude the customary pastoral perception of communal land ownership, it does not only affect livelihoods of pastoral people, but also generate severe conflicts, immortalize existing conflicts and frustrate sustainability objectives. This case emphasizes the urgent need to reform the land tenure laws of Cameroon so as to recognize customary pastoral land tenure system.

 no. 27

Migratory small ruminant management system in Nepal: current status and interventions for improvement

B.S. Shrestha and B.R. Joshi, Nepal Agricultural Research Council, P.O. Box 1950, Kathmandu, Nepal

Migratory management of small ruminants is a long tradition and means of livelihood in the high hills and mountains of Nepal, which is practiced in almost all districts adjoining to the southern flanks of Himalayan massifs. Small ruminants are an important source of cash generation with their multidimensional uses and serve as liquid assets especially during the time of famine, illness and emergencies for poor farming families and women. The current paper deals with the present status of migratory management system in three hilly and mountain districts of Nepal and discusses on the responses of interventions made to mitigate the existing constraints. The migratory system is in rapid decline due to many constraints such as diseases, parasites, predation, shrinkage of grazing areas, lack of shepherds and plant poisonings. Solar lighting and nylon net during night was found to be effective for predation control, as none of the animals was predated during night with this intervention. Use of anthelmintic at monthly intervals during wet summer months (June-Oct)has resulted significantly ($P<0.01$) higher weight gain in lambs and kids (3.8kg and 2.2kg respectively) as compared to the weight gain in non treated lambs and kids. Oral administration of sodium thiosulphate has been found to be effective for treating plant poisoning cases with recovery rate of almost 98%. With these simple interventions, the productivity of the animals and income of farmers have increased significantly, which is essential for sustainability of the system.

---

no. 28

Livestock systems and livelihoods: responding to competing claims for land at the tourism/livestock interface

P. Chaminuka[1,2], A. Van Der Zijpp[1], C. Mccrindle[3], H. Udo[1], K. Eilers[1], E. Van Ierland[1] and R. Groeneveld[1], [1]WUR, P.O. Box 338, Wageningen, Netherlands, [2]UL, Private Bag X1106, Sovenga, South Africa, [3]UP, Private Bag X04, Onderstepoort, South Africa

Rising demand for livestock products increases opportunities for improved livestock production in Southern Africa at a time that new transfrontier parks present new opportunities for rural communities to generate incomes from ecotourism. These multiple opportunities intensify competing claims on grazing land, which influences the nature of livestock systems and their related cultural and capital functions. At this tourism/livestock interface, livestock production is constrained by restricted livestock movement, livestock loss to predation and diseases and emerging preferences for tourism-related land uses. The objective of this study is to determine the contribution to household incomes and employment of livestock and to analyse the influence of factors peculiar to this interface on livestock systems. In estimating economic value of livestock systems we utilise bio-physical and socio-economic survey data on biological productivity, tangible and intangible livestock products to estimate the values of marketed and non-marketed livestock products. We consider spillover effects of livestock incomes in the local economy. Costs related to tourism/livestock interactions are incorporated in the input-output analysis. Results highlight unique aspects of communal grazing systems and how they have adapted to increased land uses related to tourism/livestock interactions.

Organic livestock production: emerging challenge for the tropical animal husbandry systems

M. Chander[1], B. Subrahmanyeswari[2], S. Kumar[1] and R. Mukherjee[1], [1]Indian Veterinary Research Institute, Extension Education Division, Izatnagar, 243 122, India, [2]N T R College of Vetrerinary science, Veterinary & A H Extension, Gannavaram, 521 102, India

The intensive livestock production methods involving agro-chemicals, allopathic drugs, antibiotics, vaccines and improved feeds significantly increased the food supply by improving the livestock productivity world over. But, in recent times, consumers around the world are increasingly seeking environmentally safe, health foods along with concern for animal welfare issues. This growing demand for good quality foods can be met using organic production methods, so, it's a new challenge for the developing country producers in particular. Additionally, organic livestock production offer new export opportunities for the developing country producers. Thus, many developing countries are now exporting organic agricultural products, but organic livestock product export from developing countries save beef from Brazil and Argentina, is yet to take off. As such, organic livestock farming is a challenge not only for the farmer but also for research, extension services and interdisciplinary work. Thus, this paper, basing on the extensive surveys done in India, explores the prospects of organic livestock production, while reviewing the status, potential, constraints, priorities, opportunities and threats of organic livestock production in context of developing countries.

---

no. 30

Bullocks vs tractors: draught animal power still has a future

N. Akila and M. Chander, Indian Veterinary Research Institute, Div. of Extension Education, Izatnagar, 243 122, India

About 2 billion people in developing countries still depend upon draught animal power, which is provided by bullocks, buffaloes, equines and camels. In India, more than 75% of the land holding is $<$2ha in size, whereas, over 50% of agricultural land is still being tilled by the bullock drawn implements though the number of tractors has increased, these till only 20% of land. Small scale Indian farmers owning about 83 million land holdings maintain bullocks for manifold farming operations including rural transport. 71.5 million bovines which also haul 14 million carts have helped save fossil fuel and helped in reducing the pollution. It is estimated that India would need about 80 million bullocks for agriculture and other uses over the next five years. These farmers, however, need training to improve the work efficiency of bullocks and also implements require to be developed to reduce the drudgery of work animals, while improving the work efficiency. This paper reviews the status of draught animal power in Indian agriculture indicating that it still has relevance amid massive efforts to mechanize farming.

Effect of incremental dietary level of *Commelina benghalensis* on intake, digestibility and nitrogen balance in sheep fed *Sorghum almum*

T.P. Lanyasunya, Kenya Agricultural Research Institute, Animal Production, KARI-NAHRC, P.O. Box 25, Naivasha, Kenya, 25, Naivasha, Kenya

Effect of increasing dietary level of *Commelina benghallensis* on intake, digestibility and N balance in sheep fed *Sorghum almum* was investigated. Twelve wethers (6 months; 18.6±1.8 kg LW) fitted with rumen canula were housed in metabolic crates and allotted to 4 treatment diets (D0, D10, D20, D30) in a randomized complete block design. The diets were constituted from fresh *Sorghum almum* and pre-wilted *Commelina benghalensis*. Control diet (D0) comprised of 3 kg fresh *Sorghum almum* (≈ 535.5 g DM/head/d about 3% LW), whereas D10, D20 and D30 were D0 + 300, 600 or 900 g of wilted *Commelina benghalensis* ((≈34.02, 68.04 or 102.06 g DM/head/d), respectively. Each diet was fed for 14 d during which time the wethers adapted to confinement and rations, followed by 7 d period of urine, faecal and rumen liquor sampling. Laboratory analyses were conducted according to standard procedures. Wethers on D10, D20 and D30 recorded 27.9, 26.25 and 36.61% higher fresh matter intake (FMI) than D0 respectively. Their mean DMI were 306.2, 348.1, 349 and 370.7 g/d, respectively. Dry matter digestibility (DMD%) and N intake increased with level of *Commelina benghalensis* in the diet ($P<0.01$ and $P<0.0001$, respectively). Results demonstrated that inclusion of *Commelina benghalensis* in *Sorghum almum* diet improved intake, digestibility and N intake, suggesting its potential for used as low quality feeds supplement.

---

no. 32

The Ankole cattle in Uganda: productivity and morphology in three production systems

D.R. Kugonza[1], A.M. Okeyo[2], M. Nabasirye[1] and O. Hanotte[2], [1]Makerere University, Faculty of Agriculture, P.O. Box 7062, Kampala, Uganda, [2]International Livestock Research Institute, P.O. Box 30709, Nairobi 0100, Kenya

Phenotypic characterisation is critical in breed improvement and conservation. To determine the performance and morphological features of Ankole cattle in three livestock production systems (LPS) of Uganda, 248 farms were surveyed. Live weight and six body measurements were taken on 35 bulls and 70 cows. The measurements were height at withers (HW), heart girth (HG), body length (BL), ear length (EL), horn length (HL) and horn spacing (HS). Data was analysed using LPS, county and sex. Mean age at sexual maturity (ASM) was 23.6 months for bulls and 22.7 months for cows. Age at first calving (AFC) was 33.2 months, while calving interval (CI) was 12.9 months. Lactation length significantly differed between LPS ($P<0.01$) and counties ($P<0.001$), unlike ASM, AFC and CI. Mean daily milk off take was 2.2 kg/cow while calf survival rate at weaning was 90%. Sex and LPS significantly influenced HW, HL and HS. Positive correlations were also observed between BW and HG, BL and HL; while those between BW and EL, and BW and HS were high but negative. Correlation coefficients in females were generally lower than in males. Results show wide variations within this breed both in performance and morphology. Selection, preferably in an open nucleus scheme would enable Ankole cattle breed improvement.

Factors influencing herd sizes, breed preference and production system of pig genetic resources from smallholder farmers in South Africa

T.E. Halimani[1], P.M. Phitsane[2], B.J. Mtileni[2], F.C. Muchadeyi[1], M. Chimonyo[3] and K. Dzama[1], [1]Stellenbosch University, Animal Sciences, P Bag X1, Matieland, 7602, South Africa, [2]Agricultural Research Council, Livestock Production, ARC Livestock Business Division, Private Bag X2, Irene, 0062, South Africa, [3]University of Fort Hare, Livestock and Pasture Science, University of Fort Hare, Post Bag X1314 Alice, 5700, South Africa

Factors influencing pig numbers, breeds and production systems were investigated in 12 South African smallholder farming communities from Limpopo and Eastern Cape. A total of 122 questionnaires were administered. Sex and marital status of the head of household and village did not influence pig numbers, sow and boar numbers, the rank of the pigs in the production system, causes of mortality, the litter size and marketing channels ($P>0.05$). Breed was dependent on village ($P<0.01$) with 40% of the households in Limpopo having crossbred pigs. The annual income was the main determinant of the number of sows per household, number of weaners and pig housing ($P<0.01$). Farmers with incomes above R12000.00/year had higher number of sows and 55% of them housed their pigs compared to the low income group (30%). The higher income group had more crossbred pigs and higher farrowings per year ($P<0.01$). The main reasons for keeping pigs were also dependent on income with more low income farmers raising pigs for sale ($P< 0.01$) and home consumption ($P<0.05$). It was concluded that the main determinant of pig production parameters in smallholder farmers is income.

---

no. 34

Effect of carob pods by-products on performance of goat

M.A.S.L. Mohamed[1], I.S. Melad[1] and A.A. Abdel-Ghani[2], [1]Fac. of Agric., Omar Al-Mukhtar University, Animal Production, Al-Beida, 218, Libyan Arab Jamahiriya, [2]Fac. of Agric., Minia University, Animal Production, Minia, 61111, Egypt

The aim of this study was to determine the effects of diets containing 15% or 30% of carob pods by-products in replacing similar levels of yellow corn on the performance of local kids. Nine of male goats at approximately 7 months of age and initial body weight (16±0.06 kg) were divided according to weight to three groups (3 each). Growth trial was lasted 60days. Results showed that carob seeds contain high percentage of crude protein, about three times as the protein in carob pods by-products and about five times as in deseeded pods (26.3, 8.4 and 5.3%) respectively. The percentages of crude fiber was 10.8, 10.4 and 11.6 for carob seeds, carob pods by-products and deseeded pods respectively. Total tannins in carob pods were two fold of that in carob pods by products (9.0% vs. 4.3%). Results cleared that the total weight gain (kg) as higher ($P<0.01$) for control and 15% carob pods by-products treatments than that for 30% carob pods by-products treatment (9.8, 9.0 and 8.1 respectively. The same trend was observed for the feed conversion (5.5, 5.9 and 6.7) respectively. This study concluded that carob pods by-products can be used in goat feeding up to 30% of the ration without any detrimental effect and this percentage was the least cost ration for goat.

Genetic analysis of carcass characteristics and indicator traits recorded using ultrasound

M.D. Macneil[1] and S.L. Northcutt[2], [1]USDA – Agricultural Research Service, 243 Fort Keogh Rd., Miles City, MT 59301, USA, [2]American Anugus Association, 3201 Frederick Av., Saint Joseph, MO 64506, USA

Our objective was to evaluate relationships of indicators of carcass merit obtained using ultrasound with carcass traits of steers. Records of intramuscular fat content (IMF), longissimus muscle area (uLMA), and subcutaneous fat depth (SQF) derived from ultrasonic imagery, and live weight from 33,857 bulls, 33,737 heifers, and 1,805 steers were used with 38,296 records of marbling (MRB), fat depth at the 12th-13th rib interface (FD), carcass weight (CW), and longissimus muscle area (cLMA) from harvested steers. Analyses were conducted using ASREML. Heritability estimates were 0.45±0.03, 0.34±0.02, 0.40±0.02, and 0.33±0.02 for MRB, FD, CW, and cLMA, respectively. Genetic correlations of carcass measures from steers with ultrasonic measures from bulls and heifers indicated sex-specific relationships for IMF (0.66±0.05 v 0.52±0.06) and uLMA (0.63±0.06 v 0.78±0.05), but not for weight (0.46±0.07 v 0.40±0.07) or SQF (0.53±0.06 v 0.55±0.06). Estimates of genetic correlations between traits of bulls and heifers measured using ultrasound were > 0.8. Prototype national cattle evaluations (NCE) conducted using estimated genetic parameters resulted in modest re-ranking of sires relative to previous analyses. Unified NCE for carcass traits using data from harvested animals and ultrasonic imagery of seedstock appropriately weights this information and provides breeders estimates of genetic merit for traits in their breeding objectives.

---

no. 2

Improving the quality of wool through the use of genetic markers

T.O. Itenge[1], J.G.H. Hickford[2] and R.H. Forrest[2], [1]University of Namibia, Animal Science, P/Bag 13301, 9000, Namibia, [2]Lincoln University, Agriculture and Life Sciences, P.O. Box 84, 7647, New Zealand

Consistency of wool fibre traits is extremely important to the wool industry. Although selective breeding based on quantitative genetic approaches has been practised for many years in order to improve the quality of wool, considerable variation still exists in wool traits. This study aimed at identifying gene markers associated with wool quality traits. Genes that code for the keratin intermediate-filament proteins and the keratin intermediate-filament-associated proteins were targeted for the investigation. Polymerase chain reaction-single strand conformational polymorphism (PCR-SSCP) analysis was used to identify sequence variation in the KAP1.3, KAP3.2, KAP8 and KRT1.2 genes, whereas PCR-agarose gel electrophoresis was used to detect length polymorphism in the KAP1.1 gene. The study has identified potential gene markers that allow selection for increased staple length, increased staple strength, higher yield, whiter and brighter wool. Results also show strong evidence to suggest that keratin genes on chromosome 11 are recombining very frequently at recombination "hotspots". A high recombination rate among loci that affect wool means that breeding for consistent wool quality may be very difficult, and therefore development of genetic markers linked to wool quality traits may be important to aid with breeding for consistent wool quality.

no. 3

Economic values for dairy production traits under different milk payment systems in South Africa

C.B. Banga[1], F.W.C. Nesser[2], D.J. Garrick[3] and J. Van Der Westhuizen[1], [1]ARC, Animal Production Institute, P/Bag X2, Irene 0062, South Africa, [2]University of the Free State, Department of Animal Science, P O Box 339, Bloemfontein 9300, South Africa, [3]Iowa State University, Department of Animal Science, Ames, IA 50011, USA

Economic values of milk production (milk volume and yields of fat and protein), live weight and longevity were derived for South African Holstein and Jersey cattle, based on milk payment systems of the four major milk buyers in South Africa. A bio-economic herd simulation model was used to calculate economic values by accounting for unit changes in marginal returns and costs arising from an independent unit increase in each trait. Economic values for milk production traits varied substantially among the payment systems; particularly protein yield which ranged from 6.31 to 21.06 ZAR/kg. Fat yield had a negative economic value (-1.10) under one payment system, for the Jersey breed. Payment systems that do not pay for volume resulted in negative economic values for milk volume. The economic value of live weight was constant across payment systems and was, respectively, -7.12 and -9.15 ZAR/kg for the Holstein and Jersey breeds. Payment system had an insignificant effect on the economic value of longevity, which averaged 5.40 ZAR/day in both breeds. Economic values of all traits were sensitive to the price of feed. Relative emphasis of production traits in the breeding objectives for South African dairy cattle should take cognisance of the diversity in milk payment systems.

---

no. 4

Expression of trypanotolerant quantitative trait loci (QTL) in cattle population under natural tsetse and trypanosomosis challenge

C.O. Orenge[1,2], C.N. Kimwele[2], A.B. Korol[3], S. Kemp[4], L.K. Munga[1], J.P. Gibson[5], O. Hanotte[4], G. Murilla[1] and M. Soller[6], [1]Kenya Agricultural Research Institute, Trypanosomosis Research Centre, P.o. Box 362, 00902 Kikuyu, Kenya, [2]University of Nairobi, Department of Veterinary Anatomy and Physiology, P.o. Box 30197, 00100 Nairobi, Kenya, [3]Haifa University, Institute of Evolution, Mount Carmel, 31905 Haifa, Israel, [4]International Livestock Research Institute, Biotechnology theme, P.o. Box 307907, 00100 Nairobi, Kenya, [5]University of New England, Institute of Genomics and Bioinformatics, Armidale, NSW 2351, Australia, [6]Hebrew University of Jerusalem, Department of Genetics, Jerusalem, 91904, Israel

African Animal Trypanosomosis poses major constraints to increasing livestock production. Recent studies in cattle artificially challenged have identified trypanotolerant quantitative trait loci (QTL). However, expression of these QTLs under natural tsetse and trypanosomosis challenge remains to be confirmed. This work reports on the first ever experiment designed to authenticate this. 192 backcrosses, 37 F1 and 25 Boran were exposed to high tsetse challenge and Body weight, Parasitemia, Packed cell volume (PCV) and 25 other phenotypes measured to asses their trypanosomosis vulnerability. F1 was the most trypanotolerant followed by the BC and least was the Boran. This confirms expression of trypanotolerant QTLs under challenge through F1 and BC, thus paving way to utilization of MAS and MAI for further trypanotolerance improvement.

Parental breed effects on carcass traits of meat goat kids from a three-breed diallel

R. Browning, Jr.[1], W. Getz[2], O. Phelps[3] and C. Chisley[4], [1]Tennessee State Univ., Nashville, TN 37209, USA, [2]Fort Valley State Univ., Fort Valley, GA 31030, USA, [3]USDA Agric. Marketing Service, Lakewood, CO 80401, USA, [4]Southern Univ., Baton Rouge, LA 70813, USA

Pasture-raised buck kids (n=275) from a complete diallel of Boer (B), Kiko (K), and Spanish (S) sires and dams were harvested at 33 weeks old. Through visual scoring, live muscling grades favored (P<0.01) kids of B and S sires over K sires. Cold carcass muscling grades favored (P<0.01) B-sired kids over K- and S-sired kids. Live grades of kids favored (P<0.01) B dams over K and S dams. Dam breed did not affect carcass grades. Sire breed did not affect live, carcass or cut weights. Live and carcass weights were heavier (P<0.01) for kids of K dams (26.5±0.6 kg; 11.1±0.3 kg) than from S (26.2; 10.2) and B dams (23.7; 9.6). Cold carcass dressing percent was lower (P<0.01) for kids of B sires and dams (40.3±0.4%; 40.4±0.4%) than from K (41.8; 41.9) and S sires and dams (41.8; 41.5). Kids of K dam had heavier (P<0.01) forelegs, hind legs, and combined boneless right fore- and hind legs (3.26±0.08 kg; 3.67±0.09 kg; 1.95±0.05 kg) than S (2.91; 3.37; 1.75) and B dams (2.76; 3.24; 1.68). Parental breed did not affect lean content (69%) of leg cuts. Loins were lighter (P<0.01) for kids of B dams (1.37±0.05 kg) than from K (1.62) and S dams (1.52). Sire and dam breeds interacted (P=0.01) for kidney-pelvic fat weight as straightbred kids (74±16 g) differed from all four crossbred kid groups (127 to 164) only within B matings. Dam breed consistently affected carcass traits.

---

no. 6

Association of single nucleotide polymorphisms in apoVLDL-II gene with serum biochemical components

H. H. Musa[1,2], G. H. Chen[3] and G. Q. Zhu[1], [1]Yangzhou University, College of Veterinary Medicine, College of Veterinary Medicine, Yangzhou University, Yangzhou, 225009, China, 225009, China, [2]University of Nyala, Faculty of Veterinary Science, Faculty of Veterinary Science, University of Nyala, Nyala, 155, Sudan, 249, Sudan, [3]Yangzhou University, College of Animal Science and Technology, College of Animal Science and Technology, Yangzhou University, Yangzhou, 225009, China, 225009, China

ApoVLDL-II is a major constituent of very low density lipoprotein (VLDL) involved in lipid transportation. Blood samples were collected from Anka and Rugao chicken breeds for DNA and serum components analysis. Total cholesterol, triglycerides and high density lipoprotein (HDL) were assayed by enzymatic kit. Very low density lipoprotein (VLDL) and low density lipoprotein (LDL) were estimated according to Friedewald equation. The polymorphism of apoVLDL-II gene in each chicken was studied by PCR-RFLP, PCR-SSCP and gene sequence methods. Single nucleotide polymorphisms (SNPs) were detected at 1950bp (G/A), 2532bp (T/A), 2706bp (A/C) and 2793bp(C/T). The results showed that breeds were significantly differed in SNPs locus. SNP (A/C) at 2706bp was significantly (P<0.05) associated with triglycerides and very low density lipoprotein levels, whereas SNP (C/T) at 2793bp was significantly (P<0.05) associated with total cholesterol and high density lipoprotein levels. Our findings suggested apoVLDL-II gene as a potential candidate for serum biochemical components.

Genetic analysis of fleece traits in South African Angora goats

C. Visser[1], M.A. Snyman[2] and E. Van Marle-Köster[1], [1]University of Pretoria, Department of Animal and Wildlife Sciences, 0002, South Africa, [2]Grootfontein Agricultural Development Institute, Middelburg (EC), 5900, South Africa

The objectives of this study were to determine the most appropriate models of analysis for fleece traits of South African Angora goats and to estimate variance components and genetic parameters for each of these traits. The data used for this study were collected on 6221 kids born between 2000 and 2006 in 11 different Angora goat studs. The number of sires and dams with progeny in the data set were 302 and 3602 respectively. Variance components and genetic parameters were estimated with the ASREML program for fleece weight (FW; kg), fibre diameter (FD; kg), coefficient of variation of fibre diameter (CVFD; %), standard deviation of fibre diameter (SDFD; µm), comfort factor (CF; %), spinning/effective fineness (SF; µm) and standard deviation of fibre diameter along the length of the staple (SDA; µm). Models including only direct additive genetic effects were the most appropriate for all fleece traits analysed. Direct additive heritability estimates of 0.24±0.03, 0.45±0.03, 0.37±0.10, 0.32±0.11, 0.63±0.10,0.61±0.10 and 0.14±0.08 were obtained for FW, FD, CVFD, SDFD, CF, SF and SDA respectively. Genetic and phenotypic correlations among the traits were also estimated with multivariate analyses. Information generated by this study will be applied for the estimation of breeding values for fleece production of animals in the SA Angora goat industry and in a QTL study for fibre diameter and related fleece traits in SA Angora goats.

---

no. 8

Genetic analysis of body weight and growth traits in South African Angora goats

M.A. Snyman[1], C. Visser[2] and E. Van Marle-Köster[2], [1]Grootfontein Agricultural Development Institute, Middelburg (EC), 5900, South Africa, [2]University of Pretoria, Department of Animal and Wildlife Sciences, 0002, South Africa

The objectives of the study were to determine the most appropriate models of analysis for body weight and growth rate at different ages and to estimate (co)variance components and genetic parameters for each of these traits. The data consisting of 13825 kid records, the progeny of 339 sires and 5370 dams, were collected on the 2000- to 2004-born kids of 11 different Angora goat studs. Variance components and genetic parameters for birth weight (BW), weaning weight (WW), body weight at 8, 12 and 16 months of age (W8, W12 and W16), daily growth rate from birth to weaning (GRb-w), weaning to 8 months (GRw-8), 8 to 12 months (GR8-12) and 12 to 16 months of age (GR12-16) were estimated with the ASREML program. Direct additive heritability estimates of 0.18±0.03, 0.13±0.02, 0.11±0.02, 0.20±0.04 and 0.34±0.06 were obtained for BW, WW, W8, W12 and W16 respectively. Maternal heritabilities of 0.14±0.05, 0.05±0.04, 0.10±0.02, 0.08±0.03 and 0.07±0.03 were estimated for the respective traits, while maternal environmental effects of 0.11±0.03 and 0.14±0.04 were estimated for BW and WW respectively. This study demonstrated the importance of implementing the correct model of analysis for the estimation of (co)variance components and genetic parameters for body weight and early growth traits in Angora goats. The importance of a sufficiently structured and related pedigree structure is highlighted.

no. 9

Use of crossbreeding for increasing domestic beef production

M. Ben Salem[1] and H. Kélifa[2], [1]INRA, Rue Hédi Karray, 2049 Ariana, Tunisia, [2]OEP, Tunis, 1002 Tunis, Tunisia

Beef in Tunisia comes from native cows, exotic Holsteins and their respective crossbreds. However, given the limited meat potential of these animals, beef production does not meet the total demand. Therefore efforts have been directed to identifying appropriate crossing breeds for increased production. First, exotic dual purpose breeds such as the Brown Swiss and the Tarantaise were introduced to determine their possible use in crossbreeding with dairy animals. However, although some improvement was made in meat production, crossbreeding with these breeds still did not help in achieving the desired production level. Meanwhile, the improved revenue of households have served to emphasize the need for more beef production. Recently, alternative crossbreeding with beef breeds was implemented. The objective of this work is to evaluate, trough field studies, the potential of using the White Blue Belgium and the Charolais breeds in industrial cross breeding programs for increasing beef production. Data on breeding, birth weight (BW), calving ease, average daily gain (ADG) and carcass weight (CW) of crossbreds were collected for the 2000-2006 period. A total of 467 producers and 24000 cows were included in the study. Results showed that BW, ADG and CW increased by 8kg, 150g and 30kg, respectively. Carcass quality and market value were also improved. This suggests that crossbreeding with beef breeds is one of the successful methods which can be used in the country to reach self-sufficiency in domestic beef and improve farmers' revenue.

---

no. 10

Opportunities to breed sheep meat with healthy fats

J.C. Greeff[1], M. Harvey[1], S. Kitessa[2] and P. Young[2], [1]Department of Agriculture and Food Western Australia, 3 Baron Hay Court, 6151 South Perth, Australia, [2]CSIRO, Livestock Industries, Wembley, 6913 Perth, Australia

The potential to change the fatty acid profile of sheep meat was studied by estimating the heritability of the different fatty acids and the genetic and phenotypic correlations between the individual fatty acids. The heritability estimates varied from zero for C18:3n6 to 0.50 for C16:0. The heritability of the three omega-3 fatty acids, ALA, EPA, and DHA were 0.11±0.061, 0.18±0.067 and 0.19±0.075. The genetic correlations show that there were favourable and unfavourable relationships between the different fatty acids. C16:0 were generally favourably correlated with the unsaturated C18, C20 and C22 fatty acids while C18:0 moderately unfavourably correlated with these unsaturated fatty acids. However, the results indicate that it should be possible to breed sheep with higher amounts of omega-3 fatty acids.

no. 11

Genetic improvement of communal wool sheep in the Eastern Cape

P.G. Marais[1], D. Swart[2] and B.R. King[1], [1]Grootfontein Agricultural Development Institute, P/bag X529, 5900 Middelburg (EC), South Africa, [2]ARC-Animal Production Institute, P/bag X529, 5900 Middelburg (EC), South Africa

Reciprocal progeny tests were done in communal and commercial areas to evaluate the genetic improvement of communal wool sheep *in situ*. Four groups of 100 commercial ewes each from 4 group breeding localities were mated to 6 commercial rams and 4 ewe groups were mated to 6 communal rams, respectively. Concurrently 4 groups of communal ewes from 4 communal grazing villages each were mated to 6 communal rams, while 4 groups of ewes were mated to improved commercial rams. After mating, the ewe flocks remained in their communal or commercial farming areas. Weaning weight, growth performance and wool traits were measured for the progeny of each group. Weaning weight and 6-month body weights of lambs bred from commercial ewes x commercial rams were higher ($P<0.05$) than progeny from communal rams. No differences occurred between the progeny of communal ewes sired by communal or commercial rams. Wool traits of communal offspring did not differ between commercial and communal rams, except for greasy fleece weight which was higher ($P<0.05$) for commercial rams. Clean wool income was 54% higher for commercial offspring sired by commercial rams and 46% higher for communal offspring sired by commercial rams, as compared to their communal sired counterparts ($P<0.05$). The introduction of commercially improved rams resulted in substantial genetic improvement of communal wool sheep. Communal grazing conditions limited the expression of growth potential.

---

no. 12

Association study of the PRL and MSTN gene polymorphisms with milk performance traits in Latvian Brown cattle breed

A. Jemeljanovs[1], I. Zitare[1], N. Paramonova[2], I. Poudziunas[2], L. Paura[3], V. Sterna[1] and T. Sjakste[2], [1]Research Institute of Biotechnology and Veterinary Medicine, 1 Instituta Street, LV-2150 Sigulda, Latvia, [2]Institute of Biology of the University of Latvia, 3 Miera Street, LV-2169 Salaspils, Latvia, [3]Latvia University of Agriculture, 2 Liela Street, LV-3001 Jelgava, Latvia

Prolactine is a hormone with multiple functions including expression of milk protein genes. Myostatin, transforming growth factor, appears to play an important role in reproductive physiology and animal welfare. The aim of this association study was to genotype PRL gene microsatellite (MS) and cSNP $A^{7490}\rightarrow G$, and MSTN gene MS in 100 animals of Latvian Brown cattle breed. PRL MS locus was presented by common 160 bp and rare 155 bp alleles with common genotypes of 160/160 and 155/160. Animals with genotype 160/160 showed the highest milk yield, fat and protein content when compared with animals of 155/160 genotype. Alleles G^{7490} and A^{7490} were detected as common and rare correspondingly. Heterozygous GA animals showed highest milk yield. MSTN MS locus was presented by two L and M alleles of different MS motif (frequency 0.82 and 0.18). Repeat number polymorphism was observed within both MS motifs. 4 animals of MM genotype showed highest milk yield, fat and protein content comparing with animals of LL and LM genotype. Results obtained in this study confirmed potential usefulness of PRL and MSTN genes polymorphisms in genetic improvement of milk performance traits. The analysis should be expanded on a larger animal population.

Genetic change in Bonsmara cattle

H.E. Theron, K.A. Nephawe, A.N. Maiwashe and F. Jordaan, ARC-Animal Production Institute, P/Bag X2, Irene, 0062, South Africa

The indigenous Bonsmara beef cattle breed currently has around 100,000 active animals in 300 herds. It is the breed with the highest number of performance tested animals in South Africa. Performance recording data has been collected on farm and central testing centers since the inception of the breed. All pedigrees are known (1,25 million animals). Performance data was electronically captured since the early 1980's (760,000 weaning weight records) and breeding values became available in 1998. Estimated breeding values are obtained from five different multivariate models. Weaning weight is included in all the models for post-weaning traits to account for selection. Heritability estimates for reported traits are as follows: birth weight direct (0.29) and maternal (0.09); weaning weight direct (0.17) and maternal (0.11); yearling weight (0.37); 18-month weight (0.41); mature cow weight (0.43); average daily gain (ADG) (0.34), daily feed intake (0.36), shoulder height (0.51), scrotal circumference (0.46); body length (0.34) and feed conversion ratio (FCR) (0.31). Farmers select primarily for weaning weight. Wean direct, twelve and eighteen month weights, ADG, FCR, scrotum circumference and body length has constantly improved while wean maternal has only increased slightly. Since the mid 1990's, birth weight direct, mature weight and shoulder height have stabilized. This indicates that Bonsmara breeders were successful in breeding more efficient animals, while body size is limited for adaptation to the South African environment.

---

no. 14

Estimation of genetic parameters of test day fat and protein yields in Brazilian Holstein cattle using an autoregressive multiple lactation model

C.N. Costa[1], J.G. Carvalheira[2], J.A. Cobuci[3], A.F. Freitas[1] and G. Thompson[2], [1]Embrapa Gado de Leite, Research & Development, Rua Eugenio do Nascimento, 610, 36038-330 Juiz de Fora, MG, Brazil, Brazil, [2]University of Porto, 2Research Center in Biodiversity and Genetic Resources (CIBIO-ICETA), Rua Padre Armando Quintas-Crasto, 4485-661 Vairão, Portugal, [3]Universidade Federal do Rio Grande do Sul, Departamento de Zootecnia, Av. Bento Goncalves, 7712 São José, 91540-000 Porto Alegre RS, Brazil

In the scenario of dairy cattle genetic selection, the possibility to improve breeding value estimation is driving the research aimed at replacing the use of cumulative lactation records by test day (TD) measurements. This study was aimed to estimate variance components and genetic parameters for daily fat and protein yields of Brazilian Holstein cattle, using an autoregressive test day multiple lactations (AR) animal model. Data consisted of test day records produced by Holstein cows under milk recording supervised by the Brazilian Holstein Association, calving from 1993 to 2004. Medium to high heritability estimates (from 0.18 to 0.30 and from 0.30 to 0.43 for fat and protein TD yields, respectively) suggest opportunities for larger genetic gain by selection.Results from this study confirm the potential of using TD yields to replace the lactation model to estimate breeding values of Holstein cows in Brazil. Further studies are needed to compare these results with other modeling approaches e.g., the RR model.

no. 15

Developmental adaptive reaction norm estimates using a multiple random regression model

N.T. Pegolo[1], L.G. Albuquerque[2], R.B. Lobo[1], L.A.F. Bezerra[1] and H.N. Oliveira[1,2], [1]University of Sao Paulo, Genetics, Av. Bandeirantes, 3900 Ribeirão Preto SP, 14049-900, Brazil, [2]Sao Paulo State University – UNESP, Animal Breeding and Nutrition, Caixa Postal 560, Botucatu SP, 18618-000, Brazil

Developmental reaction norm (DRN) is considered an ideal framework for the understanding of evolution of phenotypes through the elucidation of the developmental programs involved in the epigenetic transition from genotype to phenotype. Here, we joined a Quantitative Genetics viewpoint isolating the additive genetic component of one character, in an adaptive DRN (ADRN). We used 462,513 progeny weights from Brazilian Nelore cattle sires to estimate the covariance function (CF) and predict the ADRNs using a random regression model with multiple control variables. The model considered linear Legendre polynomials along an environmental gradient and cubic ones along an age gradient. We separated male progenies data analyses (MP) from female ones (FP). Correlations among FC coefficients indicated important connections among ADRN attributes, as between plasticity and growth curve slope in males. The MP heritability surface showed lower levels (between 0.11 and 0.39), with a depression in unfavorable environments. In FP analysis, heritabilities were higher, with a limited depression in a dislocated position, confirming a genotype-environment-age-sex interaction. The methodology permitted the prediction of phenotypic selection acting in the CF coefficients and correlated response along all environment and development gradients.

---

no. 16

Multi-trait and random regression mature weight breeding value estimates in Nelore cattle

L.G. Albuquerque[1], A.A. Boligon[1], M.E.Z. Mercadante[2] and R.B. Lôbo[3], [1]FCAV/UNESP, Via de Acesso Prof. Paulo Donato Castellani, 14884-900, Brazil, [2]IZ, Sertãozinho, 13460-000, Brazil, [3]USP, Ribeirão Preto, 14049-900, Brazil

Mature weight breeding values were estimated using a multi-trait animal model (MM) and a random regression animal model (RRM). Data consisted of 82,064 weight records from 8,145 Nelore females, recorded from birth to 8 years of age. Weights at standard ages were considered in MM. All models included contemporary groups as fixed effects, and age of dam (linear and quadratic effects) and animal age, as covariates. In the RRM, mean trends were modeled through a cubic regression on orthogonal polynomials of animal age and genetic maternal and direct and maternal permanent environmental effects were also included as random. Legendre polynomials of orders 4, 3, 6 and 3, were used for animal and maternal genetic and permanent environmental effects, respectively, considering 5 classes of residual variances. Mature weight (5 years) direct heritability estimates were 0.35 (MM) and 0.38 (RRM). Rank correlation between the sires breeding values estimated by MM and RRM was 0.82. However, selecting the top 2% (12) or 10% (62) of the young sires based on the MM predicted breeding values, 71% and 80% of the same sires, respectively, would be selected if RRM estimates were used instead. The RRM modeled the changes in the (co)variances with age adequately and larger breeding value accuracies can be expected with this model. Financial support from CNPq and FAPESP.

Estimation of genetic (co)variance functions for weights from birth to adult age in Nelore cows using random regression models

A.A. Boligon[1], L.G. Albuquerque[1], M.E.Z. Mercadante[2] and R.B. Lobo[3], [1]FCAV-UNESP, 14884-900, Jaboticabal SP, Brazil, [2]IZ, 13460-000, Sertãozinho SP, Brazil, [3]FMRP-USP, 14049-900, Ribeirão Preto SP, Brazil

A total of 82,064 weight records from 8,145 Nelore females, measured from birth to adult age, were used to estimate (co)variance functions, using random regression on Legendre polynomials of age. The models included direct and maternal additive genetic effects and animal and maternal permanent environmental effects as random, and fixed effects of contemporary group (herd, year and month of birth and of recording and sex of calf), and age of cow at calving (linear and quadratic effect) and orthogonal polynomials of animal age (cubic regression) as covariables. Residual effects were modeled using 1 to 5 classes of variances. Orders of polynomial fit varied from 1 to 6. A model including fourth- and sixth-order polynomials for direct additive genetic and animal permanent environmental effects, respectively, and third-order polynomials for maternal genetic and permanent environmental effects, and with 5 classes for residuals, was found to be the best according to the criteria adopted. Heritability estimates of direct genetic effects ranged from 0.32 to 0.48. Maternal heritability estimates were higher close to 240 days of age. Correlation estimates between weights from birth to 8 yr of age were positive and decreased with increasing distance between ages. Financial support from CNPq and FAPESP

---

no. 18

Incidence of dark skin spots and pigmentation in commercial Corriedale flocks

J.I. Urioste, F. Penagaricano, R. Lopez, J. Laporta, F. Llaneza, C. Lafuente and R. Kremer, Univ. de la Republica, 18 de Julio 1968, Montevideo, Uruguay

The presence of pigmented fibres (PF) in Uruguayan wool prevents access to high quality markets. One strategy of PF reduction is by genetics means. Pigmentation traits are thought to be genetically correlated with PF. Population studies are needed to evaluate their variation. The aim of this study was to assess the incidence of dark skin spots and pigmentation scores of mouth and hooves of sheep from 13 randomly sampled commercial Corriedale flocks. During 2006-2007 data from 2478 animals (10-15% of each flock) with different age (1-6 yr) and genetic origins were taken at shearing. Pigmentation in non-fleece areas was assessed using a subjective scale of 1(light) to 5 (dark). Spots were identified in fleece areas and percentage PF on its surface determined using a subjective scale of 0 to 5. Distribution of mouth pigmentation scores was symmetric and unimodal, whereas it was skewed to high score values in hooves. In both cases, a reduction in pigmentation with age of animals was observed. 62% of animals presented spots; 20% had spots without PF but 19% showed spots with more than 40% PF. Presence of spots, particularly those with high percentage of PF, increased with the age of animals. Important differences between flocks were detected for all pigmentation traits ($p<0,0001$). This first population study on pigmentation traits of Corriedale sheep showed that there is an important phenotypic variation among animals and flocks, suggesting the existence of an underlying genetic variation.

no. 19

Association between polymorphism of bovine lactoferrin gene and somatic cell count in milk

G. Sender, A. Korwin-Kossakowska, M. Sobczyńska and J.M. Oprządek, Institute of Genetics and Animal Breeding, Jastrzebiec, 05-552 Wolka Kosowska, Poland

Currently, somatic cell count (SCC) in cow milk is the primary trait used to evaluate susceptibility to mastitis. Identifying candidate genes could be used to introduce selection for resistance to mastitis into breeding programs. In this study the gene coding for bovine lactoferrin is targeted as a candidate gene, because of its role in immune response during udder infections. The objective of this study was to identify an association between polymorphisms at the bovine lactoferrin locus and SCC in milk of Polish dairy cattle. Lactoferrin alleles were determined according the PCR-RFLP method. The genotyping of 479 Holstein cows from two Polish dairy herds was carried on. Statistical analyses were performed on test day SCC using the GLM analysis of variance within the SAS software. The results showed that the significantly ($P<0.01$) lowest test-day SCC characterized cows with lactoferrin genotype BB compared to AA and AB animals. The variance components for test-day SCC were estimated by the REML method using animal model. The estimated breeding value (EBV) of SCC was calculated according to the BLUP method. Linear contrast between model adjusted EBV of SCC for all animals was used to test for differences between genotypes. In Polish Holstein cows population no significant association was observed between lactoferrin genotypes and EBV of SCC.

---

no. 20

Some important random and non-random effects affecting growth of two beef cattle breeds in Botswana

K. Raphaka[1] and K. Dzama[2], [1]Agricultural Research, P/Bag 0033, Gaborone, Botswana, [2]Stellenbosch University, P/Bag X1, Matieland 7602, South Africa

Over 8000 records of Tswana (5923) and Botswana Composite (2257) cattle breeds collected between 1988 and 2006 were analysed for important fixed and random effects. It was found that non-genetic effects of breed of calf, sex of calf, month and year of birth, previous parous state, weight of cow at parturition, age of dam, and age of calf at weaning significantly affected BW, WW, 18MW, pre-weaning average daily gain (ADG1) and post-weaning average daily gain (ADG2) in both breeds. The Composite breed had higher BW, ADG1 and WW whereas the Tswana had higher ADG2 and 18MW. Single-trait analysis was used in the estimation of (co)variance components by fitting an individual animal model (AM) and the animal maternal model (AMM) for the two breeds. Direct heritabilities for BW, WW, ADG1, 18MW and ADG2 in the Tswana were 0.45, 0.32, 0.37, 0.31 and 0.31, respectively from the AM analysis. Fitting the AMM resulted in direct heritabilities of 0.31, 0.20 and 0.16 for BW, WW and ADG1, respectively, while the maternal heritabilities were 0.11, 0.15 and 0.21, respectively. For the Composite the direct heritabilities for BW, WW and ADG1 were 0.58, 0.32 and 0.30, respectively with single-trait AM. Partitioning using the AMM resulted in the direct heritabilities for BW, WW and ADG1 of 0.55, 0.17 and 0.14, respectively, while corresponding maternal effects were 0.09, 0.15 and 0.15, respectively. The genetic correlations between direct and maternal effects were positive and ranged from 0.20 to 0.89.

Use of random amplified polymorphic DNA (RAPD) markers for detecting genetic similarity between and within Zulu (Nguni) sheep populations

N. Kunene[1], C. Bezuidenhout[2] and I. Nsahlai[3], [1]University of Zululand, P/b X1001, Kwadlangezwa,3886, South Africa, [2]North West University, P/b X60001, Potchefstroom,2520., South Africa, [3]University of Kwazulu-Natal, Private bag X01, Scottsville,3200, South Africa

The objective of this study was to assess genetic relationships among three populations of Zulu sheep (*O. aries*) using RAPD markers. The sheep were at Makhathini, Kwamthethwa community and the University of Zululand (UNIZULU) in Kwazulu-Natal, (S.A). The two areas are 250 km and 40 km away, respectively, from UNIZULU. Genetic relationship between this breed, the Merino and Zulu goats (*C. hircus*) was studied for comparison. The DNA samples were isolated from 100 animals, 21 random primers were screened and 6 produced clear reproducible results. 1654 scorable bands were generated of which 824 were polymorphic. Estimation of genetic relationship for the data comprised of *C. hircus*, Merino and Zulu sheep revealed three distinct groups: one which consisted of the goat as an out-group, one of Zulu sheep with subgroups of the community and the UNIZULU sheep and the third group was the Makhathini sheep which formed a cluster with the Merino. The second set of data consisting of Zulu sheep only, formed three main clusters with each cluster depicting each population. The genetic similarity within each population of sheep ranged from 77% to 95%. It was concluded that RAPD markers can be used as a tool for classifying and estimating genetic similarity among sheep populations, between sheep breeds and between the two species

---

no. 22

Genetic polymorphisms associated to beef tenderness in Nellore cattle

J.B. Ferraz, F.M. Rezende, F.V. Meirelles, J.P. Eler, R.C.G. Silva, M.N. Bonin, M.E. Carvalho and J.C. Balieiro, University of Sao Paulo, Av. Duque de Caxias Norte, 225, 13635-900 Pirassununga, SP, Brazil

Longissimus dorsi samples aged for 7, 14 and 21 days of 630 Nellore young bulls, genotyped for 18 molecular markers (CAPN316, CAPN530, CAPN4751, UOGCAST1, WSUCAST, DGAT, EXON2FB, FABP4, LEPA1457, LEPA252T, LEPA59V, LEPC963T, LEPT945M, TFAM1, TFAM2, UASMS1, UASMS2 and SCD1), were analyzed to verify the association of those SNP markers to beef tenderness. This research is part of a project of validation of genetic markers in Nellore, the most important *Bos indicus* beef breed in Brazil, lead by the Breeding, Genetic and Biotechnology group of the University of Sao Paulo. Beef tenderness varied from 2.30 to 9.74, 1.38 to 8.39 and 1.61 to 7.39 kg for 7, 14 and 21 days of ageing. Polymorphisms associated to CAPN316, FABP4, LEPA252T, LEPA59V, LEPC963T, SCD1, UASMS1 and UASMS2 had one of the alleles only in heterozygosity. Analysis of variance found a significant effect for CAPN4751, DGAT and UASMS2 on TENDER7 ($P<0.05$) and a suggestive effect for LEPA252T and LEPC963T ($P<0.016$). On TENDER14 a significant effect for CAPN4751, DGAT, LEPA252T, UOGCAST and WSUCAST was observed, while a suggestive effect was showed for CAPN316 and UASMS2. CAPN4751, UOGAST and WSUCAST presented a significant influence on TENDER21, and, DGAT and TFAM2 showed suggestive effect. Based on the results, it is possible conclude that besides of molecular marks previously related with beef tenderness, others marks can be have its effect associated with this phenotype in Nellore cattle.

no. 23

Estimative of (co)variance components for growth traits in broilers in Brazil

J.L.B.M. Grosso[1], J.P. Eler[1], J.C.C. Balieiro[1], J.B.S. Ferraz[1], E.C. Mattos[1], A.M. Felício[1] and T. Michelan Filho[2], [1]Faculdade de Zootecnia e Engenharia de Alimentos da Universidade de São Paulo, Grupo de Melhoramento Animal e Biotecnologia, Departamento de Ciências Básicas, Avenida Duque de Caxias Norte, 225, 13635-900, Pirassununga, São Paulo, Brazil, [2]Aviagen do Brasil Ltda., Avenida 23, 1120, 13500-280, Rio Claro, São Paulo, Brazil

The current research was conducted to estimate the (co)variance components and the heritability coefficients for growth traits in broilers. Data of 69,267 chickens from Aviagen do Brasil, company that has a broiler local selection program, were used to estimate genetic parameters for body weight at 7 (BW7), 30 (BW30), 38 (BW38) and 42 (BW42) days of age. (Co)variance components were estimated by restricted maximum likelihood method, using MTDFREML program. Estimates of heritability for direct effect were: 0.26 (BW7), 0.36 (BW30), 0.31 (BW38), and 0.30 (BW42). Maternal heritability estimates were 0.12 for BW7, and 0.04 for traits BW30, BW38 and BW42. The results suggest that analyzed traits can be selected and significant response to selection can be achieved. Maternal effects were important for BW7 and had little effect of BW30, BW38 and BW42.

---

no. 24

Genetic relationships between linear type traits in first lactation of Iranian Holsteins

S. Toghiani, A.A. Shadparvar, M. Moradi Shahrbabak and M. Dadpasand, graduate student(M.S.) of guilan university(Iran), #3 Bastan Alley-Shikhalikhan Lane-Malek St -Esfahan, 8154748741-Iran, 8154748741, Iran

The main objective of this study was to estimate genetic parameters and relationships for 10 linear type traits, 10654 records for type traits of Iranian Holstein cows which were collected during 1980 to 2004 at Animal Breeding Center of Iran, were used. The estimations were performed using restricted maximum likelihood method under an animal model. Variance components were estimated from single-trait analysis using MATVEC software and covariance components were obtained from four-trait analysis using DF-REML software. Linear type traits were Angularity, Stature, Body depth, Rump width, Rear leg side view, Foot angle, Fore udder attachment, Rear udder height, Suspensory ligament and Udder depth Heritability estimates for type traits were low to moderate, from 0.075 for rear leg side view to 0.376 for rump width. Genetic correlations between type traits were estimated from 0.72 for udder depth with fore udder attachment to -0.75 for foot angle with leg side view. The results of this study showed favorable and high genetic correlation between mammary system traits, therefore the response for selection of these traits are directly similar.

no. 25

Genetic and phenotypic parameters for test day milk yield of Friesian cattle in Kenya

T.K. Muasya[1], E.D. Ilatsia[1], T.M. Magothe[2] and M.G. Githinji[1], [1]Kenya Agricultural Resrearch Institute, Animal Production, P.O. Box 25 Naivasha 20117, Kenya, [2]Ministry of Livestock, Livestock Recording Centre, P.O. Box 257, Naivasha, 20117, Kenya

A total of 78493 test day milk records (TDMY) from the first three lactations of 6508 Friesian cows were used to estimate variance components and genetic, phenotypic parameters for TDMY and to determine correlation between TDMY and lactation milk yield (LMY) in the Friesian cattle in Kenya. Variance components were estimated through univariate and multivariate animal models that defined contemporary groups based on herd, year and season of calving (HYSC), or on herd, year and season of test day milk sampling (HYSTD). The models were run using derivative free restricted maximum likelihood procedures. Variance components were influenced by contemporary group and models based on HYSTD resulted in higher additive genetic variances and lower residual variances. Heritability estimates for TDMY based on HYSTD and HYSC were 0.46, 0.32 and 0.18; and 0.17, 0.11 and 0.15 in the first, second and third lactations respectively. Genetic and phenotypic correlations between TDMY and LMY across lactations were high and positive and ranged from 0.65 to 0.99 and 0.45 to 0.89 respectively. The high genetic correlations indicate that the two traits are influenced by similar genes and that TDMY records can be used in genetic evaluations of the Friesian cattle breeding programme.

---

no. 26

Genetic analysis of reproduction traits in Bonsmara beef cattle

R.R. Van Der Westhuizen, Agricultural Research Council, Privatebag X2, Irene, 0062, South Africa

Although reproduction is economically the most important trait in beef cattle production, limited selection tools for reproduction are available in most countries. The aim of this study was to estimate the genetic parameters for and between age at first calving (AFC) and inter calving period (ICP) using data on South African Bonsmara cattle. Two separate analyses were conducted. The first analysis was a multi-trait analysis between AFC and ICP with ICP treated as a repeatable trait. The number of records were 19 731 for AFC and 34 849 for ICP. Heritability estimates for AFC and ICP were 0.25 and 0.03, respectively. The genetic and phenotypic correlations between them were -0.50 and -0.21, respectively. In the second analysis ICP between different calvings were treated as different traits (ICP1 is the ICP between the 1st and 2nd calving; ICP2 between the 2nd and 3rd calving ect.). Estimates of heritability for AFC, ICP1, ICP2, ICP3 and ICP4 were 0.23, 0.09, 0.12, 0.10 and 0.04, respectively. The genetic correlations between ICP1 to 4 ranged from 0.56 (between ICP2 and 3) to 0.97 (between ICP2 and 4) while the phenotypic correlations were all below 0.05. The genetic correlation between AFC and ICP1 to 4 ranged from -0.12 to -0.21. Results from the two analysis indicates that different ICP are genetically different traits except for ICP 2 and 4 and thus a multi-trait model should be more appropriate than a repeatability model for genetic analysis of ICP. The results from this study provide the basis for development of genetic predictions for reproductive traits in Bonsmara cattle.

Extent and implications of incorrect offspring-sire relationships in pastoral production system in Kajiado district, Kenya

M.W. Maichomo[1], J.M. Gathuma[2], G.K. Gitau[2], J.M. Ndung'u[1], W.O. Kosura[2] and O.H. Olivier[3], [1]Kenya Agricultural Research Institute, Box 362, 0 Kikuyu0902, Kenya, [2]University of Nairobi, College of Agriculture and Veterinary Sciences, P.O Box 29053, Nairobi, Kenya, [3]International Livestock Research Institute, P.O Box 30709, Nairobi, Kenya

Pastoral farmers have embraced livestock improvement programs in an effort to increase production and ensure food security. Use of bulls of superior genotypes is the common feature in a communal grazing system. The aim of this study was to evaluate accuracy of farmer's paternity identification which determines success of future breed selection and hence genetic gain. Paternity of 269 Orma/zebu and Sahiwal/zebu calves was evaluated using genetic markers and the likelihood based method. Results indicate that only 6.7% farmer alleged paternities were confirmed, 88% parent-offsring relationships were rejected and 18 parent-offsring relationships were undetermined. However, 82% of offsprings were assigned at least 80% confident paternities to one of the sampled candidate males. These results suggest that there is need to institute proper breeding program in the pastoral area if farmers are to benefit from their current efforts of breed improvement.

---

no. 28

Development of a sustainable beef cattle breeding program for low input smallholders of Southern Brazil

F.F. Cardoso[1], T.S. Palma[1], B.B.M. Teixeira[1], M.M. Oliveira[1], M.F.S. Borba[1] and F.E. Schlick[2], [1]Embrapa South Cattle & Sheep, P.O. Box 242, 96401970 Bage, Brazil, [2]Emater/RS, Av. Gen. Osorio 1000, 96400100, Brazil

Beef cattle breeding is one of most important economic activities in the lower half of the State of Rio Grande do Sul in Southern Brazil and low input smallholders represent the majority of people surviving from this activity. These farmers manage cow-calp operations and their cattle lack consistency of production and quality and organized recording, breeding and selection systems. This project aimed to develop methodology and strategies for performance recording, genetic improvement and sustainable breeding, through a multidisciplinary team work of researchers, extension specialists and the farmers. Initially, we surveyed 32 smallholders of three locations and performed community meetings to describe their breeding, production and market systems and to identify traits of economic relevance. These traits were fertility, adaptation and calf weight, which then were used to derive breeding objectives. A total of 1,892 cattle of 32 producers were identified and performance recording was encouraged by training; however, this initiative failed due to low literacy and tradition. Most successful genetic improvement strategies were based on financed access to proven bulls, community joining and artificial insemination, using breed complementarity to produce crosses with ¾ British ¼ Zebu composition, which were identified as suitable for their production and market systems.

Comparison of reproductive performance of friesian and jersey crosses at Kingolwira dairy farm, Morogoro, Tanzania

Z.C. Nziku[1] and S.W. Chenyambuga[2], [1]West Kilimanjaro Livestock Research center, Animal breeding, P.O. Box 147, Sanya Juu, Moshi-Tanzania, 255, Tanzania, [2]Sokoine University of agriculture, Animal Science and Production, P.O. Box 3004, Morogoro-Tanzania, 255, Tanzania

This study was carried out to assess the reproductive performances of Friesian and Jersey crosses and determine the non-genetic factors that affect the reproductive performance. Data were collected on age at first calving, calving interval, lactation length, dry period and calf mortality between 1997 and 2004. Friesian and Jersey crosses did not differ significantly ($P \geq 0.05$) in age at first calving, calving interval, dry period and calf mortality, but differed significantly ($P \leq 0.01$) for lactation length. Mean age at first calving was 38.06 and 37.68 months in Friesian and Jersey crosses. Mean calving interval was slightly longer 452.91 days in Friesian than in Jersey crosses 421.48 days. The dry period averaged 190.47 and 168.43 days in Friesian and Jersey crosses. Mean lactation length was 255.79 days in Friesian and 304.34 days in Jersey crosses. Calf mortality was lower 19.91% in Jersey than in Friesian crosses 24.53%. Year of calving significantly influenced ($P \leq 0.001$) by calving interval and lactation length while season of calving had significant effect ($P \leq 0.05$) only on lactation length. Parity significantly affected ($P \leq 0.05$) by lactation length and dry period. It is concluded that Jersey crosses are relatively better in reproductive performance compared to Friesian crosses.

---

no. 30

Trypanotolerance in Ethiopian cattle breeds

J. Stein[1], W. Ayalew[2], W. Mulatu[2], E. Rege[3], T. Dessie[2], B. Malmfors[1], H. Lemecha[4] and J. Philipsson[1], [1]Swedish University of Agricultural Sciences, Dept of Animal Breeding and Genetics, Box 7023, S-750 07 Uppsala, Sweden, [2]International Livestock Research Institute, P.O. Box 5689, Addis Ababa, Ethiopia, [3]International Livestock Research Institute, P.O. Box 30709, Nairobi, Kenya, [4]National Animal Health Research Center, P.O. Box 04, Sebeta, Ethiopia

Trypanosomosis (T) is considered the single most important livestock disease in Sub-Saharan Africa and causes enormous economic losses through reduced production and death of livestock. One way of managing T is enhanced use of naturally existing trypanotolerance in indigenous livestock. As part of a bigger study, an extensive field survey was conducted in the tsetse infested home areas of four indigenous cattle breeds of Ethiopia – Abigar, Gurage, Horro and Sheko. Cattle keepers in each area were interviewed to explore their perceptions about T and animal performance. In addition 100 animals from each breed were blood-sampled during the expected peak season to investigate parasitemia and PCV. About 59% of the cattle keepers ranked T as the most important disease, but Sheko cattle keepers did not in general consider diseases to be a main constraint. Sheko and Abigar cattle had highest reported milk production and reproduction rates. Sheko cattle had lower trypanosome prevalence (Sheko 5%, Abigar 23%, Gurage 20%, Horro 18%), higher PCV (25.1) than Abigar (20.0) and Gurage (22.7), and got fewer trypanocidal treatments per year (1.0) than Abigar (3.8), Gurage (24.1) and Horro (3.6). Thus, the Sheko breed showed a higher level of trypanotolerance than the other breeds.

Genetic evaluation of growth traits in the crosses between two ecotypes of Nigerian local chicken

O.M. Momoh[1] and C.C. Nwosu[2], [1]University of Agriculture, Department of Animal Breeding and Physiology, PMB 2373, Makurdi, Nigeria, [2]University of Nigeria, Department of Animal Science, Nsukka, Nsukka, Nigeria

Genetic evaluation of the Nigerian local heavy chicken ecotype (HE) and its F1 crosses with the light ecotype (LE); HE X LE as the main cross (MCX) and LE X HE as the reciprocal cross (RCX) was carried out at the Poultry Farm of the Department of Animal Science, University of Nigeria, Nsukka. The study was designed to provide estimates of heritability of body weight (BW), daily weight gain (DG) and feed conversion ratio (FCR) at different ages and the genetic correlations between them in HE and its crosses. A total of 214 pedigree hatched day old chicks of HE, 190 of MCX and 185 of RCX from 15 sires was used in the study. BW in HE, MCX and RCX were on the average lowly to moderately heritable. Heritability estimates of DG were generally low in HE and RCX with a range of h_2=0.04±0.13 to 0.12±0.14. FCR had moderate h2 estimates in all the genetic groups. In all the groups, genetic correlation estimates between BW and DG were high, positive and highly significant ($P<0.001$). The low to moderate heritability of growth traits in the Nigerian local heavy chicken ecotype and its crosses as well as the variations in additive genetic effect reflect genetically unimproved populations.

---

no. 32

The possibility of a genotype x environment interaction in the South African Jersey population with reference to age at first calving

D.J. Van Niekerk[1], F.W.C. Neser[1] and J. Van Der Westhuizen[2], [1]University of the Free State, Animal, Wildlife and Grassland Sciences, P.O. Box 339, 9300 Bloemfontein, South Africa, [2]ARC, Animal Production, P Bag X2, 0062 Irene, South Africa

The possibility of a genotype x environment interaction for age at first calving in the South African Jersey population were investigated by grouping 54 864 lactation records completed between 1980 and 2002 into different production or environmental levels. Two different sets of criteria were used to group the animals. Firstly a cluster analysis, using both environmental and genetic factors was done which resulted into four different clusters. Secondly the herds were divided into four categories according to feeding systems and geographic location. All the herds using a Total Mixed Ration (TMR) were placed in one group while the pasture herds form a second group. The herds that are situated in the warmer areas were placed into a third group and the herds in the Overberg area that do not have access to irrigation water formed a fourth group. Bivariate animal models were used to determine the genetic correlations between the different production and/or environmental scenarios. The correlations vary between 36% (Cluster 1 and 3) and 79% (TMR and pasture herds). It was concluded that there are a GXE effect for age at first calving within the South African Jersey population.

no. 33

Skin spots and pigmentation scores as indicator traits for presence of pigmented fibres in Corriedale fleeces

J.I. Urioste[1], F. Penagaricano[1], R. Lopez[1], C. Lafuente[1], J. Laporta[1], H. Naya[1,2], I. Sienra[1] and R. Kremer[1], [1]Univ. de la Republica, 18 de julio 1968, Montevideo, Uruguay, [2]I. Pasteur, Mataojo 2020, Montevideo, Uruguay

Presence of dark, pigmented fibres in sheep wool is an economically important quality fault. Direct selection against dark fibres is expensive and cumbersome. An alternative approach is through indirect selection with indicator traits. The aim of this study was to estimate genetic parameters (heritability, repeatability, genetic correlations) for the incidence of skin spots and fibres and a pigmentation score in nose-lips in Corriedale sheep. During 2002-2007, 5183 records from 2400 animals on two experimental flocks, aged 1 to 6 years, were taken at shearing for spot presence (SP) and a pigmentation score (1-5) on nose-lips (PS). In 2005-2006, presence of pigmented fibres (PF) in fleeces of 471 animals was assessed in a wool lab. Model for PF included flock-year, animal and residual effects. For SP and PS, age of animal and permanent environmental effects were included also. SP and PF were threshold traits. Heritability estimates (standard deviations) were 0.25 (0.05), 0.21 (0.10) and 0.33 (0.03) for SP, PF and PS. Repeatability was 0.50 (0.03) and 0.47 (0.02) for SP and PS. A strong, positive genetic correlation (0.72 ±0.24) was found between SP and PF. Correlation estimates of PS with SP (0.52±0.08) and PF (0.47±0.23) were positive and medium. These preliminary results suggest that traits are moderately heritable and highly repeatable. SP could be a useful indicator trait for PF.

---

no. 34

Characterisation of the indigenous sheep of Kenya using microsatellite markers

A. Muigai[1], O. Mwai[2], A.K. Kwallah[1] and O. Hanotte[2], [1]Jomo Kenyatta University, P.O. Box 62000, 00200,Nairobi, Kenya, [2]ILRI, P.O. Box 30709, 00100,Nairobi, Kenya

Indigenous small ruminants play a major role in the resource-poor indigenous people and pastoralist set-up. In Kenya, indigenous sheep can be classified as either fat-tailed or fat-rumped they include the Red Maasai, East African fat-tailed and Somali Blackhead Persian. In this study the genetic diversity, differentiation and population structure was determined in nine sheep populations using 15 autosomal microsatellite markers. High genetic diversities were observed for all the populations. Most of the genetic variation was observed within the populations rather than among the populations. Populations from the same geographic regions showed small genetic distances. Phylogenetic and Principal Component Analyses separated the sheep into two main groups, indigenous and exotic sheep. However the Red Maasai sampled from Kadjiado clustered with the exotic Dorper sheep unlike the Red Maasai sampled from Naivasha which clustered with the indigenous populations. Our results show that there is rampant crossbreeding of the Red Maasai with the Dorper thus putting the breed at risk of becoming extinct.

no. 35

Reproductive performance of two sow lines under desert climatic conditions

E. Lutaaya, A. Nakafeero and S. Nemaire, University of Namibia, Animal Science, Private Bag 13301, Windhoek, Namibia

Data from a pig farm in Namibia recorded over the period 2002 – 2006 were analysed to: (i) determine factors influencing Age at First Farrowing (AFF), Number Born Alive (NBA), Weaning-to-conception interval (WCI), Farrowing Interval (FI) and; (ii) estimate the nonproductive sow days (NPSD) and replacement rates in two sow lines. The mating scheme involved a within herd grandparent programme where purchased line A females were mated to line B boars to generate line C females, which were then mated to terminal line D or line E boars. Sows born during winter months had the lowest AFF; Line C had lower AFF than line A possibly due to greater heterosis. NBA was influenced by year of service, parity, service season and service season x parity interaction. The NBA peaked at about fifth parity and then declined; females served during the hot summer months had lower litter sizes, compared to those in other seasons. WCI was shortest for sows weaning their piglets in winter months. FI was shortest for farrowings in winter months. NPSD was estimated to be 76 d in line A and 52 d in line C; replacement rate was 50.5% in line A and 42.1% in line C. Component traits of reproduction indicate depressed performance of gilts and sows during the summer period. Under desert conditions, a wide array of interventions are required to mitigate temperatures during the summer months, because of increased heat stress that depresses reproductive performance.

---

no. 36

Prospects of performing multiple-country comparison of dairy sires for countries not participating in Interbull international genetic evaluations

A. Loberg, H. Jorjani and F. Fikse, Interbull Centre, Animal Breeding and genetics, SLU/Hgen Box 7023, 75007 Uppsala, Sweden

The international trade with genetic material requires genetic evaluations to be performed internationally. Today 42 countries are member of Interbull, but only 26 countries participate in the international genetic evaluations. The minimum requirement for a country to participate is that it can provide breeding values of their own bulls. The aim of this study was to investigate the possibility to include a country in the international genetic evaluations without including any data from the country in question. In this study Argentina exemplified a member country not yet participating in the international evaluation. Predicted genetic merits for Argentina were estimated as a linear function of breeding values for traits with records (i.e. all other countries). This model was used with six different sets of correlations between Argentina and the other countries. Argentinean data was added in a regular genetic evaluation, as performed by Interbull, providing a reference for comparison. The results showed that it is possible to perform an international genetic evaluation for a country without submitted data. For the top 100 bulls the alternative methods found 37 to 67 co-selected bulls compared with the reference situation in which Argentinean data was included. When looking at the country of origin for the top 100 ranked bulls, the different alternatives did with some exception include the same countries as the reference situation.

no. 37

Genetic relationships between calving interval and linear type traits in South African Holstein and Jersey cattle

M.L. Makgahlela, B.E. Mostert and C.B. Banga, ARC-Animal Production Institute, Private Bag X2, Irene, 0062, South Africa

Genetic correlations between first calving interval (CI) and linear type traits, in South African Holstein and Jersey cattle, were estimated to assess the possibility of using type information as selection criteria for CI. Data were obtained from the national dairy recording scheme and comprised records of 30 503 Holstein cows in 640 herds and 27 360 Jersey cows in 460 herds. Multiple-trait animal models were used to estimate parameters, based on the restricted maximum likelihood methodology. Fixed effects in the model varied depending on the individual trait. Linear type traits reflecting body size generally had much higher correlations with CI than udder traits. Genetic correlations of CI with body size traits were mostly positive in both breeds, ranging from 0.04±0.16 with bone strength to 0.51±0.08 with body depth. Correlations between CI and body depth (0.51±0.08), angularity (0.32±0.08) and rump angle (0.32±0.12) indicate that cows with deep, more angular bodies and low pins have longer CI. Genetic correlations with udder type traits were generally low, ranging from -0.01±0.20 with udder width to 0.25±0.11 with rear teat placement. The highest genetic correlations with CI were found for rear leg rear view (-0.70±0.34), body depth (0.51±0.08), dairy strength (0.51±0.09), rear leg set (0.51±0.06), foot angle (-0.44±0.04), rump angle (0.32±0.05) and angularity (0.32±0.08). These traits may therefore be used to predict CI in South African Holstein and Jersey cattle.

---

no. 38

Breeding objetives for beef cattle cow-calf operations of low input smallholders in Southern Brazil

F.F. Cardoso[1,2], B.B.M. Teixeira[1], C.H. Laske[1,2], T.S. Palma[1] and M.F.S. Borba[1], [1]Embrapa South Cattle & Sheep, P.O. Box 242, 96401970 Bage, Brazil, [2]Universidade Federal de Pelotas, P.O. Box 354, 96010-900 Pelotas, Brazil

The objective of this work was to identify breeding objectives, selection criteria and their economic weights for cow-calf operations managed by low input smallholders in the lower half of the State of Rio Grande do Sul in southern Brazil. From data collected by 32 interviews with smallholders of three different locations and participatory action research meetings, we described their breeding, production and market systems, based on selling male calves and culling cows at weaning. Fertility, calf weight and adaptation, perceived as the ability to produce a healthy and heavy calf in a harsh environment, were recognized as the most important traits for this system. A bio-economic model for the breeding objective was derived and weaning weight (WW), weaning rate (WR) and cow weight (CW) were the chosen selection criteria. Economic weights, derived as the partial derivative of the bio-economic model with respect to each trait evaluated at the average value of the remainder criteria, were US$ 13.13 and US$ 2.53 for increasing 1 kg in WW and CW, respectively, and US$ 61.59 for 1% increase in WR. A selection index for joint selection of these three traits would place 63% emphasis on WR, 19% on WW and 18% on CW, showing that despite its low heritability, WR is the most important trait for genetic improvement targeting increased economic return of low input cow-calf operations.

Accuracy values associated to estimated breeding values for mature weight in Nelore cows obtained by different models

A.A. Boligon[1], L.G. Albuquerque[1], M.E.Z. Mercadante[2] and R.B. Lobo[3], [1]FCAV/UNESP, Jaboticabal, 14884-900, Brazil, [2]IZ, Sertãozinho, 13460-000, Brazil, [3]FCAV/USP, Ribeirão Preto, 14049-900, Brazil

Records of 18,770 Nelore animals, born from 1975 to 2002 in 8 herds participating in the Nelore Cattle Breeding Program, were analyzed to estimate genetic parameters and accuracy values associated to breeding values for mature weight. The mature weights were analyzed as a single weight taken closest to 4.5 years of age for each cow in the data file, considering weights starting from 2 (W2Y_S), 3 (W3Y_S), or 4 (W4Y_S) years of age or as repeated records, including all weights starting from 2 (W2Y_R), 3 (W3Y_R), or 4 (W4Y_R) years of age. The variance components were estimated by restricted maximum likelihood, fitting univariate and bivariate animal models, including weaning weight. The heritability estimates were 0.29; 0.34; 0.36; 0.41; 0.44; and 0.46 for W2Y_S; W3Y_S; W4Y_S; W2Y_R; W3Y_R; and W4Y_R, respectively. The accuracy values associated to the estimated breeding values ranged from 0.52 to 0.94. Larger accuracies were obtained in repeated records models. Selecting the top 2% or 10% of the sires based on the predicted breeding values for W4Y_R, 84% and 91% of the same sires, respectively, would be selected if W3Y_R were considered as selection criterion using the same intensity. The most appropriate model for genetic evaluation of mature weight would be a repeatability model, including weights starting at 3 years of age. Financial support from CNPq and FAPESP

---

no. 40

Residual correlations between body measurements and carcass traits of *Bos indicus* and tropically adapted *Bos taurus* breeds

J.N.S.G. Cyrillo[1], R.F. Nardon[2], A.G. Razook[1], M.E.Z. Mercadante[1], L.A. Figueiredo[1] and S.F.M. Bonilha[1], [1]Instituto de Zootecnia, Estação Experimental de Zootecnia de Sertãozinho, CP 63, 14160-000 Sertãozinho, Brazil, [2]APTA, Polo Regional Médio Paranapanema, CP 263, 19802-970 Assis, Brazil

Records from 144 Nellore, Guzerah and Caracu males selected for post-weaning weight from the Estaçao Experimental de Zootecnia de Sertãozinho-SP-Brazil were analyzed. During three years the animals were finished in feedlot and slaughtered averaging 19 months of age, at three different points of maturity. The multivariate procedure (MANOVA) was used to estimate the residual correlation between body measurements and quantitative carcass traits. The model included genetic groups, order of slaughter and year of birth effects. Estimates of correlations between hip height (HH) and carcass weight (CW) was 0.37, and it was antagonist with carcass fat percentage (FAT%) (-0.15), and kidney heart fat (KF) (-0.14). High correlations were estimated between chest girth (CG) and CW (0.77),chuck (0.57)and fore ribs (0.67) and positive correlations were estimated between CG and FAT% (0.13), KF (0.38) and carcass fat (0.44). Estimates of correlations between body length (BL) and CW and FAT% were 0.38 and 0.15, respectively. Correlations between dorsal line length, rump length and the distance between pin bonesand all the cuts of commercial interest analyzed ranged from 0.24 to 0.56. The results indicate that hip height and chest girth may be useful indicators of carcass composition in beef cattle.

no. 41

(Co)variance components and breeding values of Holstein cows, using a random regression model

H.R. Mirzaei[1], H. Mehraban[1] and H. Farhangfar[2], [1] Zabol University, Zabol, Iran, [2]Birjand University, Birjand, Iran

(Co)variance components, heritability and breeding values were estimated from 38475 test-day records of Holstein cows, using a random regression test day model. The model included fixed effects; herd-test date, milking time and sire's origin and random effects; additative genetic and permanent environmental effects. The additive genetic variance was the less in the early and the highest in the late lactations. The permanent environmental variance intended to increase in early and late lactation. Heritability increased in the second half of the lactation, so that the most value of coefficient was in the 19th stage of lactation (.216). Because of high correlation between breeding value and the total breeding value (0.927) and also high rank correlation between bull's breeding value and the total breeding value (0.907) at the 19th stage of lactation, it was suggested to select bulls based on that stage.

---

no. 42

Estimation of parameters and genetic trends for production traits of Iranian Holstein cattle in Markazi province

H.R. Mirzaei[1], M. Razavi[1] and M. Vatankhah[2], [1]Zabol University, Zabol, Iran, [2]Research inst. of Kord city, Kord city, Iran

(Co)Variance components and genetic trends of15539 records collected 31 herds during 1990- 2004 were estimated for milk, fat yields and fat percentage, using an animal model. The model was consisted of herd-year-season of calving and parity as fixed effects, and additive genetic effects, permanent environmental effects due to repeated records and residuals as random effect. Heritability of milk yield, fat and fat percent were found to be 0.20, 0.23 and 0.32 respectively. Repeatability of milk yield, fat and fat percent were estimated as 0.46, 0.39 and 0.40, respectively. Phenotypic trends of traits were 22.29 kg/year for milk, 0.23 kg/year for fat and 0.05 percent/year for fat percent. Genetic trends for those traits were 3.75 kg/year, 0.06 kg/year and -0.002 percent/year and environmental trend for them were 19.79 kg/year, 0.23 kg/year and 0.007 percent/year, respectively.

no. 43

Estimation of genetic parameters for body weight traits in Iranian Lori-Bakhtiari lambs using random regression model

H.R. Mirzaei[1], H. Moradian[1] and M. Vatankhah[2], [1]Zabol University, Zabol, Iran, [2]Research Ins. of Kord City, Kord City, Iran

Genetic parameters for body weights of Lori-Bakhtiari lambs were estimated, using random regression model. Data consisted of 24056 recors of 5633 lambs collected from 1989 to 2005. The model included fixed effects (year of birth, sex of lamb, birth type of lamb and age of dam), and random effects (additive genetic effect, permanent environmental effect and residual effects). Results indicated that the 5th order of polynomial was more appropriate than others. The heritability at birth, 1, 2, 3, 4, 5, 6, 7, 8, 9, 10, 11 and 12 months of age were estimated as 0.18, 0.16, 0.23, 0.26, 0.25, 0.22, 0.21, 0.21, 0.23, 0.25, 0.27, 0.28 and 0.26, respectively. The proportion of permanent environmental variance to phenotypic variance was increased from 0.01 at birth to 0.74 at 7 months of age and decreased afterwards. Genetic correlations between body weights at closer ages were higher than farther. In general the lamb body weights affected by different factors as additive genetic and permanent environment.

---

no. 44

Genetic characteristics of growth traits and yearling wool weight in Baluchi sheep breed

H.R. Mirzaei[1] and A. Kamali[2], [1]University of Zabol, Animal science, Zabol, Iran, [2]University of Zabol, Animal science, Zabol, Iran

The genetic (co)variances, parameters and correlations were estimated for growth traits; birth weight, weaning weight, average daily gain from birth to weaning and yearling wool weight, collected from 4099 flock of Baluchi Sheep. Restricted maximum likelihood procedure conducted using univariate and bivariate animal models. Heritability estimates for birth weight, weaning weight and average daily gain from birth to weaning and yearling wool weight in the univariate analyses were 0.074, 0.03 and 0.172 and 0.075, respectively. In the bivariate analyses genetic correlations between growth traits ranged from 0.069 to 0.995 for birth-weaning weights and for weaning weight-average daily gain. Genetic correlations were 0.899, 0.647 and 0.737 between yearling wool weight and birth weight, weaning weight and average daily gain, respectively. Results indicated that yearling wool weight might be expected to increase if selection were practiced on any of those growth traits.

no. 45

Relationship between ultrasonic measures of reproductive and fat thickness traits in pre-pubertal Nelore heifers

M.E.Z. Mercadante[1], F.M. Monteiro[1,2], J.M. Garcia[2] and A.G. Razook[1], [1]Instituto De Zootecnia, CP63, 14160970-Sertãozinho, Brazil, [2]UNESP, FCAV, 14884-900-Jaboticabal, Brazil

The objective of this study was to assess the relationship between reproductive and fat thickness traits and body condition score. Three non invasive ultrasonic evaluations (14, 16 and 19 months of age, SD=0.7) of reproductive (transrectal probe 7.5MHz) and fat thickness (carcass probe 3.5MHz) traits were made in 61 Nelore heifers born in 2006. These animals were from two lines (control and selection) selected by yearling weight since 1981. Mean of ovaries area (OV), maximum follicular diameter (FOL), backfat thickness (BFAT), rump fat thickness (RFAT) and body condition score (BCC) were analyzed. The observed means were: 3.4cm^2, 8.1mm, 2.1mm, 6.3mm, 5.8 and 299.7kg for OV, FOL, BFAT, RFAT, BCC and body weight, respectively. The OV and FOL repeated records were analyzed by proc mixed (SAS), with a model including the selection line, measurement (1 to 3), and BFAT or RFAT effects. For OV and FOL the effects of selection line and measurement (only for OV) were significant, however the effects of BFAT and RFAT were not. The residual correlations were estimated using the same procedure, fitting the selection line and the measurement effects for all traits. OV showed low correlation with FOL (0.08). The correlations between OV and FOL with BFAT (0.12 and -0.02), RFAT (0.24 and 0.13) and BCC (0.25 and 0.30) were also not high. The fat thickness traits can not be a good predictor for pre-pubertal reproductive traits.

---

no. 46

Genotype-environment interaction for eighteen month weight of Canchim cattle in São Paulo State, Brazil

M. Mattar[1], M.M. Alencar[2], F.F. Cardoso[3], A.S. Ferraudo[4], L.O.C. Silva[5] and A.C. Espasandin[6], [1]Graduate Student, FCAV/Unesp, Via Prof. Paulo Donato Castellane, Jaboticabal, SP, Brazil, [2]Southeast Embrapa Cattle/ CNPq's fellow, Rod. Washington Luiz, km 234, São Carlos, SP, Brazil, [3]South – Cattle & Sheep, BR 153, km 603, Bagé, RS, Brazil, [4]FCAV/Unesp, Via Prof. Paulo Donato Castellane, Jaboticabal, SP, Brazil, [5]Embrapa Beef Cattle, BR 262, km 4, Campo Grande, MS, Brazil, [6]UDELAR, Av. 18 de Julio, Paysandú, Uruguay

Genotype-environment interaction (GEI) in beef cattle can be characterized by the change of the set of genes which express a trait in function of the environment production. Genetic correlation between the trait studied in different environments measures the extent in which this change occurs. The aim of this study was to evaluate the existence of GEI on eighteen month weight (EMW) in Canchim cattle, through genetic correlations obtained by bayesian inference. Three different clusters of cities in the State of São Paulo, homogeneous with respect to environment variables, were considered as the environments. The statistical model included the fixed effects of contemporary group and age at weighing (covariate), and additive and residual random effects. The heritability estimates of EMW were low to moderate in magnitude [0.20-0.35] for the three clusters. The genetic correlations between EMW in the three clusters varied from 0.13 to 0.34, suggesting the existence of GEI. The results showed that phenotypic expression of EMW depended on the environment in which it was measured.

Performance of beef cattle bulls tested under extensive rangeland conditions

G. Mukuahima, W.A. Van Niekerk, N.H. Casey, R.J. Coertze and J.H. Dreyer, Department of Animal and Wildlife Sciences, University of Pretoria, Pretoria, 0002, South Africa

Comparative bull performance recording under extensive rangeland conditions allows for direct comparisons between and within genotypes. The study quantified growth performance and feed conversion efficiency of 444 bulls, of the Angus, Beefmaster, Bonsmara, Drakensberger, Nguni and Simbra breeds on rangeland in the Eastern Highveld, over a four year period. Bulls were selected according to the entry standards for the National Beef Cattle Performance Testing Scheme. Parameters recorded were live weight (LW), ADG, Kleiber ratio (KR) and FCR on rangeland. An AOV was applied to determine significance ($P<0.05$) between breeds, year, breeders within breed and the interactions for the dependent variables. Linear regressions established the relationships between variables. ADG and FCR differed ($P<0.05$) within breeds. Bulls with low initial LW had higher ADG than bulls with high initial LW, though they did not necessarily have higher final weights. ADG, FCR and KR were influenced ($P< 0.05$) by breed, year and breed x year interactions. High FCR occurred in years of high rainfall whereas high growth rates occurred in average rainfall years. Under the test circumstances the Simbra and Beefmaster showed better performance compared to the other breeds. The study revealed a substantial variation in growth and efficiency traits within beef breeds grazing rangeland under these test conditions.

---

no. 48

Effect of sire-cluster interaction on genetic values for eighteen month weight of Canchim cattle

M. Mattar[1], M.M. Alencar[2], F.F. Cardoso[3], A.S. Ferraudo[4], L.O.C. Silva[5] and A.C. Espasandin[6], [1]Graduate Student, FCAV/Unesp, Via Prof. Paulo Donato Castellane, Jaboticabal, SP, Brazil, [2]Southeast – Embrapa Cattle/ CNPq's fellow, Rod. Washington Luiz, km 234, São Carlos, SP, Brazil, [3]South – Embrapa Cattle&Sheep, BR 153, km 603, Bagé, RS, Brazil, [4]FCAV/Unesp, Via Prof. Paulo Donato Castellane, Jaboticabal, SP, Brazil, [5]Embrapa Beef Cattle, BR 262, km 4, Campo Grande, MS, Brazil, [6]UDELAR, Av. 18 de Julio, Paysandú, Uruguay

The extent to which a genotype is expressed can be determined by the environment, therefore, it is possible that the best genotype in an environment may not be in another one. The aim of this study was to evaluate the existence of genotype–environment interaction (GEI) in eighteen month weight (EMW) in Canchim cattle, using Pearson's and Spearman's rank correlations of breeding values (BV) of sires in different environments. Three different clusters of cities in the State of São Paulo, homogeneous with respect to environment variables, were considered as the environments. The three-trait statistical model used to predict BV by bayesian inference included the fixed effects of contemporary group and age at weighing (covariate), and additive and residual random effects. Pearson's and Spearman's correlations varied from 0.39 to 0.78 and 0.42 to 0.75, respectively, suggesting that the genetic potential of sires for EMW depended on the clusters environment conditions.

no. 49

The use of polymerase chain reaction (PCR) for sex typing of ostrich (*Struthio camelus*) in Iran

M. Faghani and A. Doosti, Islamic Azad University of Shahrekord, Department of agriculture, Department of agriculture, Islamic Azad University of Shahrekord, P. Box 166, Shahrekord, Iran, 0098, Iran

Ostrich farming has been dramatically increased in the past few years in Iran. At this bird, it is very difficult to distinguish between males and females based on an analysis of their external morphology, especially young birds. The objective of this study was to develop DNA markers that can be used for sex identification in the ostrich using polymerase chain reaction (PCR) technology. Two methods were used for this study that was included PCR and random amplified polymorphic DNA (RAPD). DNA isolated from blood and feather of 45 birds (20 male and 25 female). We used feather because feather sampling requires less training for field workers, results in shorter handling times for the organism, generates no hazardous wastes, and requires simpler storage procedures. Polymerase chain reaction was performed with using from one pair w-linked primer and one w-linked primer was used for RAPD analysis. At results in PCR method one band approximately 650bp identified in female birds that in male birds don't observed. Also in RAPD method female birds have one band, approximately 1300bp, which in male birds don't observe. Comparing the two methods it was found that PCR method was better than RAPD method Because the results of PCR method could be obtained faster and easier than RAPD method. Therefore PCR is a fast, accurate and inexpensive procedure for sex typing of ostriches that can use for breeding of this bird.

---

no. 50

A genetic analysis of the population of Polish Red-and-White cattle

J.M. Oprzadek, G. Sender and M. Łukaszewicz, Institute of Genetics and Animal Breeding Polish Academy of Sciences, of Animal Sciences, Jastrzebiec ul. Postepu 1, 05-552 Wolka Kosowska, Poland

The work aimed at estimating relations between chosen loci (CSN3, BLG, PRL, GH, PIT-1, LEP), production and functional traits (calving progress, calf vitality, somatic cell count in milk, number of inseminations per pregnancy, health status, calving interval, culling reasons–barrenness, diseases of the udder, parturient paresis, leg diseases, low productivity, accidents, sex and number of calves born). The animal material consisted of 498 Polish Red-and-White cows. Data on reproduction, semen doses per effective insemination, length of the calving interval, culling reasons and length of productive life referred to the years 1999–2003. Among the 82 sires 26 were represented by at least four daughters. Heritability coefficients were low for all the traits examined and ranged from 0.05 for calving for which the assistance of a veterinary surgeon was necessary to 0.38 for life milk production and 0.23 for the survival rate of calves until week 2 of life. The heritability coefficient for calf mortality was low and amounted to 0.01. Among the seven polymorphisms analysed significant differences in breeding value for the production of milk and its components during a 100 day lactation of primiparas were observed only in the case of the variants of genes CSN3, PRL-HD and GH.

Study on effect of non-genetic factors on body weight of tellicherry goats

A.K. Thiruvenkadan, M. Murugan, K. Karunanithi, J. Muralidharan and K. Chinnamani, mecheri sheep research station, pottaneri, tamil nadu, 636 453, India

Data on 566 Tellicherry goats, recorded between 1988 and 2007 were used to study non-genetic factors on body weight and daily gain from birth to 12 months of age. The least-squares means for body weight at birth, and at 12 months of age were 2.17±0.03 and 18.78±0.44 kg, respectively. The pre- and post-weaning average daily weight gains were 72.41±1.68 g and 37.46±1.49 g respectively, and the associated growth efficiencies were 3.11±0.08 and 1.34±0.05 respectively. Significant differences associated with the year of kidding were observed in body weight, weight gain and efficiency in weight gain at different stages of growth. Age of the dam influenced body weights at birth and weaning and early growth rate. Growth rate of kids born between December to February was relatively slower than those born in other months and this can result from seasonal changes and suggests that it is necessary to plan the kidding season rationally by controlling the estrus and mating time. The kids born as twin had lower birth weight and slower early growth rate than those born as single but had a higher post-weaning growth rate. The heritabilities of different traits were moderate to high, except for birth weight, which was of low heritability. The phenotypic and genetic correlations among the different body weights were positive and high, except for between birth weight and 12 months of age.

---

no. 52

Height growth curve parameters of Nellore, Angus x Nellore, Canchim x Nellore and Simmental x Nellore females

M.M. Alencar[1], F. Barichello[2], A.R. Freitas[3] and P.F. Barbosa[3], [1]Southeast-Embrapa Cattle/CNPq's fellow, Rod. Washington Luiz, km 234, São Carlos, SP, Brazil, [2]FCAV/UNESP -FAPESP's scholarship, Via Prof. Paulo Donato Castellane, Jaboticabal, SP, Brazil, [3]Southeast-Embrapa Cattle, Rod. Washington Luiz, km 234, São Carlos, SP, Brazil

The objective of this study was to estimate the height growth measured on the hip from eight to 90 months of age (t) of females of four genetic groups (GG): Nellore (NEL), Angus x Nellore (AN), Canchim x Nellore (CN) and Simmental x Nellore (SN). The animals were born in two seasons (E: autumn and spring) and were submitted to three levels of supplementation (Treat), depending on the season (0.0 and 3.0 kg of concentrate for animals born in autumn, and 0.0, 1.5 and 3.0 kg of concentrate for animals born in spring). The estimate of height growth, as a function of t, was realized by Brody's non-linear model with parameter A (height at maturity), b (parameter that shaped the curve) and k (rate of growth, t^{-1}). Using the MIXED procedure of SAS, the effects of GG, E, GG x E, and Treat(GG x E) on A, b and k were studied. The effect of GG on A was significant ($P<0.05$), indicating that the genetic groups showed different height at maturity. For the parameters b and k, the effect of Treat(GG x E) was significant ($P<0.05$), suggesting that the effect of Treat depended on GGxE.

Investing in genetics no. 53

Analyses of breeding values for simulated discrete visual score data with different distributions and different genetic parameters

F. Barichello[1], M.M. Alencar[2] and R.A.A. Torres Júnior[3], [1]FCAV/Unesp – FAPESP's scholarship, Via Prof.Paulo Donato Castellane, Jaboticabal, SP, Brazil, [2]Southeast Embrapa Cattle – CNPq's fellow, Rod. Washington Luiz, Km 234, São Carlos, SP, Brazil, [3]Embrapa Beef Cattle, BR 262, km 4, Campo Grande, MS, Brazil

The aim was to evaluate the effect of the form (Y) of assigning discrete visual scores (VS) based on a continuous underlying scale (US) on the estimates of the breeding values (BV) for two heritability values (H: 0.25 and 0.49) and two contemporary group variance values (GC: 0.04 and 0.16). Herds with 40 bulls and 1,200 cows, mated at random, were simulated for 20 years. Direct and maternal BV, maternal permanent environmental, contemporary group and age of dam effects were generated and combined with an independent error term to form the phenotype in the US. The VS data were assigned according to symmetric relative and fixed and asymmetric relative distributions. The BV was estimated using a linear model with the Gibbs Sampler. The procedure was repeated five times for each situation. Correlation (R) between estimated and true BV was obtained for each animal class (sires, dams and offspring). Significant effects of H on R for all animal classes were found (0.49 presented greater R). For bulls, significant effects of H x GC interaction on R were found (greater H and smaller GC presented greater R). Y had no effect on R. Larger samples may be needed for better evaluating the effects of the factors on BV estimates for VS.

Human nutrition and livestock products no. 1

Strategic use of naturally selenium (Se)-rich milling coproducts to eliminate Se deficiency and create Se-enriched foods

J.B. Taylor, ARS, USDA, 60 Office Loop, Dubois, ID 83423, USA

Selenium (Se) is essential for sustaining a healthy life. When dietary Se was marginally deficient, populations experienced impaired reproduction and growth rates and increased disease rates. Selenium-rich grains, harvested from regions with seleniferous soils, were natural sources of bioavailable Se. These grains contained 3 to 60 times as much Se as grains grown in soils with adequate Se availability. We isolated a specific wheat-milling coproduct that had 37% greater Se content than the parent grain. Through strategic feeding of Se-enriched coproducts, we enhanced the Se status of livestock. Muscle foods harvested from lambs fattened on Se-enriched coproducts for 14 days provided more than 70% of the daily Se requirement for humans. Milk collected from ewes fed Se-enriched coproducts during the first 28 days of lactation had 7 times as much Se as milk from ewes fed sodium selenite. Because of the degree of Se enrichment achieved, these ewes and their nursing young grazed range without the need for supplemental Se for almost 1 year. Ultimately, we used a naturally occurring Se-rich feed source to add nutritional value to food-animal products and enhance the long-term Se status of livestock. Based on our research, Se-enriched feeds can be prepared from high-Se grains that were harvested from regions with seleniferous soils. Furthermore, these feeds can be strategically and safely used to eliminate the risk of Se deficiency in livestock while simultaneously creating Se-rich food products, such as meat and milk, for human consumption.

no. 2

Validity of using a constant value for heme iron

H.C. Schonfeldt and N. Gibson, University of Pretoria, School of Natural and Agricultural Sciences, Pretoria, 0002, South Africa

It is well documented that iron from animal sources are more readily available than those from plant sources, relating to their higher content of heme iron. According to the Monsen model a constant average of 40% of total iron in meat, fish and poultry is in the heme form, while the remaining 60% is in the lesser bioavailable non-heme form. The Monsen equation has been modified for use in populations with the amount of heme iron in meat, fish and poultry calculated as 45% of the total iron content. From these constant values derived from estimation, the amount of iron that will be absorbed is calculated. However, new evidence suggests that the type of meat, species and the specific meat cut appear to play a significant role in the amount of heme iron present, thus directly affecting the amount of iron that will be available for absorption in the human body. The heme content of red meats seems to be more than 45%, while poultry and fish might have lower heme iron contents, questioning the validity of using a constant value in calculations. The implications of different heme values (applied vs. determined) to calculate the iron bioavailability of beef, mutton, fish and poultry will be presented, and the impact thereof on treating iron deficiency anaemia from a human perspective discussed.

---

no. 3

Traceability and the eating-out-paradox

C.B.E. Rogge, L. Lichtenberg and T. Becker, University of Hohenheim, Institute for Agricultural Policy and Agricultural Markets, Schloss, 70593 Stuttgart, Germany

Various studies have analysed consumer behaviour concerning food, covering most retailing sectors in great detail. However, they have neglected that the food service industry market grew in importance over the last few decades, due to changes in modern society. We present a first time study examining the demand for information on meat in selected sectors of the food service industry compared to retailing. In order to define consumer expectations towards a traceability and information system for the entire food supply chain, we conducted two different surveys in the federal state of Baden-Wurttemberg, Germany. In the second half of 2007, more than 600 people were surveyed in fast food restaurants, restaurants in shops/stores, staff canteens, supermarkets and shopping centres. Based on these results, a second survey was conducted in the first half of 2008, and around 60 people were interviewed in depth at their homes. Compared to consumers in retailing, significant differences appear in information seeking behaviour, as well as desired information. These differences are reflected by a lower willingness to pay for additional information in the food service industry. Since the product is the same we call this the Eating-Out-Paradox. Further discrepancies show in consumer trust and it can be shown that heterogeneous criteria are being used when selecting meat (products) in the food service industry and in retailing. The establishments have to adapt their consumer communication in order to meet these different requirements.

no. 4

Determination of German consumer attitudes towards meat traceability by means-end chain theory

L. Lichtenberg and T. Becker, Institute for Agricultural Policy and Agricultural Markets, University of Hohenheim, Department of Agricultural Markets and Marketing, Institute 420 b, 70593 Stuttgart, Germany

The attitudes of German consumers towards the traceability of meat are analysed by the means-end approach using laddering interviews to identify means-end chains, in order to understand consumer decision making. The "standard" means-end chain model links attributes to functional and psychological consequences which are linked to values or goals. Between February and May 2008, 126 laddering interviews were conducted with people who are responsible in general for their households´ food purchases, as well as for meat in particular. The interviews were carried out in two major German cities and their rural areas in order to compare consumer attitudes between the two cities and between urban and rural areas. The results indicate a difference concerning the information demand towards meat traceability between Germany's urban and rural areas. However, as a whole meat traceability is of particular importance for both. In total, there exists a willingness-to-pay for meat traceability at an average of 10%. With the help of the empirical study results, better marketing and advertising strategies can be developed to the benefit of consumers as well as food retailers. Food retailers should adjust their communication and price policies in order to satisfy the information demand of consumers concerning traceable meat and to take advantage of the higher WTP for traceable meat exhibited by consumers.

---

no. 5

Consumer perception of livestock production systems and animal welfare in Chile

B. Schnettler and N. Sepulveda, Universidad de La Frontera, Prod. Agropecuaria, Av. Fco. Salazar 01145, Temuco, Chile

Given the importance of animal welfare (AW), a survey was carried out among 384 consumers in Temuco, Chile, to establish their knowledge and perceptions about animals handling during production, to detect preferences for meat produced under AW principles, their willingness to pay a higher price and to distinguish different consumer segments. 60% of people surveyed knew about livestock management practices, half of them considered that had a negative effect on the animals, but only 32.1% have changed their meat consumption habits due to this. 70% of the people surveyed had over 50% of knowledge about AW aspects. There is a strong preference and willingness to pay a higher price for meat produced under AW principles. Consumers have a positive perception of the fact that the meat that they consume comes from pasture-fed animals, raised in the open, and raised, transported and slaughtered following humane principles. Three segments were identified by using cluster analysis: the most numerous (58.6%) considers confinement and feeding with concentrates as positive; the second group (25.5%) showed a strong rejection of the use of hormones, feeding with broiler litter and concentrates, and places a positive value on raising animals in the open. The smallest segment (15.4%) placed the highest value on humane treatment of the animals, plus a positive value on concentrates use and rejected confined raising. The conclusion is that a large part of the population perceives AW as a desirable condition when purchasing beef.

no. 6

Levels of copper, chromium, manganese, potassium and sodium in cow's whole milk from Addis Ababa, Ethiopia

E.A. Engda, Bahir Dar University, Department Of Chemistry, Bahir Dar University, P.O Box. 429, Bahir Dar, Ethiopia

Bulk milk samples representative of 309 cows were collected in two periods from four different farms, each with population of 78 Jersey, 125 Holstein, 59 Holstein and 47 Crossbreeds cows, which supply milk to Addis Ababa, Ethiopia. In each sampling period, one bulk milk sample was collected from each farm. Milk samples were analyzed for various elements (K, Na, Cr, Cu and Mn) levels. Known weights of freeze dried cow's whole milk samples were digested with 3.5 mL HNO_3 and 2.0 mL $HClO_4$ on a hot plate for 4.5 h. The contents of nutrients in the digests were then analyzed employing flame atomic absorption spectrometer. The following mean levels were recorded (mg/L): K 907.22, Na 316.45, Cu 0.14, Mn 0.045 but Cr was not detected in any of the farms. Despite topographical and environmental differences that prevail, the mineral concentrations in cow's whole milk were found within normal intervals described in the literature. Results further indicated positive correlations between animals' feeding habits and milk moisture content. Ethiopian mean consumption of cow's milk supplies smaller amounts of Na, K and Cu to the daily intake of the analyzed minerals for Addis Ababa population due to the consumption of cow's whole milk was observed. The accuracy of results was checked by analyzing the NIST reference material SRM 8435 and good agreement was achieved with certified values.

---

no. 7

Production of dairy products using undomesticated fruits in Swaziland

A.M. Dlamini[1], S. Mahlalela[2] and R. Mamba[2], [1]University of Swaziland, Animal Production and Health, Private bag Luyengo, M205, Swaziland, [2]Ministry of Agriculture and Cooperatives, Home Economics, Private bag Mbabane, H100, Swaziland

Fruits are used for coagulating and enhancing organoleptic properties of dairy products. The objective of this research was to determine the role undomesticated fruits in dairy technology. Fruits tested were tincozi, tineyi, umfomfo, emahlala and ematfundvuluka and the control were strawberry and lemons. Sensory evaluation was done by a panelist of 40, assessing appearance, texture and general acceptability. Data were analyzed using MSTAT-C. Fruits with high acidity, Umfomfo (pH 3.2), emtfundvuluka (pH 2.8) and emahlala (3.8) were used for cheese making. Tincozi (pH 4.8) and tineyi (pH 5.1) were suitable for yoghurt flavouring. All indigenous fruits flavoured yoghurts (30%) had lower ($P<0.05$) syneresis than the plain (47%) and strawberry yoghurts (60%). Appearance was ranked highest for strawberry (7.53) and lowest in tincozi (5.57). Cheese produced using ematfundvuluka had highest crude protein, (25.6%) whereas ematfundvuluka + umhlala cheese was lowest (23.8%). Cheese made from ematfundvuluka was ranked highest, 6.81 in general acceptability whereas ematfundvuluka + umhlala cheese was the least preferred (5.43). It is concluded that tineyi, tincozi and umfomfo fruits, can be used to enhance yoghurt flavours, while ematfundvuluka produced cheese of highest quality. It is recommended that teneyi and ematfundvuluka should be included in local yoghurt and cheese, making procedures respectively.

no. 8

Effect of breed and diet on the composition of milk fat fatty acids

Y.C. Tesha, L.R. Kurwijila and A.E. Kimambo, Sokoine University of Agriculture, Animal Science and Production, P.O. Box 3004, Chuo Kikuu Morogoro, Tanzania

The effect of breed and different oil seed cake supplementation on milk fat fatty acid composition was studied using twelve (12) Friesian and 12 Ayrshire lactating cows. The animals were divided into three groups of eight animals each and allocated to three dietary treatments. Diet 1 (control) contained maize bran (MB) plus minerals, in diet 2 and 3 some of the maize bran in the control diet was replaced with cotton seed cake (CSC) and sunflower seed cake (SSC) respectively. Each diet was fed for 21 days in a change over design. Animals were milked twice per day and milk samples were collected for butter making. The butter oil was methylated and the composition of fatty acids methyl esters were analysed by gas chromatograph. The proportions of short chain fatty acids were higher ($P<0.05$) for Ayrshire breed (5.33%) than for Friesian (4.46%). The proportion of medium (12.56%) chain fatty acids for Friesian was lower ($P>0.05$) whilst those of long chain (82.99%) were higher ($P>0.05$) than those of Ayrshire (14.85 and 79.81) % respectively. Diet 3 led to higher ($P>0.05$) short chain (5.95%) and lower long chain (79.62%) fatty acids than diet 2 (3.62 and 84.13) % and diet 1 (5.12 and 80.12) % respectively. The proportion of unsaturated fatty acids was slightly higher for Friesian (57.92%) than Ayrshire (46.33%) and for those fed diet 2 (57.42%) than those fed diet 3 (55.15%) and diet 1 (43.8%). It is concluded that both breed and diet have slight influence on the proportion of butter fat fatty acids composition.

---

no. 9

Effect of linsed supplementation on beef quality traits using different forages and cattle breeds

G. Hollo, I. Repa, J. Seregi and I. Hollo, Kaposvár University, Guba Sándor street 40, 7400 Kaposvár, Hungary

The effect of diet (extensive vs intensive diet, forage to concentrate ratio, feeding concentrates are rich in n-3 fatty acids) and the breed (old: Hungarian Grey (HG), dual purpose: Hungarian Simmental (HS), dairy: Hungarian Holstein-Friesian (HF)) was investigated on the fatty acid composition of beef in order to produce beef with healthy fatty acid composition. Findings reveal that the extensive diet with linseed supplemented concentrate influenced the n-6/n-3 ratio and the CLA content of longissimus muscle more advantageous concerning human nutrition. The meat from HG contained more CLA and less n-6 fatty acids than that of HF bulls. The different forage to concentrate ratio with/without linseed supplementation did not significantly affect the performance and slaughter traits in HS young bulls. The wider forage to linseed concentrate ratio caused slightly higher dressing percentage, meat and fat proportion and lower bone in carcass. The effect of muscle type on chemical composition of muscles is more significant than that of the diet. SFA and MUFA were affected by muscle type, n-3 fatty acids and n-6/n-3 fatty acid ratio by diet, the linseed supplementation caused higher n-3 fatty acids accumulation in all examined muscles. The level of cis-9 trans-11 CLA (mg/100 g) was influenced by muscle type, but not by diet.

no. 10

Effect of replacing maize with palm oil sludge on performance and meat quality of broiler chickens

A.O. Bobadoye[1], G.N. Onibi[2] and A.N. Fajemisin[2], [1]Federal college of forestry, Agricultural extension and management, pmb 5054, 234 ibadan, Nigeria, [2]federal university of technology akure, animal production and health, pmb704, 234 akure, Nigeria

This study assess the effect of substituting 10, 20, 30 and 40% of energy supplied by maize in a control diet with energy from palm oil sludge on performance and meat quality of broiler chickens. A total of 200 shaver sturbo day old broiler chicks were randomly allocated to 5 treatment at 10 birds/replicate and 4 replicate/treatment. The feeding trial lasted 8 weeks during which weekly feed intake and group live weight were measured. Three birds per replicate were sacrificed and sample of the thigh, drumstick and breast muscles were taken for meat analysis and oxidative stability. Results showed that the final live weight, weight gain and feed conversion ratio were not significantly ($P>0.05$) influenced by the diets. The total feed intake increased from 5492.10+20.30g to 5778.70+11.80g with increasing levels of palm oil sludge in the diets ($P<0.05$). Total cost of feed and feed cost/kg weight gain decreased as levels of dietary palm oil sludge increased. The fat content of selected muscle fat increased with increasing levels of palm oil sludge in the diets ($P<0.05$). Oxidative deterioration of the meat measured as malonaldehyde (MDA) concentration was lowest for birds on the control and increased with increasing levels of palm oil sludge ($P<0.01$).

---

no. 11

The analysis of the changes in the functioning of rural farming households brought about by the assistance of Heifer International

R.E. Laski, Heifer International, International Programs, One World Avenue, 72202, USA

The research study carried out examined the changes and other effects in the functioning of rural farming households through the assistance of the American charity, Heifer International. The assistance of American charity-Heifer International was directed in particular to this group of rural farming households. This assistance produced positive and noteworthy outcomes. Principally, the effect of this assistance was seen in the improvement in agricultural equipment and in the quality of life of these farming households. The most significant improvement in the life of the farming families was in their economic situation and general quality of life as well as an improvement in their diet and nutrition. To a lesser degree, the help of Heifer International fulfilled other needs than nutritional, but respondents expected that these needs would be met in the near future. Those farming families who received the support and resources of Heifer International assessd the role of Heifer International to be highly beneficial to them. The assistance of Heifer International was also beneficial in creating a more pro-market attitude amongst the rural farming households.

no. 12

Does restricted feeding affect development of intramuscular fat across the musculature in Hanwoo carcasses?

S.K. Hong[1], S.H. Lee[1], W.M. Cho[1], I.H. Lee[1], D. Perry[2] and J.M. Thompson[2], [1]National Institute of Animal Science, RDA, Suwon, 441-706, Korea, South, [2]Beef CRC, UNE, Armidale 2351, Australia

Korean consumers place great emphasis on the level of intramuscular fat in beef cuts. Korean native cattle (Hanwoo) are generally fed a restricted concentrate ration up to 18 months and then ad-libitum concentrate. This study investigated the development of intramuscular fat in the body and whether restricted feeding up to 18 months affected intramuscular fat percentage in the muscle. 196 Hanwoo steers were allocated to *ad libitum* and restricted treatments, and to slaughter at 2 monthly intervals from 6 months to 30 months. The restricted steers were fed a concentrate ration daily at 1.5% of live weight from 6 to 12 months, 1.75% of live weight from 12 to 18 months and *ad libitum* thereafter. At slaughter the left side was broken into 10 joints and each joint dissected into muscle, bone, and fat. The muscle tissue from each joint was chemically analysed for fat. The relative growth of intramuscular fat in the 10 joints was analysed using the allometric function. Chemical fat was average maturing across most of the joints. At the same total chemical fat weight there was little difference between nutritional treatments in the weight of chemical fat in the joints. There was approximately a 1.5% decrease in chemical fat in most joints of restricted steers at all slaughters. Thus early nutritional restriction reduced chemical fat percentage and this persisted despite later *ad libitum* feeding. This has implications for optimising marbling in Hanwoo cattle.

---

no. 13

Effect of age on carcass and cut composition of South African beef carcasses

H.C. Schonfeldt[1], P.E. Strydom[2] and M.F. Smith[2], [1]University of Pretoria, School of Agricultural and Food Sciences, Pretoria, 0002, South Africa, [2]Agricultural Research Council, Private Box X2, Irene, 0062, South Africa

The physical composition (proportion subcutaneous fat, meat and bone) of 15 primal cuts from beef animals (n=122) of three different age groups (as defined in the current South African classification system) and representing the full variation in fatness within each age group, was assessed. Furthermore, proximate analysis (percentage total moisture, fat, nitrogen and ash) were performed on the meat including the subcutaneous fat of each cut. Total fat and muscle content for each cut were then calculated. To ensure that the effect of differences in carcass fatness over the age classes were accounted for in the analyses, percentage chemical fat of the carcass was included as a covariant in the statistical analysis. The bone and meat content of the different cuts within the same carcass varied considerably. The meat content decreased and bone content increased with increasing age. A large variation in compositional and chemical characteristics for the various cuts for the three age groups was observed. On average the hind and fore shins contained the lowest amount of chemical fat, followed by the fillet and thick flank.

no. 14

Karoo sheep meat as a potential Geographical Indication in South Africa

H. Vermeulen, H.C. Schonfeldt, J.F. Kirsten and C. Leighton, University of Pretoria, School of Agricultural and Food Sciences, Preroria, 0002, South Africa

Geographical Indications (GIs) are a prominent alternative food quality movement globally. Despite a rich diversity of traditional knowledge, indigenous resources and agro-food production based on local resources, such as Karoo sheep meat, South Africa does not have any registered GIs. It is widely argued that Karoo sheep meat is imbued with the subtle, fragrant flavours of the Karoo bush grazing, being 'spiced on the hoof'. This paper investigates the reputation of Karoo sheep meat, as a potential GI, through sensory evaluation to establish product specificity and the investigation of consumers' perceptions. No sensory detectable difference exists between the two main sheep breeds, within a region implying consistency in the context of a potential GI. Furthermore, the Karoo region consistently produces a similar type of sheep meat product, with similar grazing plants. Namibian sheep meat grouped separate from meat from all the other Karoo-like regions. Principal Component Analysis was applied to identify the attributes that differentiate the most between the sheep samples. The consumer survey results presents potential market segments developed through cluster analysis. There is relatively strong evidence of the reputation of Karoo sheep meat among consumers, especially among consumers that is aware of Karoo sheep meat. However, definite scope exists to increase the awareness of the 'romantic' Karoo region and Karoo sheep meat and its unique qualities among sheep meat consumers.

---

no. 15

A comparison of the nutrient content of South African mutton and lamb

H.C. Schonfeldt[1], J. Sainsbury[1] and S.M. Van Heerden[2], [1]University of Pretoria, School of Agricultural and Food Sciences, Pretoria, 0002, South Africa, [2]Animal Nutrition and Animal Products Institute, Agricultural Research Council, Private Box X2, Irene, 0062, South Africa

The aim of the study was to determine the nutrient content of South African mutton (C2) and lamb (A2). The meat samples, incorporated in the study, comprised of the most commonly consumed sheep breeds in South Africa, namely Dorper and SA Mutton Merino. The carcasses were subdivided into nine wholesale cuts, which were then dissected into meat, bone and subcutaneous fat for determining physical composition. Proximate analyses, sodium, potassium, iron, magnesium, zinc, cholesterol, fatty acid profile, thiamin, riboflavin, niacin, vitamin B6, and B12 were determined on a double blind basis in South African National Accreditation Service accredited laboratories. South African mutton and lamb are lower in fat and cholesterol, than the values that was extracted from other tables.

no. 16

Fatty acid analysis and CLA content as related to flavour of mutton from selected regions in Southern Africa

H.C. Schonfeldt, C. Leighton and H. Vermeulen, University of Pretoria, School of Agriculture and Food Sceinces, Pretoria, 0002, South Africa

In South African mutton is mostly produced on natural pastures and in arid areas. According to literature this has an impact on the fatty acid profile and sensory attributes of the meat so produced. The specific flavours are unique flavours that are located in the lipid-soluble fraction and are appealing to consumers. However, the correlation between frequently consumed natural grazing plants and its effect on the fatty acid profile of the meat so produced is not known within the South African context. Consequently, the fatty acid profile and conjugated linoleic acid (CLA) content of selected grazing plants and mutton samples from five regions were analysed. Both the fatty acid profiles and CLA content of cooked mutton from five regions as captured form the cooking losses (separated fat only) and five Karoo shrubs (leaves and thin twigs) were analysed by an accredited analytical laboratory. Although present in significant amounts in all the mutton fat and grazing plants studied, no direct significant link could be found between a particular fatty acid (including CLA) in a grazing plant with a flavour attribute in the mutton from a particular region.

---

no. 17

Dietary n-3 fatty acids source supplementation with an ionophore could alter milk composition in lactating ewes

F. Mirzaei, A. Towhidi, M. Rezaeian, A. Nikkhah and K. Rezayazdi,

16 lactating Chall ewes were assigned to 4 groups and received following dietary treatments for 10 weeks: control diet (CON), diet contained fish oil (20g/kg DM; FO), diet contained monensin (15mg/kg DM; as a ionophore, MON), diet contained fish oil and monensin (FM). Amounts of the feed were offered and orts for ewes as individually were daily recorded. Milk samples were weekly collected. At the end of trial, rumen fluid and plasma samples were obtained. All of the diets significantly reduced DMI compared to CON. Milk yield was lower in MON than the other groups. Milk fat percentage of FM significantly decreased. Milk fat yield was significantly reduced in MON and FM groups .Milk protein percentage were not significantly different among the diets. The MON increased significantly milk urea nitrogen concentration. Fish oil supplementation decreased protozoa population and the acetate-to-propionate ratio in rumen fluid compared to CON. Concentration of plasma glucose and urea were not affected by the treatments, but plasma concentrations of triglycerides, total cholesterol, HDL-cholesterol were higher in FO than the other groups. The results suggested that n-3 fatty acids source supplementation with an ionophorecould change milk composition especially reduce fat percentage in sheep so that make it more suitable for human consumption.

Carnosine and anserine as biomarkers of beef meat quality

A.M. Giusti[1], R. Monticolo[2], F. Perer[1], L. Mosca[2], A. Macone[2] and C. Cannella[1], [1]University Sapienza of Rome, Medical Physiopathology- section of Food Science, pl. Aldo Moro, 5, 00185 Rome, Italy, [2]University Sapienza of Rome, Biochemical Sciences, pl. Aldo Moro, 5, 00185 Rome, Italy

Meat quality is influenced by pre-slaughter events such as breed, age, environment, stress and post-slaughter events such as carcass pH and temperature, electrical stimulation. All these factors can facilitate the interaction of prooxidants with the cellular macromolecules resulting in the generation of free radicals and the propagation of oxidative reactions. The oxidation is a major deterioration reaction which often results in a significant loss of meat product quality. Carnosine and anserine are naturally occurring skeletal muscle dipeptides. Their function is not completely understood, but they are thought to act both as buffering agents and as antioxidants *in vitro* e *in vivo*. The aim of this study was to evaluate the carnosine and anserine content in muscles of conventionally processed young bulls of two different breeds (Charolais and Limousine) raised in 4 different intensive-farms. All young bulls of both races were divided in two categories based on the animal health conditions :"Maximum" and "Minimum" and following the bulls slaughtering muscles were analyzed after 3 and 7 days of storage at 2±0.5 °C. Data indicate that the category "Maximum", in both breeds, had a higher content of the two peptides compared with the "Minimum" category. The length of the ageing period and the breeding condition showed little or no influence on carnosine and anserine content.

---

no. 19

Feeding habits of pastoral and agro pastoral communities in the southern rangelands of Kenya

E.N. Muthiani, A.J.N. N Dathi, J.N. Ndungu, J.K. Manyeki and W.N. Mnene, Kenya Agricultural Research, Range management, P.O Box 12, Makindu 90138, +254, Kenya

A longitudinal survey was carried out to establish the feeding habits of pastoral Masai and agropastoral Kamba and Taita of Kajiado, Makueni and Taita-Taveta districts respectively. Thirty households stratified in three classes base on what the site communities considered were good livestock husbandry practices were selected. Data on the frequency of consumption of cereals, meat, pulses and milk among other in week was collected using a semi-structured questionnaire. Contrary to the expected, the average proportion of households consuming meat was lowest (36.3%) among the Maasai of Kajiado District while Taita-Taveta District recorded the highest percent of 63.1%. Makueni recorded 43.3% of the households as consuming meat at least once in a week. More households in Makueni and Taita- Taveta had meat in their diets in December 2005 than in other months probably due to the festivities as compared to Kajiado where 73% did not eat meat in the same month. The highest proportion of households consuming meat of 64% in Kajiado was recorded in November 2004 which could have been due to slaughter of the weak animals towards the end of the drought. Cereals were consumed by all households across the district on daily basis. However, Makueni District reported a higher proportion of households without pulses in their weekly diets than the other districts due to drought in the area

no. 20

Nutritional manipulation of livestock products and their impact on human nutrition

Z. Hayat[1,2], T.N. Pasha[2], F.M. Khattak[2] and G. Cherian[3], [1]University College of Agriculture, Deapartment of Animal Sciences, University of Sargodha, Sargodha, 40100, Pakistan, [2]University of Veterinary & Animal Sciences, Out Fall Road, Lahore, 54000, Pakistan, [3]Oregon State University, Department of Animal Sciences, 122, Withycombe Hall, Corvallis, 97331, OR, USA

Nutritional manipulation of livestock products especially chicken eggs has been the topic of much interest. There are a number of livestock products with various health claims are available in the market, however information on their impact on human nutrition is scanty. Present study was therefore planned to examine the effects of incorporating designer eggs in human diet on their lipid profile, blood glucose and blood pressure. The designer eggs were produced by feeding diets containing 10% flaxseed. This provided a marked progressive increase in egg n-3 fatty acid content in the egg ($P<0.05$). However there was no change in the cholesterol concentration, and egg quality parameters. Blood glucose and blood pressure of volunteer were independent of dietary treatment. There were also non significant effects of designer eggs on serum total, HDL and LDL cholesterol, however, a mild HDL cholesterol elevating effect has been observed. Serum triglyceride concentrations showed a significant ($P<0.05$) decrease due to consumption of n-3 enriched eggs. These data demonstrate that livestock products such as nutrient enriched eggs which have beneficial effect on human nutrition can be generated by minor diet modifications without affecting production and egg quality parameters.

---

no. 21

Goat milk acceptance and promotion methods in Japan

T. Ozawa[1], J. Nishitani[1] and H. Blair[2], [1]Nippon Veterinary and Life Science University, School of Animal Science, 1-7-1 Kyonan-cho, Musashino-shi, 186-8602, Tokyo, Japan, [2]Massey University, I.V.A.B.S., Private bag 11222, Palmerston North, New Zealand

A consumer questionnaire conducted with the purpose of ascertaining the acceptability of goat milk and related products in Japan was carried out on 345 guarantees of NVLU in 2006. 275 effective sponses representing middle class urban households were returned. The results revealed that: 1) 30% of respondents have experienced drinking goat milk and only 10% are aware of the current retail situation of goat milk and related products; 2) over 70% of goat milk drinkers raised goats by hand at some point in their past and their first experience drinking goat milk was in infancy; 3) those with experience drinking goat milk expressed a vague evaluation and minimal understanding of drinking goat milk; 4) respondents who were inexperienced goat milk drinkers expressed a strong desire to taste and a weak desire to purchase goat milk; and 5) low goat milk characteristics recognition, but high evaluation is obtainable those who knows. Goats are perceived as being "mild and familiar." It is necessary for those who manage goat husbandry to present goat milk and related product tasting opportunities to consumers. The key point is to make the functional differences between cow and goat milk clear and present the advantages of goat milk at the fore of this promotion. Goat milk should not be promoted merely as a drink that is similar to cow milk, but must be positioned as a functional drink or health food in order to expand the market.

no. 22

Indigenous milk products of Nepal

N. Pandey, Nepal Agricultural Research Council, Bovine Research Programme, P.O. Box 8504, Kathmandu, Nepal

Nepal is a landlocked country with 147,181sq.km. area located in southern Asia between India in the south and China in the north. Livestock is an integral component of agricultural farming system of Nepal. Consumption of milk is a part of Nepali culture. Presently the total milk production of Nepal is 1312140 MT. Based on this figure, per capita milk consumption is about 51 liters per annum. Milk supply channel is mainly through the informal market. Around 10% goes through the formal sector. Out of which 95% is sold as pasteurized milk and only 2-3% is used for product diversification. Indigenous technology occupies bigger place than modern technology, which are very typical ones. Those products are ghee, chhurpi, khuwa, dahi, kheer, gudpak, buttermilk and yak cheese. Among these products ghee, chhurpi, khuwa, dahi,gudpak and yak cheese are the commercial products and kheer, butter milk are non commercial products. Ghee is the main traditional product in Nepal. It is one of the major exportable dairy products. Ghee made from cow is believed to have medicinal value. Dahi is second major traditional milk product and has a religious value as well as the medicinal value. Khuwa is used as a base product for making various types of sweets. Chhurpi is a hard type of cheese, which has very long self-life. In eastern Himalayan region, Serkum (similar to Chhurpi) is produced. Gudpak is made from Khuwa by mixing with other ingredients and have high market value especially in Nepal. Yak cheese is a typical type of cheese which is produced only in Nepal and export to foreign countries. Kheer is more used by vegetarian peoples.

---

no. 23

Effect of season, breed and anatomical location on palatability and chemical composition of beef in Uganda

M.H. Kwizera, D. Mpairwe and F.B. Bareeba, Makerere University, Animal Science, P.o. Box 7062 Kampala, 256 Kampala, Uganda

The study was conducted on a total of 77 adult bulls: 45 Ankole (AK), 14 Ankole X Friesian (AF) and 18 Karamojong (KJ) cattle breeds. Meat samples [Small Intestines (SI), *Longissimus dorsi* (LD), *Psoas major* (PM) and *Semitendinosus* (ST)] were collected from three abattoirs during dry and wet seasons. Cholesterol and fat contents were higher in the dry than wet seasons while palatability did not differ among breeds and seasons. Breed x season interaction effect was significant ($P<0.05$) for T. cholesterol, fat and protein. Protein in AK and KJ was similar (21 g/100g) in the wet season and higher ($P<0.001$) than the dry season. Fat in the AK and KJ was higher ($P<0.001$) in dry than wet seasons (4 & 10 vs 2 and 3 g/100 g) for the breed and season respectively. AK and KJ had similar T. cholesterol (59 vs 57 g/100g) but higher than AF (52 g/100 g). Fat was lowest ($P<0.05$) in AK and highest ($P<0.05$) in KJ. Protein was highest in KJ ($P<0.05$) followed by AK and AF cattle (21, 15 and 19 g/100 g) respectively. SI had higher cholesterol (77 mg/100 g) and fat (5.4 g/100g) but lower protein and palatability rating than muscles. Cholesterol and protein content was higher ($P<0.05$) in PM than ST and LD. The study revealed that beef from Karamajong and Ankole had lower cholesterol levels than the recommended level of 76 mg/100 g. This attribute can be exploited to market the two breeds in Uganda. Health-conscious consumers would prefer wet season meat since it possessed low cholesterol and fat levels.

no. 1

Optimum growth rate of Belgian Blue double-muscled heifers

L.O. Fiems and D.L. De Brabander,

Belgian Blue double-muscled (BBDM) heifers (n=341) were used to investigate the effect of body weight (BW) at first calving at an age of two years (BWC). Females with a birth weight (BWB) <40 kg realized a lower BW-gain up to an age of 4 months and they also had a lower BWC ($P<0.05$). Because intra-uterine growth retardation may slow down growth throughout life, heifers with <40 kg BWB (n=26) were eliminated. Dams with a higher BWC gave birth to calves with a higher birth weight. Milk yield during the first four months after calving linearly increased with a higher BWC ($P<0.001$; $R^2=0.34$). Females with a daily BW-gain from 0-4 months of age <0.6 kg have a lower BWC than females with a higher growth rate. A higher rate of gain from 0-4 months did not affect heifer growth rate. With a mean adult BW of BBDM cows of ±710 kg, heifers should gain more than 0.8 kg/d from birth to conception, to achieve 55-60% of their adult BW at the conception at 15 months of age. A lower BWC was only partly compensated from the first up to the third calving, but there was no clear effect on the interval from the first to the third calving. From the present study we can conclude that it is desirable to strive for a pre-partum BWC of about 600 kg in BBDM heifers when their first calving occurs at an age of 24 months. BW gain during the last months of gestation was negligible, because of a reduced intake capacity and a heavy BWB of the offspring. Therefore, extra attention should be paid to the nutrition of BBDM heifers to realize an accelerated growth rate before they become pregnant.

---

no. 2

Effect of cold shock on frozen-thawed spermatozoa of Nili-Ravi buffalo (*Bubalus bubalis*) and Sahiwal bulls

A. Ijaz[1], S.A. Mahmood[1], M.I.R. Khan[1] and Z.U. Rahman[2], [1]University of Veterinary and Animal Sciences, Lahore, 54000, Pakistan, [2]University of Agriculture Faislabad, Faisalabad, 38040, Pakistan

Effect of cold shock on frozen-thawed bovine sperm was measured in terms of motility (%), viability(%), plasma membrane integrity(%) and acrosomal integrity(%). A total of 14 ejaculates, seven from each species, were processed for freezing(-196 °C). Three straws(0.5ml each) from each species of the bulls were thawed and pooled at 37 °C and were evaluated before and after cold shock(4 °C for 2 min). Motility was observed using phase contrast microscopy before and after cold shock. Viability, plasma membrane and acrosomal integrity were evaluated by supravital stain, hypo-osmotic swelling test and normal acrosomal reaction before and after cold shock, respectively. Two hundred sperms were counted for each parameter. Results revealed that cold shock treatment had no effect on motility, viability, plasma membrane and acrosomal integrity of buffalo bull sperm. However, cold shock treatment significantly affected motility (before;59.2±1.0 vs after;41.9±1.1), viability (before;70.07±1.4 vs after;45.2±1.3) and plasma membrane integrity (before;62.3±4.2 vs after;47.24±3.7) of Sahiwal bull sperm. There was no effect of cold shock treatment on acrosomal integrity (before;73.9±1.1 vs after;70.3±1.2) of buffalo bull sperm. In conclusion, cold shock treatment has detrimental effects on Sahiwal bull sperm with out affecting buffalo bull sperm.

no. 3

Evaluation of quality and quantity of fresh and frozen-thawed semen of Golapayegany native bulls

M. Borji and S.R. Masihi, Research Center of Agriculture and Natural Resources, Animal science, Bigining of Mobarak Abad road-North belth way, 38135-889, Iran

This study was conducted to evaluate the quantity and quality of Golpayegany bull semen. Two ejaculates were obtained by artificial vagina on one day per wk for 24 wk from 8 bulls aged 24 and 30 months. Two successive ejaculates collected were collected at 15 min intervals and diluted in an egg yolk-citrate medium. Samples were examined of fresh and after freezing and storage in liquid nitrogen for 1 month. Results indicate that two years old bulls were better than tree years old in terms of semen characteristics. Mean volume, pH, motility, spermatozoa concentration, dead spermatozoa and abnormal spermatozoa were: 3.56, 6.55, 60%, 718*106, 10.3% and 8.7% respectively. The effect of freezing on all factors other than sperm concentration was significant ($P<0.05$). Correlation between ejaculation and other factors with the exception of pH, spermatozoa concentration and dead spermatozoa were significant.

---

no. 4

The effect of face colour and lactation status on reproductive performance in Awassi ewes

R. Kridli, M. Husein and A. Abdullah, Jordan University of Science and Technology, Faculty of Agriculture, Irbid, 22110, Jordan

This study was conducted to evaluate the effect of face color and lactation status on reproductive performance of Awassi ewes treated with fluorogestone acetate sponges (FGA) and equine chorionic gonadotropin (eCG). A total of 282 ewes were utilized in the study. Intravaginal FGA sponges were inserted for 14 days in early July (breeding season). A 600 I.U. dose of eCG was administered to each ewe (i.m.) at the time of sponge removal. Fertile rams were allowed with the ewes at the time of sponge removal. Estrus expression and the interval from sponge removal to onset of estrus were neither affected by face colour nor by the lactation status of ewes. Lambing rate was similar between brown- and black-faced ewes, while the percentage of multiple births was greater ($P<0.01$) in black-faced ewes. Both the fecundity and prolificacy were greater ($P<0.05$) in the black-faced ewes. Lambing rate and the percentage of multiple births were similar between lactating and dry ewes. Fecundity tended to be greater ($P<0.10$) in lactating than in dry ewes (1.24 ± 0.05 and 1.03 ± 0.1 lambs/ewe, respectively) while prolificacy was similar regardless of lactation status. Results of the present study indicate that black-faced Awassi ewes respond better to hormonal treatment than brown-faced ewes while the lactation status dose not appear to influence the ewes' response to hormonal treatment.

Sexual development in SA Landrace and Black Indigenous gilts under two dietary treatments

P.C. Coetzee, N.H. Casey, S. Greaves and H.L. Lucht, University of Pretoria, Department of Animal and Wildlife Sciences, 0002 Pretoria, South Africa

In a 2x2 factorial design, sexual development of 32 Black Indigenous (BI) and 60 SA Landrace (SAL) gilts was studied under the effects of two dietary treatments fed individually *ad lib*. Treatment 1 contained 16% CP, 13.6MJ DE/kg DM. Treatment 2, a 25% dilution of Treatment 1 using wood shavings, contained 12% CP, 10.2MJ DE/kg DM. Target weights set for BI gilts were from 10 kg to 50 kg and for SAL, 15 kg to 120 kg, because of the large difference in their respective mature weights. A sample was slaughtered at each target weight. Reproductive organs were removed, measured and the number, size and physiological state of follicles recorded. Carcasses were analysed for fat. Treatments effected growth ($P<0.05$). The allometric equation, $\ln(\text{tissue mass kg})=\ln(a)+b\ln(\text{EBM kg})$, was applied to calculate relative growth of components. Cumulative mass (lnkg) was regressed against cumulative ME intake (lnME) in an autoregressive model. Treatments caused no autoregressive difference, though genotypes differed in the pre-pubertal phase, but not the pubertal phase. Age and weight at puberty differed by treatment for each genotype. Differences in component growth in the two phases were non-significant between treatments within genotypes. In both genotypes, 12% CP delayed ovarian development in terms of live weight and age. Treatments had no significant effect on ovarian development in terms of carcass fat. Genotypes differed marginally with the BI gilts having 29.3% and the SAL gilts 31.6% carcass fat, respectively.

---

no. 6

Meta-analyses of the effect of a mixture of 28% eugenol and 17% cinnamaldehyde in dairy cows

D. Bravo[1] and P. Doane[2], [1]Pancosma Research, Research & Development, Voie-des-Traz 6, 1218 Geneva, Switzerland, [2]Archer Daniels Midland Company, 10203 North 200 East Decatur, 46733 Indiana, USA

This abstract presents a meta analysis study using the trials carried out with a mixture of 28% eugenol and 17% of cinnamaldehyde (XT) which is a rumen active additive dedicated towards milk production. When added in a diet, XT improves lactation performance within the trials available. Milk production (34.2 vs. 33.1 kg/day, N=36, $P<0.001$, $R^2=0.45$) and dry matter intake (22.1 vs. 20.6 kg/day, N=16, $P<0.001$, $R^2=0.96$) are significantly improved with a constant feed efficiency (1.44 vs. 1.46, N=16, $P=0.441$, $R^2=0.10$). Milk protein content is decreased and milk fat content is unchanged, but protein or fat yields are increased. Between trials, the search for correlation between nutrients and the response to the product clearly indicates that XT interacts with the nutrition of the cow and with the nutrient composition of the diet. Dietary NEL serves as an example, the higher NEL was in this data set, the greater the extent to which milk production was increased by XT. The results indicate that the definition of the efficacy of a feed additive should not be based only on the presentation of individual and consecutive trials. A feed additive acts with a consistent mechanism leading to an overall response. The definition of the efficacy should look for global responses to the product and this is why meta analysis is a relevant tool. Moreover, it provides an initial framework for how best to use this type of additive.

Estrogenic effects of gibberellic acid on reproductive and production characteristics of immature female fowl

A.E. Elkomy[1], A.A. Elnagar[2], E. El-Ansary[3] and G. El-Shaarrawi[3], [1]Arid Lands Cultivation And Development Research Institute, Livestock department, Mubarak City for Scientific Research, 21934 Alex, Egypt, [2]National Research Center, Animal Production Department, National Research Center, 11787 Gizza, Egypt, [3]Faculty of Agriculture, Poultry Production Department, Alexandria University, 21545 Alex, Egypt

The objective of this study was to investigate if gibberellic acid had an estrogenic effect on productive and reproductive performance of immature female fowl. One hundred hens (18 weeks old) were injected subcutaneously with 0, 100, 200, 400 and 800 µg GA3/kg B.W./week for 4 weeks as an experimental period. After that, all groups were allowed without treatment for 8 weeks as a recovery period. GA3 resulted in a significant decrease in age at first egg laid (sexual maturity) compared with the control group. Egg production and egg weight were non-significant increased due to GA3 treatment, Feed conversion was improved but not significantly. Higher GA3 doses (400 and 800 µg) reduced eggshell thickness and relative eggshell weight. A non-significant increase in relative egg albumin weight was observed in the GA3 treated groups, while, the relative egg yolk weight was decreased. Egg yolk total lipids was increased by GA3 treatment and this effect was GA3 dose-dependent. GA3 at any dose resulted in a significant increase in hen's serum 17- β Estradiol concentration when compared with the control group during the two experimental periods.

---

no. 8

Study of the serum protein profile of cows that aborted after artificial insemination in Senegal

M.M.M. Mouiche and G.J. Sawadogo, Inter-States School of Sciences and Veterinary medicine of Dakar, Deprtement of Animal Production, EISMV de Dakar/, BP 5077, Senegal

In cattle, the economic losses due to embryonic mortality observed after artificial insemination remain important. The proportions and causes of this remain unknown in Senegal. This study assessed the causes of embryonic mortality in cattle through the exploration of metabolic indicators associated with the effectiveness of the defense system. To do this, serum of 21 cows that aborted were collected, 40 pregnant cows and 20 inseminated unpregnant cows, from the regions of Dakar and Thies (Mbour). Serums samples were analysed for serum proteins and tested for brucellosis by the rose bengal test. Results show that the ratio of albumin to globulin (A/G) was significantly different ($P<0.05$) between the unpregnant cows and those that aborted. Almost three quarters of cows that aborted, had high gammaglobulin levels. Thus, abortions related to pathological causes can be suspected. Brucellosis testing revealed one positive case of 81 cows tested. Since the cost of artificial insemination remains high for the African farmer, particular emphasis should be placed on these deaths to limit economic losses.

no. 9

Effects of season, breeds and environmemtal conditions on cow milk protein composition

A. Jemeljanovs, V. Sterna and B. Lujane, Research Institute of Biotechnology and Veterinary Medicine, 1 Instituta Street, LV-2150 Sigulda, Latvia

Nutritive value of milk is determined primarily by total protein content. Coagulation properties are largely affected by stage of lactation, breed and season. The aim of the present study was to compare milk protein content and composition in Latvia between season, breed and environmental factors. In the present research, 416 milk samples of Latvian Brown and Holsteins Black and White breed cows were evaluated. Measurements and statistical analyses were based on milk samples collected from 2006 through 2007 in outdoor and indoor seasons from different regions of Latvia. Protein and casein contents were analyzed using Milkoscan 133. Results show that milk samples of HB&W cows had 3.27% protein and milk samples of LB breed cows3.55%, while casein content was 2.56% and 2.80% respectively. Protein and casein contents (3.71% and 2.83%) where higher in outdoor season, than in indoor season when protein content was 3.44% and 2.61% respectively And differed significant ($P<0.01$). Somatic cell count is one of measure of environmental factors, but no differences were observed ($r=0.06$).

---

no. 10

Effect of heat stress on six beef breeds in the Zastron district: the significance of breed, coat colour and coat type

L.A. Foster[1], P.J. Fourie[1], F.W.C. Neser[2] and M.D. Fair[2], [1]Centrally University of Technology Free State, Agriculture, Private Bag X20539, 9300, Bloemfontein, South Africa, [2]University of the Free State, Animal Wildlife and Grassland Sciences, P.O. Box 339, 9300, Bloemfontein, South Africa

A study was done to ascertain which factors had the greatest influence on an animal's susceptibility to heat stress. Factors tested for susceptibility to summer heat stress were breed, coat colour, coat type, hide thickness, weight gain and body condition score. Only breed, coat colour, coat type and body weight were included in the test for winter susceptibility. A system of subjective scoring of cattle coats, ranging from extremely short to very woolly and coat colours ranging from white to black has been described. Sixty heifers from six breeds (10 from each breed) participated in the study. The breeds were Afrikaner, Bonsmara, Braford, Charolais, Drakensberger, and Simmentaler. Sampling for coat score and coat colour was carried out twice – in August (winter) and December (summer). Rectal temperature (RT) was used as a parameter to determine heat stress. During both periods significant differences ($P<0.0001$) in RT were observed between breeds. None of the factors tested had any significant influence on RT within breeds in winter. However, the following factors had a significant influence ($P<0.1$) on RT in the following breeds in summer: Afrikaner–body condition score and hide thickness, Bonsmara–hide thickness, Braford-none, Charolais-coat score, Drakensberger–hide thickness, Simmentaler-none.

Heat tolerance of Nelore, Senepol x Nelore and Angus x Nelore heifers in the southeast region of Brazil

A.R.B. Ribeiro[1], M.M. Alencar[2], A.R. Freitas[3], L.C.A. Regitano[2], M.C.S. Oliveira[3] and A.M.G. Ibelli[4], [1]Post-doc Embrapa – CPPSE (Scholarship by FAPESP), Rod. Washington Luiz, km 234, 13560-970 São Carlos SP, Brazil, [2]Embrapa – CPPSE (CNPqs researcher), Rod. Washington Luiz, km 234, 13560-970, Brazil, [3]Embrapa – CPPSE, Rod. Washington Luiz, km 234, 13560-970, Brazil, [4]PhD Student – Federal University of São Carlos, Rod. Washington Luiz, km 235, 13560-970, Brazil

The Brazilian beef production chain has experienced an increase in the utilization of adapted and non-adapted taurine breeds in crossbreeding systems. In spite of this, little is known about the adaptability of these groups and of their crossbred products when raised in tropical climate. The aim of this study is to evaluate the physiological responses related to adaptability of Nelore (NE) and crossbred Angus x Nelore (TA) and Senepol x Nelore (SN) cattle submitted to a heat tolerance trial. The study was conducted in the Southeast – Embrapa Cattle (CPPSE), São Carlos, Brazil. A total of 45 heifers, 15 of each genetic group, were evaluated in three days, at 7:00 a.m. (resting measure), at 1:00 p.m. (after 6 h under the sun with no access to water and shade) and at 4:00 p.m. (after 2 h under the sun with access to shade), during the summer 2008. Rectal temperature and sweating rate were measured and the data were analyzed by the least squares method. The effect of genetic group for the rectal temperature was significant ($P<0.05$) only at 4:00 p.m. and for the sweating rate in all three measurements ($P<0.01$). The SN group had the lowest values.

---

no. 12

Assessment of buffalo (*Bubalus bubalis*) semen by tetrazolium salt reduction assay

M. Iqbal, M. Aleem, A. Ijaz and T.N. Pasha, University of Veterinary and animal Sciences, Lahore, 54000, Pakistan

The aim of the study was to use the tetrazolium salt reduction assay (MTT) as an indictor to evaluate sperm viability of fresh buffalo semen. Semen samples from twenty Nili-Ravi bulls of two age groups; Group A (5 years; n=13) and Group B (6-8 years; n=7) were collected once a week. The MTT reduction assay was standardized and then compared with sperm motility, eosin and nigrosin (E&N) staining and hypo-osmotic swelling (HOS) for each semen sample. Results revealed a similar correlation ($r=0.995$) between the MTT reduction assay and viability of sperm in both groups. There was a wide variation when sperm motility was compared with MTT assay. However, the sensitivity of MTT was > 90%. Similarly, MTT reduction assay was more sensitive than E&N staining when the sperm viability was > 80%. Likewise, compared to HOS, MTT showed higher sensitivity when sperm viability was > 70%. In conclusion, MTT reduction assay is a reliable assay to measure semen quality and it is not influenced by the experience of the individual worker.

no. 13

Effect of feeding regime on growth and carcass yield and quality of Ankole, Nganda and Ankole x Friesian crossbred bulls in Uganda

D. Mpairwe, C.B. Katongole, F.B. Bareeba and E. Mukasa-Mugerwa, Makerere University, Animal Science, P.O. Box 7062 Kampala, 256 Kampala, Uganda

The study involved seventy eight growing bulls (27 Ankole 212±26 kg; 27 Ankole x Friesian 112±27 kg, and 24 Nganda 118±32 kg); and three feeding regimes: grazing alone (GZ), grazing supplemented with a concentrate of 16% CP (GZ+MD) and 20% CP (GZ+HG) in a 3 x 3 factorial arrangement. The bulls were slaughtered after 400 days, carcasses weighed, graded and dressing percentages determined. Samples for chemical composition were taken from four sites [small intestine (SI), *Longissimus dorsi* (LD), *Psoas major* (PM) and semi-tendinosus (ST)]. Average daily gain differed significantly ($P<0.05$) between the dietary treatments, but did not differ between the breed-types. Daily gains were higher ($P<0.05$) for GZ+HG, than for GZ+MD and grazing alone (351, 291 and 193 kg/day, respectively). Carcass weight was higher ($P<0.05$) for GZ+HG (149 kg) and GZ+MD (129 kg) than for grazing alone (101 kg). Similarly, dressing percentage was higher ($P<0.05$) in the supplemented animals compared to grazing alone. Chemical composition was similar across the breed types, but differed between the sites sampled. The SI had higher ($P<0.05$) in cholesterol (83 mg/100g) and fat (5 g/100g) levels but lower in protein than the muscles. Cholesterol and fat levels were higher ($P<0.05$) for LD than the PM and ST muscles. The results of this study showed that beef production of grazing indigenous cattle can be improved by supplementation.

---

no. 14

Assessing feed efficiency in beef steers through infrared thermography, feeding behaviour and glucocorticoids

Y. Montanholi[1], K. Swanson[1], F. Schenkel[1], R. Palme[2], D. Lu[1] and S. Miller[1], [1]Univ. of Guelph, Guelph, N1G2W1, Canada, [2]Univ. of Vet. Medicine, Vienna, A-1210, Austria

There is demand for reliable low-cost alternatives for predicting feed efficiency. Thus, the potential of infrared thermography (IR; °C) traits: hind area (HA), eye (EY) and snout (SN); feeding behaviour (FB) traits: time at bunk (TB;min/d), time per meal (TM;min), meal size (MS; kgDM), eating rate (ER;gDM/min), number of meals per day (NM) and daily visits to the bunk (VB); glucocorticoid (GL) levels: faecal cortisol metabolites (FE;ng/g) and plasma cortisol (PC;ng/ml) as predictors of residual feed intake (RFI) were evaluated in 91 steers (436+39 kg) in 2 years (Y1=46; Y2=45). The individual daily feed intake of a corn silage and high-moisture corn-based diet was measured using an automated feeding system. Body weight and thermographs were taken every 28d. Correlations of 0.25 (HA), 0.20 (EY) and 0.35 (SN) were observed between RFI and IR traits. FB traits had correlations of 0.23 (MS), 0.42 (ER) and 0.37 (VB) with RFI. RFI was correlated (-0.45) with FE, but not with PC. Steers were also classified into 3 RFI-groups: LOW (-0.7+0.1 kg/d, more efficient), MED (-0.3+0.2) and HIG (0.2+0.2). LOW steers tended to have lower NO (30.1 vs. 31) than the HIG group ($P<0.1$). LOW steers had lower ER (59 vs. 68), VB (49.2 vs. 57.3) and tended to have lower MS (0.84 vs. 0.93) than HIG steers. LOW steers had greater FE than HIG steers (51.1. vs. 31.2), but similar PC (40.9 vs 41.3). Thus, IR, FB and GL may have application in the assessment of RFI in cattle.

no. 15

Udder health traits as related to economic losses in dairy cattle

H.G. El-Awady[1] and E.Z.M. Oudah[2], [1]Department of Animal Production, Faculty of Agriculture, Kafr El-sheikh University, Kafr El-sheikh, Egypt, [2]Department of Animal Production, Faculty of Agriculture, Mansoura University, 35516, Mansoura, Egypt

A total number of 4752 lactation records of Friesian cows from 2000 to 2005 were used to determine the relationship between somatic cell counts (SCC), udder health status (UHS) and losses in milk production. Studied traits were milk yield traits (i.e., milk yield, (MY), fat yield (FY) and protein yield (PY); udder health traits (UHS) (i.e. SCC, mastitis (Mast), udder quarter infection(UQI)). MTDFRML software was used to estimate genetic parameters. The effects of SCC, UHS and UQI on milk traits and clinical mastitis were also studied. Unadjusted means of MY, FY, PY and SCC were 3936, 121, 90 kg and 453,000 cells/ml, respectively. All fixed effects were significantly ($P<0.01$) affect all studied traits except the effects of month of calving on both fat and protein yields. The SCC, MAST, UHS and UQI increased during winter and summer than spring and autumn. Additionally, SCC and MAST noticeably increased with advancing in parity, while UHS and UQI were increased in the first and second parities then decline with advance of parities. Increasing rate of UHS and UQI lead to increasing SCC and MAST and decreased in monthly and lactationally milk yield. Likewise, losses in monthly and lactationally milk yields per cow ranged from 14 to 89 and from 105 to 921 kg, respectively. It could be concluded that the SCC can be used as a perfect tool for quarter infections, udder health status, milk quality and judging dairy farm profit.

---

no. 16

Effect of butylated hydroxy toluene on post-thaw semen quality parameters of Nili-Ravi buffalo bulls (*Bubalus bubalis*)

A. Hussain, A. Ijaz, M. Aleem, M.S. Yousaf and T.N. Pasha, University of Veterinary and Animal Sciences, Lahore, 54000, Pakistan

The study was to evaluate the potential impact of butylated hydroxy toluene (BHT) on the post- thawed semen quality when added to an egg yolk-Tris-citrate semen extender at varying concentrations (0.0, 0.5, 1.0, 2.0, 3.0 mM). Semen samples were collected from buffalo bulls (n=16), once a week and 2 ejaculates were collected at each collection. The semen straws were cooled, equilibrated at 4 °C and freezed (-196 °C) for 15 days. Thereafter, five semen straws from each treatment were thawed at 37 °C to assess the semen quality in terms of sperm motility (%), viability (%), plasma membrane integrity(%) and acrosomal integrity(%). Post-thawed sperm motility was determined using phase contrast microscopy. Viability, plasma membrane integrity and acrosomal integrity were evaluated by supravital stain, hypo-osmotic swelling test and normal acrosomal reaction, respectively. Two hundreds sperms were counted for each parameter. Results showed that BHT significantly improved motility, plasma membrane integrity, viability and acrosomal integrity at the concentration of 1.0mM and 2.0mM compared to the control. However, higher concentration of BHT failed to affect the viability of post-thawed spermatozoa. In conclusion, addition of BHT in egg yolk-Tris-citrate semen extender can improve the post-thawed semen quality of buffalo bull.

no. 17

Effect of All-lac xcl 5x, Acid-pak 2x, Bio-mos® and Zinc Bacitracin on nutrient digestibility and gastrointestinal morphology of broiler chickens

A.M. Ngxumeshe and R.M. Gous, Tsolo Agriculture and Rural Development Institute, Agriculture, Private Bag X1008, 5170, Tsolo, South Africa

An experiment was conducted with Ross broiler chickens from day-old to 42d of age. A prebiotic (Bio-Mos®, probiotic (All Lac XCL 5x), organic acid (Acid pak 2x), individually or in combination were used to supplement an antibiotic growth promoter (Zinc bacitracin). The chickens were challenge with *Clostridium perfringens* (CP) at 21, 22 and 23 days to determine the efficacy of these additives for replacing antibiotics in hindering the effects of CP on the villus surface area. Feed additives in this experiment prevented the negative effects of CP as the treated birds did not have lesions on their villus surfaces.

---

no. 18

Haematology and plasma biochemistry of the African wild grasscutter (*Thryonomys swinderianus*, Temminck): a zoonosis factor in the tropical humid rainforest of southeastern Nigeria

M.N. Opara[1] and B.O. Fagbemi[2], [1]Federal University Of Technology, Animal Science And Technology, PMB 1526,Owerri Imo State, PMB 1526 Owerri, Imo State, Nigeria, [2]University Of Ibadan, Vet. Microbilogy And Parasitology, Ibadan,Oyo State, Nigeria

Haematological and plasma biochemical values of wild grasscutters, their potential to transmit zoonotic pathogens were evaluated. Three 5ml blood samples were collected from each of 1,000 grasscutters caught in the wild, for haematology, biochemical and parasitological tests. Haematological and biochemical values were compared with those from captive-reared. There were significantly ($P<0.05$) higher lymphocyte, eosinophil and basophil values for wild grasscutters, than the captive-reared. Parasitological examination revealed 15% prevalence of blood protozoa in the wild grasscutters. Blood pathogens encountered were Trypanosoma sp, 66.7% and Plasmodium sp.33.3%, with 20.7% mixed infection. Sex did not ($P>0.05$) affect blood protozoa infection while season did. It is concluded that wild grasscutters serve as efficient reservoir hosts for agents of African trypanosomiasis and malaria in the tropical humid rainforest region of Nigeria.

no. 19

Feeding n-3 fatty acid source improves semen quality by increasing n3/n6 fatty acids ratio in sheep

A. Towhidi[1], F. Samadian[1], K. Reza Yazdi[1], F. Rostami[2] and F. Ghaziani[1], [1]University of Tehran, Department of Animal Science, University College of Agriculture and Natural Resources, P.O. Box 4111, Karaj, 3158777871, Iran, [2]AREEO, Tehran, 19395-1113, Iran

The fatty acid composition of spermatozoa plays an important role for successful fertilization. The aim of this experiment was to study the effect of dietary fish oil as a n-3 fatty acids source on semen quality and sperm fatty acid composition in sheep. Eight Zandi fat-tailed rams were assigned in two groups and fed either a control diet or a supplemented diet with fish oil. Both of the diets were isocaloric and isonitrogenic and formulated according to AFRC (1998). Semen samples were collected weekly and characteristics of samples were evaluated by standard methods. At the end of experiment, one sample of all rams collected and total lipids were extracted by Folch method. Phospholipids were separated by thin layer chromatography. Fatty acid composition of total and phospholipid fractions of sperm was determined by gas chromatography. Feeding fish oil improved the proportion of spermatozoa with progressive motility, percentage of motile sperm and sperm concentrations. The dietary fish oil increased the proportion of docosahexaenoic acid (C22:6, n-3) in sperm fatty acid composition. The ratio of n-3/n-6 fatty acid was also increased in total lipid and in phospholipids of sperm. It was concluded feeding fish oil as n-3 fatty acids source can improve semen quality by changing n-3/n-6 fatty acids ratio.

---

no. 20

Effect of FSH-P and PMSG administration on superovulatory response in Sahiwal cows

A. Ijaz[1], S. Mehmood[1], M. Hussain[1] and Z.U. Rahman[2], [1]University of Veterinary and Animal Sciences, Lahore, 54000, Pakistan, [2]University of Agriculture, Faisalabad, 38040, Pakistan

The effects of gonadotropins, FSH-P and PMSG, were studied to induce a superovulatory response in Sahiwal cows. Cows were divided into 3 groups: T1(FSH-P), T2(PMSG) and T3(normal saline) with 12 cows in each group. T1 cows were administered FSH-P(40mg;in eight divided doses in a tapering fashion) starting on day 9 of estrous cycle for 4 days. T2 and T3 cows were administrated(I/M) PMSG(6IU/kg body weight) and normal saline on day 12 of the estrous cycle, respectively. PGF2α(Cloprostenol;1000mcg/cow) was administered (I/M; morning and evening)to T1, T2 and T3 cows to induce heat after 48 hrs. The cows were inseminated with frozen-thawed semen on the day of estrus. Superovulatory response in each group was confirmed by rectal palpation on day 7 post insemination and embryos were recovered by a non-surgical method on the same day. Time(hrs) from PGF2α administration to exhibition of estrus was similar in T1(48.20±0.15), T2(48.17±0.12) and T3(48.10±0.73). The number of corpora lutea palpated on both ovaries were significantly higher in T1(7.16±2.37) and T2(8.08±1.19) compared to T3(1.00±0.00). Viable embryos recovered for T1(1.5±0.3) and T2(2.1±0.4) were significantly higher compared to T3(0.6±0.4). Similarly, degenerated embryos were significantly higher in T1(0.5±0.3) and T2(0.9±0.4) compared to T3(0.1±0.1). It was concluded that gonadotropins had no effect on the superovulatory response and the number of viable embryos in the Sahiwal cows.

no. 21

Effect of long term use of bovine somatotrophic hormone (bST) on reproductive parameters and health status in Nili-Ravi buffaloes

M.A. Jabbar, I. Ahmad, T.N. Pasha, M. Abdullah and M.E. Babar, University of Veterinary and Animal Sciences, Lahore, 54000, Pakistan

The study was conducted to determine the effect of long term use of bovine somatotropic hormone (bST) on reproductive parameters and health status of Nili-Ravi buffaloes. Thirty Nili-Ravi lactating buffaloes with similar milk production and stage of lactation were selected and randomly divided in two equal groups; A (control) and B (given injection of bST @ 250 mg/animal after an interval of 14 days). Nutritional requirements of experimental animals were met through green forage (45-50 kg/day) supplemented with concentrate ration. The composition of concentrate ration was 17.0% CP and 2.20 Mcal/kg, ME. The calving interval, dry period and lactation length were shorter by 14.0, 26.0 and 2.7 percent in treated group compared with the control. The postpartum estrous period, service period and services per conception were 98.2±76.40 vs.160±56.9 days, 115.10±107.0 vs. 207.04±85.0 days and 1.31±0.51 vs. 1.47±1.11 in the group B and A respectively. Statistical differences were significant for postpartum estrous and service period but for services per conception the difference was non-significant which reflects the positive effect of bST on reproductive parameters. Prevalence of mastitis was 57.14% higher in treated animals. There were variations in body weights for animals in group A and B but these changes over the time were non-significant.

---

no. 22

Plasma haematochemical profile of recombinant bST treated Nilli ravi buffalos

Z. Rahman, T. Khaliq, I. Javed, I. Ahmad, A. Malik and R. Ullah, University Of Agriculture, Physiology And Pharmacology, Department of Physiology and Pharmacology, University of Agriculture, Faisalabad, 38040, Pakistan

The effect of rbST (500 mg) on plasma biochemical and hormonal profiles were studied in two groups, with eight buffaloes in each group. Buffaloes in group I were administered 500 mg of rbST subcutaneously, twice during the experimental period of 32 days at an interval of 16 days, while those in group II acted as control. Blood was collected to determine DLC, ESR and PCV. Serum glucose, cholesterol, urea, protein, albumin, globulin and malondialdehyde, total antioxidant status paraoxonase, AST and ALT were determined. Plasma hormones (T3, T4 and cortisole) were measured using ELISA kits. Overall mean ESR (5.74%), neutrophils % decreased and lymphocytes % increased significantly in bST treated as compared to control buffalo. Regarding plasma biochemical profile there was an overall increase in plasma glucose (15.11%) and triglycerides of bST treated buffalos as compared to control group, while significant decrease in overall plasma total proteins, albumin (0.87%) globulin and urea (4.02%) was observed in bST injected buffalos. Plasma ALT, paraoxonase, malondialdehyde (MDA) were significantly high and Alkaline phosphatase and the antioxidant status decreased significantly in bST injected buffalos. Thyroxin (T4) did decrease significantly in bST injected buffalo as compared to control animals.

Effect of gonadotrophin releasing hormone (GnRH) on postpartum ovarian activity in Nili-Ravi buffaloes (*Bubalus Bubalis*)

A. Tabassam, M. Aleem, A. Ijaz and K.R. Chohan, University of Veterinary and Animal Sciences, Lahore, 54000, Pakistan

Sixteen pluriparus Nili-Ravi buffaloes on 30-35 days postpartum were used to study the effect of GnRH on the initiation of ovarian activity. The buffaloes were randomly divided in groups A and B (A; GnRH-treated, B; normal saline-control; n=8 in each). Group A was administered (I/M) 200ug GnRH and group B normal saline (2ml). Blood samples were collected on day 0, 4 and 8 post-treatment. The average diameter (cm) of the largest follicle on days 4 and 8 in group B (1.1±0.2, 1.9±0.2) was significantly larger than group A (0.6±0.2, 0.37±0.17). The average number of follicles in group A on days 0 and 8 (1.5±0.17, 1.5±0.21) was significantly higher than group B (0.62±.0.32, 0.37±0.17). Two animals from group A responded to GnRH and showed a rise in serum progesterone concentration (ng/ml) 0.81, 3.40, 3.44 and .101, 2.93, 3.00 on day 0, 4 and 8 respectively. However, only one animal in group B showed a rise in serum progesterone on day 4. The number of days from parturition to first estrus in group A were significantly shorter (76.5±6.31) compared to group B (89.7±7.32). Pregnancy rate on first insemination was significantly higher in group A (60.7%) compared to group B (24.3%). In conclusion, GnRH treatment at 35 days postpartum can initiate ovarian activity by luteinizing the larger follicle and can reduce the postpartum interval to estrus.

---

no. 24

Electrolytes and trace mineral status in recombinant bST treated Nilli ravi buffalo

Z. Rahman, T. Khaliq, I. Javed, I. Ahmad, A. Malik and H. Anwar, Department Of Physiology And Pharmacology, University Of Agriculture, Faisalabad, 38040, Pakistan

Two groups, with eight buffaloes in each group were selected from Livestock Experiment Station, University of Agriculture, Faisalabad. The buffaloes in group I were administered 500 mg of rbST subcutaneously, twice during the experimental period of 32 days at an interval of 16 days, while those in group II acted as control as similar amount of sterilized normal saline was injected into those animals. Blood was collected on different days of experimental period and serum was harvested and stored at -20 °C for further analysis. Likewise milk samples were also collected and stored at -20°C. Electrolytes including sodium, potassium, calcium, magnesium, chloride and phosphorus were determined. In trace elements zinc, iron, copper and manganese were estimated in the milk as well as in plasma. In the milk of bST injected sodium, magnesium, chloride, phosphorus and copper increased while non significant difference in other minerals were observed. The plasma sodium and calcium were high and potassium, magnesium, chloride and phosphorus were significantly low in bST treated buffalos as compared to the normal.

Comparison of ethanolic and aqueous extracts of *Tribulus terrestris* on androgenic activities

M.S. Yousaf[1], M.A. Sandhu[2], Z.U. Rahman[2], M. Numan[3], N. Ahmad[2] and A. Ijaz[1], [1]University of Veterinary and Animal Sciences, Lahore, 54000, Pakistan, [2]University of Agriculture, Faisalabad, 34800, Pakistan, [3]Veterinary Research Institute, Lahore, 54000, Pakistan

The experiment was undertaken to compare the androgenic activities of ethanolic extract (EE) and aqueous extract (AE) of Tribulus terrestris fruit in young rats for veterinary usage. A number of thirty, 30 day old, Wister-Albino rats were divided into three groups viz; Group-I (control), Group-II (EE) and Group-III (AE), of 10 in each. The extract was supplemented at 250 mg/kg body weight for two weeks. Rats were sacrificed and the relative weights of liver and testis were found to be significantly lower in EE rats. The epididymal sperm count and WBC count was significantly enhanced in AE. The EE and AE significantly increased the hemoglobin and cholesterol concentrations. The concentrations of total proteins, globulin, testosterone and hepatic paraoxonase were significantly decreased in EE rats. Triiodothyronine, aspartate aminotransferase and homocysteine concentrations were significantly higher in EE rats. Moreover, *In vitro* EE treated testis tissue revealed significant reduction in testosterone synthesis. Testis cultures treated with either extracts produced more nitric oxide than control. In conclusion, AE of Tribulus terrestris can be used in veterinary practice to improve the sexually performance of compromised animals. However, EE should be avoided owing to its untoward effects.

---

no. 26

Comparison of fasting-induced and high dietary zinc-induced molting: trace minerals dynamic in tissues at different production stages in laying hens (*Gallus domesticus*)

J.A. Khan[1], Z.U. Rahman[1], M.S. Yousaf[2] and A. Ijaz[2], [1]University of Agriculture, Faisalabad, 34800, Pakistan, [2]University of Veterinary and Animal Sciences, Lahore, 54000, Pakistan

Two hundred and fifty two (252), 70 week old, commercial single comb white leghorn hens at the end of their first production cycle were divided into two groups: group-I; molted by feed withdrawal and group-II; by high dietary zinc supplementation (3 g/kg feed)to obtain the second and third production cycle. Sampling was carried out at 5%, peak and end of the second production cycle and at 5% and peak of the third production cycle. Plasma, liver and kidneys were collected for electrolytes and trace mineral determination. Plasma sodium and potassium concentrations did not differ significantly between the two groups during different production stages. Calcium concentration was significantly higher at 5% of 2nd and peak of 3rd production cycle while lower at 5% of 3rd production in zinc treated as compared to corresponding stages in fast group. Plasma zinc and iron concentration were higher at 5% of 3rd production cycle in zinc treated group. Liver sodium and potassium concentrations were higher at peak of 2nd production while sodium and copper were lower at peak of 3rd production in fast as compared to that in zinc treated group. Zinc concentration was high at 5% of 2nd and 3rd production in liver and kidney while concentrations of iron and copper decreased and manganese increased in kidneys at 5% of 2nd production in zinc treated as compared to that in fast group.

Physiological limits to production efficiency no. 27

Effect of a short duration feed withdrawal followed by full feeding on marbling fat in beef carcasses

P.S. Mir[1], K.S. Schwartzkopf-Genswein[1], T. Entz[1] and K.K. Klein[2], [1]AAFC, 5403, 1st Ave S., P.O. Box 3000, Lethbridge, AB. T1J4B1, Canada, [2]U. of Lethbridge, Department of Economics, Lethbridge AB. T1K3M4, Canada

Effect of feed withdrawal for 48h, prior to initiation of the fattening period on carcass marbling fat (MF) was studied in 120 European cross-bred heifers (585±39 kg). In a 2X2 factorial design experiment, half the heifers were denied feed (water was available) for 48h prior to the initiation of the 75d fattening period, where heifers were provided either a TMR with 85% rolled barley and 15% barley silage or the feed components free choice. At slaughter samples of the skirt muscle (pars costalis diaphragmatis; PCD) were procured for determination of chemical and dissectable fat content. No differences due to feed withdrawal were observed for carcass weight, percent lean (saleable) meat yield, rib eye area, average fat cover, or fat content of PCD, but the US marbling score was increased to 545 from 502±18 (P=0.048) and the amount of dissected fat from the muscle tended to be higher (P=0.107). Thus 81% of the carcasses from feed denied heifers displayed a "small" amount of MF as compared (P=0.0807) to 68% of the heifers not denied feed. Based on weekly prices of carcasses that displayed a "small" amount of MF, it was found that the price of a carcass could increase by $C4.61, if a 48h feed withdrawal was imposed prior to initiation of the fattening phase. The odds ratio was 1.84 times for carcasses to display "small" amount of MF if heifers were denied feed prior to fattening than if fed continuously.

Technological surges and livestock production no. 1

Integrating farm animal recording with research and development: the South African model

J. Van Der Westhuizen, Agricultural Research Council, Animal Recording and Improvement, Private Bag X2, 1675 Irene, South Africa

The National Farm Animal Recording and Improvement Schemes in South Africa are governed by means of legislation. The Animal Improvement Act also refers to the centralised recording database, called INTERGIS and conduct of participation by breeders' societies by means of registering authorities. The ARC manages the Schemes and INTERGIS on behalf of the Department of Agriculture. The ARC and other research institutions also use recording data and information as sources for research and development. Linkages are also built between the INTERGIS and the Agricultural Geo-reference Information System (AGIS) for the management of natural resources and genetic biodiversity of farm animals. Research and development products are applied by means of a web based portal (called Logix) for use by primary and secondary beneficiaries. Logix also serves as the primary means of electronic data exchange from source (farm level as well as other contributions of data such as milk recording laboratories, bull and boar testing centres or fleece testing laboratories). The development of farm based personal computer software programs has lead to hands-on management based on the research products generated by the national system and resulted in an upswing in electronic data exchange and reduction of response times to users, making it possible to apply management interventions immediately. The system also assists in the uptake of livestock farmers excluded as a result of the previous political dispensation.

no. 2

The relationship between a satellite derived vegetation index and wool fibre diameter profiles

M.B. Whelan[1], K.G. Geenty[2], D.J. Cottle[2], D. Lamb[2] and G.E. Donald[3], [1]Southern Cross University, P.O. Box 157, Lismore, 2480, Australia, [2]University of New England, Armidale, 2351, Australia, [3]CSIRO, Division of Animal Production, Armidale, 2351, Australia

Satellite data provide a systematic and reliable method for recording pasture availability. Pastures from Space (PfS) provides estimates of pasture growth on a weekly basis throughout Australia. The wool fibre diameter profile (FDP) changes in response to feed intake and offers a mechanism for recording the nutritional status of sheep. The aim of this project was to establish whether a relationship exists between FDP and a normalized difference vegetation index (NDVI) from PfS. The FDPs and weekly NDVIs were scaled to the same time period for 38 FDPs from 21 flocks throughout Australia. FDP and NDVI were graphed together and correlations were categorised as strong when rises and falls in NDVI and fibre diameter aligned over the entire time period, neutral when rises and falls matched half the time period and poor when a fall in NDVI corresponded with an increase in fibre diameter or vice versa. Results were as follows; poor 3/38, neutral 9/38 and strong 26/38. Further statistical analysis is underway. Supplementary feeding and temperature were factors that may contribute to poor correlations. Strong relationships exist between FDP and NDVI. The correlation could be used to predict tenderness of the wool clip in a region and improve predictions of the wool clip's average diameter. The relationship could be used to predict the impacts of climate change on wool production.

---

no. 3

Wattle tannins have the potential to control gastro-intestinal nematodes in sheep

I.V. Nsahlai and M.A.A. Ahmed, SASA, University of KwaZulu-Natal, Animal and Poultry Science, Scottsville, Pietermaritzburg, 3209, South Africa

Nematode resistance to anthelmentic drugs is affecting small ruminant production in South Africa, so complementary treatments are needed. This study determined the effect of tannin (in solution) on the gastrointestinal nematode population. Initial egg per gram (EPG) counted in rectal faeces and liveweight (W) (of 24 sheep, 16 females & 8 males) aided to place animals into six groups of 4 animals each, within each of which animals were randomly assigned to four tannin treatments (0, 0.8, 1.6 & 2.4 g tannin/kg W) drenched in solutions of 300 ml water per sheep during three consecutive days. Rectal faeces were taken on days 0, 5, 8, 12, 15, 18 & 21 to determine EPG. L3 Larvae were cultured for 15 days on samples collected on days 0, 5, 15 & 21 to identify larvae and determine population dynamics. For all tannin treatments, EPG decreased ($P<0.05$) over time to a trough on day 18. Though differences among tannin levels varied ($P<0.05$) over time, EPG consistently decreased with increasing tannin level; the tannin effect only attained significance ($P<0.05$) from day 15. The cumulative effect of tannin on EPG was determined by integrating, for each animal, curves of the form: $f(x)=a+b.x+c.x^2+d.x^3$, where f(x) is a polynomial function of EPG on time x; a, b, c and d are coefficients derived using SAS. The area under the curve decreased linearly with increasing level of tannin. Thus 2.4 (g tannin/kg W) was the most efficacious drench.

Camel (*Camelus dromedarius*) breeding guidelines for improved performance and profitability among pastoral communities of northern Kenya

S.G. Kuria and I. Tura, Kenya Agricultural Research Institute, Marsabit, Research, P.O. Box 147, 60500, Kenya

A study was conducted among the Rendille, Gabbra, Turkana and Somali camel keeping communities of northern Kenya to document their traditional camel breeding strategies and practices with a view to formulating improved breeding guidelines. The study was necessitated by reports of declining camel performance by the pastoralists. The study took the form of a survey where 60 Rendille, 49 Gabbra, 46 Turkana and 48 Somali camel owners were randomly selected from their respective communities and individually interviewed. The findings suggested that these four communities predominantly practiced family selection, paying more attention to the traits of the bull's parents. Further, inbreeding was common among the four communities. The breeding practices employed by these communities allowed breeding of aged males and females. These practices could be partly blamed for the 20-40% camel calf mortality reported in the pastoral camel herds in Kenya. Breeding guidelines aimed at controlling inbreeding and promoting breeding of young males and females were formulated and disseminated among the Kenyan camel keepers. This study concluded that it was difficult for many Kenyan pastoralists to conceptualize breeding related problems. The tendency was to attribute productive and reproductive problems observed in camel herds to conventional diseases and nutrition.

---

no. 5

Influence of lipids on the productivity of first calving cows

A. Gonzalez, C. Dominguez, E. Dorta and O. Colmenares, Romulo Gallegos University, IDESSA, Guarico, 2301, Venezuela

The focus of this research was to assess the effects of feeding crossbred primiparous cows with African Palm Crude Oil (APCO) (Elaeis guineensis), poultry litter (PL) and molasses (M), on variables: body weight (BW), body condition (BC), conception rate (CR), calving conception interval (CCI), progesterone (P4) in skimmed milk, and luteal activity (LA). Study was in the hillside area of Guárico State in Venezuela. A sample (n=60) of crossbred Zebu cows were randomly assigned into two treatments: T1 included 2 kg/animal/d PL, 300 g/animal/d M plus 300 g/animal/d APCO; and T2: 2 kg/animal/d PL plus 300 g/animal/d M. All animals were fed this supplement in a continuous grazing system (Cynodon nlemfuensis) during the first 3 months (mo) post-calving. Grass disponibility was 2050, 2300, 1800 kg DM/ha (T1) and 2060, 2350, 1800 kg DM/ha (T2) for mo 1, mo 2, mo 3, respectively, without statistical differences. *In vitro* fractional rate of degradation of NDF was higher ($P<0.05$) during mo 2. BW during mo 1, mo 2 and mo 3 was higher ($P<0.01$) in T1 cows. BC was higher ($P<0.05$) in T1 cows during mo 1, mo 2, and mo 3. CR showed significant differences ($P<0.01$; Chi-sqr=28.23) for T1 cows. CCI was different ($P<0.01$) for T1 (97 days) vs. T2 (278 days). Daily P4 was 1.51 ng/ml, and 0.70 ng/ml ($P<0.01$) for T1 and T2. LA was higher ($P<0.01$; Chi-sqr: 14.19) in T1 (93%) vs. T2 (60%). It was concluded that supplementation with APCO improves reproductive and productive performance in first calving cows.

no. 6

Single tube guiding in conventional milking parlours

R. Brunsch and S. Rose, Leibniz-Institute for Argicultural Engineering, directorate, Max-Eyth-Allee 100, 14469 Potsdam, Germany

Where problems occur with milking techniques it has been shown that automatic milking systems (AMS) have advantages when compared to conventional milking systems (CMS). The construction with individual teat cups distributes forces more regularly to all teats in AMS with individual quarter milking than in CMS. Individual quarter milking and automation is one possibility to improve milking conditions. In quarter individual systems it is possible to analyse udder health parameters for each teat and to better control udder health and milk quality. Especially for large farms it is useful for exact analyses and automation of the milking process. In addition, the workload for the milking person can be reduced. A new single tube milking system is "MultiLactor" (produced by Siliconform GmbH, Türkheim, Germany). To investigate the influence of the single tube milking cluster in CMS, different forces for different udder formations were measured. In addition, vacuum behaviour was tested. The regular allocation of the vertical force to all teats is important for the milking process and udder health. As a first result it can be ascertained that in the "MultiLactor" all teats are loaded with nearly the same force, which means there is good adaptability to the udder by irregular udder formation that can help to reduce udder damage. The problem with wrong positioning of the teat cups could therefore be only asserted by milking clusters without claws. Vacuum level by different flow rates will also be presented.

---

no. 7

The fibrolytic activity of microbial ecosystems in three hindgut fermenters

F.N. Fon and I.V. Nsahlai, SASA, UKZN, Animal and Poultry Science, PMB, 3209, South Africa

This study evaluated the fibrolytic competence of microbial ecosystems in feces collected *in situ* from three hind gut fermenters (dwarf horse (dH), horse (H) and Zebra (ZB)) grazing in their natural environment in summer. The feces crude protein extract for all 3 species was obtained and analyzed for carboxymethyl cellulose (CMCase) and hemicellulase actvities. *In situ* fecal fluids (33 ml) squeezed through four layers of cheese cloth were incubated with 1 g dry maize stover (DM) in salivary buffer for 72h at 39 °C. Gas production (GP), apparent degradability (APD) and true degradability (TD) were determined. CMCase and hemicellulase differed among the three ecosystems ($P<0.05$). CMCase activity was highest in the H followed by the ZB and then dH. Hemicellulase activity was higher ($P<0.05$) in the ZB compared to dH and H. The overall gas production was higher ($P<0.05$) for dH than for ZB and H. APD and TD were highest ($P<0.05$) in dH (431 g/kg and 818 g/kg DM, respectively). TD was higher ($P<0.05$) in the H than ZB, both of which had similar APD. The results obtained shows that the dH as well as the H could be harbouring microbes that have evolved through millions of years of selection to become more competent in fibre breakdown. Therefore a further examination of these anaerobic ecosystems might be beneficial for selecting fibrolytic microbial feed additives.

Technological surges and livestock production no. 8

Bio-processing of some agriculture wastes by gram-negative aerobic bacteria

M. Borji, Research Center of Agriculture and Natural Resources, Animal science, Bigining of Mobarak Abad road-north belth way, 38135-889, Iran

This study was conducted to evaluate the ability of three species of gram-negative aerobic bacteria for improvement of lignocelluloses quality. The bacteria were grown on barley straw, wheat stalk and saw dust in a 6-week growth period and lignocellulose decomposition was followed by monitoring substrate weight loss, lignin loss and carbohydrate loss and improvement of protein and digestibility (*in vitro*) over time. Bacillus sphaericus was clearly the superior and most effective species of those tested that significantly decreased lignin and increased protein and digestibility ($P<0.05$). Improvement of chemical composition and nutritive value of grass lignocelluloses was more than wood lignocelluloses.

Intellectual capital for livestock production no. 1

A model for achieveing sustainable improvement and innovation in agricultural research and development interventions for maximizing socio-economic service delivery

J. Timms and R.A. Clark, Queensland Government, Primary Industries, 40 Ann Street, 4001, Brisbane, Australia

The challenge for governments is to achieve sustained prosperity and improved human, social and natural capital in a resource limited world. Agricultural research and development (R&D) projects are being called into question for less than desired achievement of outcomes, rate and scale of outcomes and ongoing improvement and innovation after the end of projects. There is a large amount of literature available about the limitations, constraints and problems of achieving outcomes and subsequently impact. Authors have put effort into the relatively less explored area of identifying key components, principles and mechanisms for the design, implementation and management of interventions to achieve sustainable improvement and innovation at a regional scale. So the question that must be answered is: How to enable people to achieve these outcomes, now and in the future? This paper presents a practical systems model of the key components that need to be included in the design and management of R&D projects to ensure impact is achieved during the project, and that ongoing improvement and innovation is sustained after completion. Examples of simple tools to operate the model are used to illustrate how it can be used to achieve and sustain outcomes, improvements and innovations in animal production industries.

no. 2

A comparative analysis of alternative methods for capturing indigenous selection criteria

M. Wurzinger[1,2], D. Ndumu[3], R. Baumung[1], A. Drucker[2], A.M. Okeyo[2] and J. Sölkner[1], [1]BOKU-University, Gregor-Mendel-Strasse 33, 1180 Vienna, Austria, [2]ILRI, P.O. Box 30709, 00100 Nairobi, Kenya, [3]NAGRC&DB, P.O. Box 183, Entebbe, Uganda

The paper compares three methods of how indigenous selection criteria can be documented, including the advantages and disadvantages. Method one is a survey where farmers rank important traits that they have previously identified. Method two is a ranking experiment, where farmers are asked to rank live animals. In addition to the phenotype, a theoretical life history (performance, reproduction) is assigned. Under the third approach, in a hypothetical choice experiment (choice cards) farmers are asked to compare theoretical profiles. All three approaches have been applied in a study of selection criteria of pastoralists keeping Ankole cattle in Uganda. Traits included in all approaches were phenotype (body size, colour, horns), production, fertility and disease resistance. Selection criteria for bulls and cows were evaluated separately. As the number of traits was low and the definition of levels was not always identical, we abstained from a formal comparison of ranks. Some traits (fertility) in cows are highly ranked in all methods. Phenotype is equally important in the selection of bulls and cows, but ranks differently. The importance of other traits varies under the different methods. A combination of a survey with one of the other two methods is recommended. All methods are applicable for situations where farmers have little formal education.

---

no. 3

A Bayesian approach to the Japanese Black carcass genetic evaluation procedure

A. Arakawa[1], H. Iwaisaki[2] and K. Anada[3], [1]Niigata University, Graduate School of Science and Technology, Nishi-ku, Niigata-shi 950-2181, Japan, [2]Kyoto University, Graduate School of Agriculture, Sakyo-ku, Kyoto-shi 606-8502, Japan, [3]Wagyu Registry Association, Research and Extension Section, Nakagyo-ku, Kyoto-shi 604-0845, Japan

The purpose of this study was to establish a Bayesian approach via Gibbs sampling (GS) to be used in the official genetic evaluation program for carcass traits of Japanese Black cattle in Japan. A total of 6 traits evaluated are carcass weight, ribeye area, rib thickness, subcutaneous fat thickness, estimated yield percent, and marbling score. As prior distributions, uniform and normal distributions and independantly scaled inverted chi-square distributions with degree of belief and scale parameters of -2 and 0, respectively, were used for nuisance parameters, breeding value and variance components, respectively. Using a dataset consisting of carcass field records of about 25,000 fattening animals with approximately 60,000 animals in the pedigree file, GS was run at varying total sample size, burn-in period and thinning interval to assess their effects on the estimated marginal distributions. The values of means of the posterior distributions calculated using every 20th sample from 90,000 GS rounds, after the first 10,000 rounds were discarded as burn-in period, were shown to be similar to the corresponding estimates using the conventional two-step procedure employing REML and EBLUP. The validity of the current approach was confirmed using variously-sized datasets including very large one.

The role of agricultural research and development: the socio-economic development nexus

N.B. Nengovhela, Agricultural Research Council, Animal Production Institute, P/Bag 2, 0062, Irene, South Africa

There are growing pressures in countries to improve the impact of their public sector research and development (R&D). The United Nations enshrines poverty reduction and the right to development as a human right, and this call is accentuated by the Millennium Development Goals. Locally, policy makers and members of the public are showing a growing concern over the cost versus performance of agricultural socio-economic intervention. For these two reasons, it is axiomatic for the South African agricultural R&D sector to improve public service quality, efficiency and effectiveness to redress poverty and inequality. There is a need to design and implement programmes and projects that demonstrate significant and lasting changes to prove their value and justify continued funding. In addition, the focus should be on the achievement of outcomes rather than the production of outputs. However, reducing poverty and inequality require targeted strategies and a concerted effort to improve service delivery. The focus of this paper will be on the role of agriculture in socio-economic development, especially regarding animal production; the South African agricultural R&D system; current problems and issues in the design, implementation, and management of interventions. This paper will provide the context and platform for other papers in the session.

---

no. 5

Capacity building in achieveing sustainable improvement and innovation: the why, what, how, implications and expriences in the agricultural reseach and development context

R.A. Clark, Queensland Government, Primary Industries and Fisheries, 40 Ann Street, 4001, Brisbane, Australia

Anybody who is involved in socio-economic development, importantly in agriculture, knows how critical human and institutional capacity building is to the development effort and the chances of its success. First, Sustainable Improvement and Innovation (SI&I) does not take place in a vacuum. People (managers, practitioners, beneficiaries) are embedded in development interventions. Second, the formal education provided to farmer support professionals and their consequent advice to farmers are technically orientated. Lastly, SI&I concepts are complex and difficult to operationalize in the design, management and evaluation of development initiatives. So, capacity development plays a crucial role in peoples development. There is an urgent need for measures to redress some of the skewedness in the formal education currently offered and an urgent need for trained people who can undertake socio-economic development. The purpose of this paper is to provide an overview of capacity development for development practitioners. The paper 1) outlines the reasons for capacity development 2) explains what is meant by the term capacity building for impact 3) share how it is implemented and 4) details some of the implications and experience when implementing the SI&I approach in the Beef Profit Partnerships project, within the animal production sector.

no. 6

The use of Internet-based tools in establishing scrapie resistant sheep flocks in Canada

D.G. Bishop and A. Farid,

Genotypes of Canadian purebred sheep at the prion protein gene from various laboratories are being assembled into a national database and linked with pedigree information. Breeders have had password protected access to their flock data through a web-interface (genenovas.ca) since early 2005. The system is designed i) to provide breeders with a fast and secure way of obtaining the genotypes of their sheep, ii) to be the depository of the official genotype of animals, iii) to help breeders to easily manage their genotype results, iv) to detect genotype inconsistencies and measure the accuracy of the data when the genotypes of parents and offspring are available, v) to allow flock owners to transfer the genotype of any number of their sheep to a 'Marketplace', which is accessible by all prospective buyers, and vi) to determine the genotype of progeny of tested parents, where possible, thus avoiding unnecessary testing and reducing the cost of establishing resistant flocks. To date (June 2008), genotypes of over 17,000 sheep of 42 breeds from 464 farms have been added to the database. The genotypes of 972 lambs (5.7% of animals tested) were predicted and added to the database. As the frequency of homozygous resistant animals to scrapie increases by selection, the number of such animals is expected to increase. The proportion of breeders who never visited their page on the website was 45.5%, while others logged into their website between one and 142 times. The results show that the use of Internet by the Canadian sheep breeders is reasonably high.

---

no. 7

Capacity building for sustainable use of animal genetic resources in developing countries

J.M.K. Ojango[1,2], B. Malmfors[3], A.M. Okeyo[1] and J. Philipsson[3], [1]ILRI, Box 30709, Nairobi, Kenya, [2]Egerton Univ., Box 536, Njoro, Kenya, [3]SLU, Box 7023, Uppsala, Sweden

Animal Genetic Resources for Food and Agriculture (AnGR) have great potential to contribute to increased food security and the improved livelihood of poor people in developing countries. However, indigenous livestock which are well adapted to local conditions are often underutilized and with few exceptions systematic breeding programs are lacking. This is due to a lack in a "critical mass" of people trained and informed in issues related to AnGR, in addition to insufficient policy support and institutional frameworks. To address these shortages the International Livestock Research Institute (ILRI) in collaboration with the Swedish University of Agricultural Sciences, and supported by Sida (Sweden), launched a project "training the trainers", for university lecturers and NARS in developing countries. So far, more than 100 scientists from 25 countries in sub-Saharan Africa and 15 countries in South and South East Asia have been trained in refresher courses and workshops. An Animal Genetics Training Resource (Web & CD) has also been developed, containing case studies from the tropics, generic information on livestock genetic improvement, breeding program design, and research and teaching methods. In general, very positive effects of the training courses and resources developed have been reported by participants on their teaching, research and on supportive activities for better use of AnGR. Despite these effects there is still a huge demand for expanded activities in this area.

Intellectual capital for livestock production no. 8

Teaching sustainable development in the animal science curriculum

A.J. Van Der Zijpp, C.H.A.M. Eilers and S.J. Oosting, Wageningen University, Animal Production Systems Group, P.O. Box 338, 6700 AH Wageningen, Netherlands

Traditionally the curricula of animal science students have moved from a comprehensive range of subjects in socio-economics, plant and soil sciences, and animal science to a focus on strictly animal science courses. System science, and sustainability science provide the framework to teach the ecological, economic and social issues of farming systems and integrate these issues to achieve the most sustainable farming system or food chain. Courses of the Animal Production Systems Group like Systems Approach in Animal Sciences (BSc), Animal Production Systems: Issues and Options and Future Livestock Systems (both MSc), part of the Animal Science curriculum at Wageningen University, prepare students for problem identification and integrative analysis. The approach is generic and case studies are derived from the tropics and developed countries. The system methodologies are used in the thesis research projects. The competence to integrate gamma and beta science and understand farming systems and value chains in their dynamic context are highly valued in research, policy and practice.

Managing resources for sustainable livestock production, ... no. 1

Managing resources for sustainable livestock production, human dignity and social security

M. Wulster-Radcliffe and L.S. Bull, North Carolina State University, Animal Science, Box 7608, Raleigh, North Carolina 27695-7608, USA

Livestock production represents a major contribution not only to the provision of significant portions of the dietary energy (about 17%) and protein (about 35%) but also critically important micronutrients and those contributions are increasing steadily. With about 30% of the global land area occupied by grass production and with much of that semi-arid and thus incapable of producing crops, food from livestock production represent the only way for individuals living in those areas to survive. That dependence on those livestock for food represents food security during times of crisis due to drought or other disasters (natural or otherwise). At the other extreme is the recent and growing trend toward high density concentrated livestock production found most often in developed countries. Issues associated with the broad spectrum of animal production practices include air and water quality impacted by animal wastes, zoonotic diseases and pathogen exposure to workers, global climate change contributions (especially greenhouse gases), and increasingly, animal welfare concerns. The importance of animals as sources of human food on a global basis dictates that the associated issues identified by society must be addressed and solutions found. At stake is the well-being of a major portion of the population in a world where food current concerns about food scarcities abound. The presentation will address these major issues and offer some suggestions for socially acceptable and sustainable livestock production into the future.

The global plan of action for animal genetic resources

I. Hoffmann, D. Boerma and B. Scherf, FAO, AGA, Viale delle Terme di Caracalla, 00153 Rome, Italy

The International Technical Conference on Animal Genetic Resources for Food and Agriculture in Interlaken, Switzerland, in September 2007, first launched The State of the World's Animal Genetic Resources for Food and Agriculture, the first authoritative assessment of global livestock biodiversity. Its main outcome was the adoption of the Global Plan of Action for Animal Genetic Resources (GPA) which is based on the fact that countries are fundamentally interdependent with respect to animal genetic resources for food and agriculture, and that substantial international cooperation is necessary to manage these resources. The GPA contains 23 Strategic Priorities, clustered into four Priority Areas: Characterization, inventory and monitoring of trends and associated risks; Sustainable use and development; Conservation; and Policies, institutions and capacity-building. The GPA provides an internationally agreed basis for the sensible management of animal genetic resources, which are increasingly at risk, and for sharing the benefits and responsibilities fairly and equitably. Animal genetic resources are of particular importance in marginal environments, where they provide the mainstay of livelihoods. Climatic change both increases the risk of the loss of these resources, and increases their importance in adapting to climatic change, as well as in meeting the food security needs of a growing world population, and in achieving the MDGs. The task before the international community now is to translate the outcomes of Interlaken into concrete and sustainable action.

---

no. 3

Development of livestock based sustainable family farms: Heifer International's approach

D.P. Bhandari and T.S. Wollen, Heifer International, Animal Well-Being, 1 World Avenue, Little Rock, 72202 AR, USA

Heifer International supports families in integrated, sustainable livestock and crop farms in order to develop strong local communities. Livestock receive protection from predators, improved feed supplies, appropriate housing and humane slaughter in exchange for the benefits they give to the families of animal source protein, draft labor, materials for clothing and household uses and the valuable manure for composting and crop growth. Heifer recognizes the need for a long-term approach to development for rural communities. Project communities practice agro-ecologically sound methods of raising animals and crops. Full participation of the community is involved in a project, recognizing the contributions of women, men, girls and boys. In Africa, Asia and Latin America, as many as 95 percent of families receive their livelihoods from agriculture every day. Meat, milk and eggs are not neatly packaged in cardboard or plastic containers, and the role of livestock is not debated. Livestock are essential for draft power; organic fertilizer; a source of nutritional protein; wool, leather and fibers; transportation and they serve as living savings banks. Most livestock are raised on small-scale holistic family farms. The health of people, land, crops, and livestock are integrated into the farm system and all family members are encouraged to participate equally in each Heifer project.

Effect of TDS plus Br on the accumulation of As and Pb from drinking water in broilers

M.C. Mamabolo, N.H. Casey and J.A. Meyer, Department of Animal and Wildlife Science, University of Pretoria, 0002, South Africa

Water quality constituents ingested through drinking water can affect the animal's physiology negatively and through bioaccumulation in tissues pose a biohazard to consumers. The study evaluated the effectiveness of a TDS (total dissolved solids) treatment as a possible alleviator of the accumulation of potentially hazardous chemical constituents in drinking water in broiler tissues. The trial design was 4 treatments x 7 replicates x 12 mixed Ross broilers per replicate. Treatments were T1=< 0.005 mg/L Br, As and Pb + <500 mg/L TDS, T2=0.1 mg/L As, 1 mg/L Br and 0.1 mg/L Pb + <500 mg/L TDS, T3=< 0.005 mg/L Br, As and Pb + 1500 mg/L TDS, T4=0.1 mg/L As, 1 mg/L Br and 0.1 mg/L Pb + 1500 mg/L TDS, administered through the drinking water from Day 1 to Day 42. Water intake and growth performance were recorded. Broilers were slaughtered, tissue samples taken and analysed for accumulation of elements. TDS had a significant effect ($P<0.05$) on the accumulation of elements in some, but not all tissues. TDS plus bromide could alleviate arsenic and lead accumulation in tissues ($P<0.05$).

---

no. 5

Women role in dairying in developing countries: a new approach

S.H.R. Raza, University of Agriculture, Faisalabad, Livestock Management, Livestock and Dairy Devlp. Board, W-76, Umar Plaza, 1st Floor, Jinnah Avn, Blue Area, Islamabad, 44000, Pakistan

Traditionally in all developing countries women play a significant role in animal production especially in dairying. Under rural social set up, a female calf is often allotted to young girl and she becomes solely responsible for her all management from feeding to breeding till she calves. This dairy animal is a part of her dowry and goes with her to new home after her marriage. Recently in Pakistan, a new project at national level has been launched to empower the landless and small dairy farmers for better prices of milk by providing them chilling units to pool their milk at one point (for better bargaining with milk collectors) and training in different aspects of improved animal production. It was observed that women farmers involved in dairying actively participated in this project. At present many milk collection units are solely operated by women organizations. They are maintaining chilling centre and earning good profit. In northern areas (Himalayan mountains range) out of 06 units two are solely run by women. Instead of selling raw milk, these women members have got training in value addition and now selling yoghurt (plain and flavoured), cheese and butter at almost 2 folds higher prices (Rs 25.0 versus 75.0/lit) than milk. This approach has provided them economic empowerment, freedom in decision making in male influenced society and social uplift. This experience must be shared with other women farmers facing the similar scenario in dairying at global levels.

The effect of heat treatment on chemical composition and nutrient apparent digestibility of *Mucuna pruriens* seeds and *Mucuna pruriens* seeds based diets by guinea fowl

M. Dahouda, University of Liège, Faculty of Veterinary Medicine, Department of Animal Production, Boulevard de Colonster B43, 4000 Liege, Belgium

Mucuna priurens, is a crop used under the tropics to improve fertility of agricultural soils owing to its good resistance to pests.However, seed production is not well valorised owing to potential toxicity due to the presence of L-Dopa. Experiments were conducted with guinea fowls to determine the effect of processing on chemical composition and nutrients digestibility of Mucuna seeds offered alone or incorporated at 4%, 12% or 20% levels in complete diets. Seeds were either cracked and toasted, cooked or used raw.Diets with 20% of seeds had significantly higher crude fiber contents.L-Dopa contents increased concordantly with increasing Mucuna inclusion levels. Cooking reduced markedly L-Dopa content while toasting had no effect. Feed intake and weight gain were not influenced by the diets with Mucuna seeds. In Mucuna seeds, cooking significantly increased the crude protein and crude fibre utilization and reduced L-Dopa content.When Mucuna seeds were fed alone, processing had no effect on feed intake which was dramatically lower than the intakes with the complete diets containing Mucuna seeds. Inclusion level of seeds in diets was therefore limited by palatability and not by toxicity.When processed adequately i.e. decorticated, heated in water and incorporated in diets at a moderate level of 12% in a complete diet the seeds of Mucuna could be profitable for guinea fowl production.

---

no. 7

The study impact of effective microorganisms (EM) on yield, chemical composition and nutritive value of corn forage (KSC 704)

M. Borji and L. Jahanban, Research center of agriculture and natural resourses, Animal science, Bigining of Mobarak Abad road-north belth way, 38135-889, Iran

The influence of Effective Microorganisms (EM) for increasing crop yields and improvement of soil properties is widely reported, but there is no very data about it effects on chemical composition and nutritive value of crops. Thus in 2006 an experiment was conducted to determine the effect of EM on yield, chemical composition and quality of corn (forage) in research center of agriculture and natural resources of Markazy province. EM as spray was including three treatments: b1=no spray or control, b2=$^1/_{200}$ and b3=$^1/_{500}$ EM concentrations, investigated by complete randomized blocks design with three replicates. Results showed that EM improved yield and content of OM, NFC, DDM, DE, and Na significantly. Because EM application increased yield from 15.9 to 17.9 ton/ha (approximately about 30%) in comparison with control (without EM), thus production of CP and DDM in unit area (kg/ha) increased significantly. Thus application of EM would undoubtedly helps to improve yield and quality of different crops as an effective way in our country.

Magnificence of locally available resources and its role in reproductive health management in livestock

R. Kumar, R. Singh and Y.P. Singh, S.V.B. Patel University of Agriculture & technology, Animal Science, Meerut, 250 110 (UP), India

An study was carried out on 2160 respondent buffaloes and observations were pooled to study the various reproductive problems viz. anestrous, repeat breeding, abortion, dystocia, utero-veginal prolapse, retention of placenta, metritis, and managemental practices using locally available resources. Therefore, responding buffalo owners had no alternatives except adoption of cost effective and easily available indigenous technical knowledge (ITK)/herbs. In case of anestrous; a significant number of respondent were using herbs in different combinations such as Methi (*Trigonella foenum graecum*) with wheat bran boiled for three days, Bajra (*Pennisetum typhoides* L.) feed ½ to 1 kg boiled with Gur (Jaggery) (250 g) for seven days, Masur Daal (*Lens culinaris moench*) ½ to 1 kg for three days at evening and for the treatment of repeat breeding, Feeding barley (*Hordium vulgare*) flour mixed with water (½ to 1 kg up to one weak), Mahandi powder (*Lawsonia alba*) 250 g mixed with 1 liter milk. Therefore, In case of Utero-veginal prolapsed, Feeding crushed 100 g Satyanasi weed seed (*Argemone mexicana*) mixed with water for a week, Feeding 250 g Mahandi (*Lawsonia alba*) with water for a three days and giving banana's root extract (*Musa paradisica*) also for the same. The above said practices give results in response. Therefore for multi-disciplinary research activities for scientific evaluation and validation of such practices are suggested.

---

no. 9

Cattle waste management practices for sustainable smallholder dairy production

F. Kabi and F.B. Bareeba, Makerere University, Animal Science, P.O. Box 7062, Kampala, Uganda

This study examined factors that influence adoption of cattle waste management practices among smallholder dairy farmers. Burying cattle excreta in trenches, topical application and control practices where farmers did not apply excreta were compared followed by adoptability analysis. 55% of the farmers kept only 2 crossbred cows producing limited excreta for farm use and 78% of the farmers produced less than 10 litres of milk from each. Burying cattle excreta produced higher ($P<0.05$) dry matter (DM) yields of 22.4 t/ha of elephant grass leaves compared to 15.1 t/ha and 10.5 t/ha from topical excreta application and control treatments, respectively. Although crude protein (CP g/kg DM) did not vary ($P>0.05$) with the different cattle excreta management practices, DM yield of elephant grass, its leaf:stem ratio and ash were superior ($P<0.05$) with excreta burying. While burying excreta improved quality and quantity of elephant grass, yields varied from farm to farm depending on the level of adoption of cattle excreta management practice. High cost, unfair division of labour among gender and lack of basic equipment such as spades and wheelbarrows to carry the excreta to gardens affected the adoption of cattle excreta management practices. Feeds were the major cost of production and accounted for 46.8% of total cost, followed by labour, veterinary services and equipment. It is concluded that high costs and unfair division of labour among gender as well as lack of basic equipment to transport excreta influence the adoption of cattle excreta use by smallholder dairy farmers.

Nitrogen flow and quality of manure from different diets fed to sheep

J.O. Ouda[1,2], A.T. Modi[2] and I.V. Nsahlai[2], [1]Kenya Agricultural Research Institute, P.O. Box 14912 Nakuru, 051, Kenya, [2]University of KwaZulu-Natal, Animal Science, Private Bag X01, Pietermaritzburg, 033, South Africa

This study investigated the influence of diets comprised of different roughages (RG) and protein supplements (PS) on partitioning of dietary nitrogen (N) ingested by sheep, manure chemical composition and mineralization of manure N. The RG were maize stovers and grass hay. The PS were lucerne hay (LH), lespedeza hay (LPZ) and sunflower oil cake (SFC). Ten diets composed of different RG:PS ratios with CP contents ranging from 55 to 230 g/kg dry matter (DM) were fed to sheep. Faeces and urine were separately collected and analysed for N content and faeces further analysed for mineral composition. Fresh and dried faeces were in addition analysed for ammonium ($NH4^+$). Fresh faeces were incubated and had net N mineralised determined over a 20 weeks period. The diets differed in DM intake, DM digested, manure output and weight gain (WtGain) by sheep. They also differed in $NH4^+$, which ranged from 653 to 4240 and 114 to 2153 mg/kg DM in fresh and dried faeces, respectively. Drying manure led to substantial loss (24 to 83%) of $NH4^+$. Net N-mineralization was maintained by a fertilizer (control) and manure from diets which had CP content above 170 g/kg DM. These diets supported sheep WtGain of 62.0 to 172 g/d. Faeces from diets containing SFC showed highest N, P and K content while diets containing LH showed high Na content. The results showed that strategic supplementation can be used to optimize maize-ruminant system productivity and N-recycling.

---

no. 11

Assessment on the availability of cattle and feeds for quality beef production in Tanzania

S. Nandonde[1], M. Tarimo[1], G. Laswai[1], D. Mgheni[1], L. Mtenga[1], A. Kimambo[1], J. Madsen[2], T. Hvelplund[3] and M. Weisbjerg[3], [1]Sokoine University, Morogoro, 255, Tanzania, [2]University of Copenhagen, Copenhagen, 45, Denmark, [3]University of Aarhus, Tjele, 45, Denmark

There has been a growing demand for quality beef in Tanzania that has motivated business entrepreneurs to enter into finishing cattle under feedlots but, the sustainability of this system in terms of feed and animals availability is yet to be documented and hence the purpose for this study. Steers available for fattening were predicted from average cattle population, while amount of available nutrients were derived from crop harvests using secondary data from different ministries and FAO. The national cattle herd was estimated to be 18 million with annual growth rate of 1.94%. The indigenous steer herds accounted for 99.6% of the total steers in Tanzania estimated at 2.5 million. The average annual concentrates that could be available for finishing cattle under feedlot were (kg, DM) maize grain (6.1×10^7), maize bran (5.6×10^8), rice polishing (8.9×10^7), Wheat bran (1.4×10^7), Wheat pollard (4.2×10^6) and Molasses (4.5×10^7), whereas available protein concentrates were (kg, DM) sunflower (2.1×10^7) and cotton (6.2×10^7) seed cakes. The nutrient outputs were 1.0×10^{10} MJ ME and 1.2×10^8 kg CP and these could support 58% of the steers and if more energy feed resources are exploited the number of steers to be fattened could be raised to 72%. The demand for quality beef could be met by proper investment in exploiting useful and crops and crop by products.

Latest development on cryopreservation of zebrafish (*Danio rerio*) oocytes

S. Tsai, D.M. Rawson and T. Zhang, University of Bedfordshire, LIRANS, 250 Butterfield, Great Marlings, Luton, LU2 8DL, United Kingdom

Cryopreservation of germplasm of aquatic species offers many benefits to the fields of aquaculture, conservation and biomedicine. Although successful fish sperm cryopreservation has been achieved with many species, there has been no report of successful cryopreservation of fish embryos and late stage oocytes which are large, chilling sensitive and have low membrane permeability. In the present study, cryopreservation of early stage zebrafish ovarian follicles was studied for the first time using controlled slow freezing. The effect of freezing medium, cooling rate, method for cryoprotectant removal, post-thaw incubation time and ovarian follicle developmental stage were investigated. Stage I and II ovarian follicles were frozen in 4M methanol in either L-15 medium or KCl buffer. Ovarian follicle viability was assessed using trypan blue and FDA+PI staining. The results showed that KCl buffer was more beneficial than L-15 medium, optimum cooling rates were 2-4 °C/min, stepwise removal of cryoprotectant improved ovarian follicle survival significantly, stage I ovarian follicles were more sensitive to freezing and there were no significant differences between ovarian follicle survival when trypan blue or FDA+PI staining was used. The highest survival obtained so far is 69.4±5.4% with stage II ovarian follicles after freeze-thawing and incubation at 28 °C for 1h when assessed with trypan blue staining. Studies are currently being carried out on *in vitro* maturation of the cryopreserved ovarian follicles.

---

no. 13

Evaluation of *Moringa oleifera* and bamboo leaves as feed supplements for ruminants using *in vitro* gas production technique

V.O. Asaolu[1], S.M. Odeyinka[2] and O.O. Akinbamijo[3], [1]International Trypanotolerance Centre, PMB 14, Banjul, No code, Gambia, [2]Obafemi Awolowo University, Animal Science, Ile-Ife, No code, Nigeria, [3]African Union Interafrican Bureau for Animal Resources, Nairobi, No code, Kenya

Moringa and bamboo leaves were evaluated as single fodders and in the following combinations; 75% moringa:25% bamboo, 50% moringa:50% bamboo, and 25% moringa:75% bamboo, using crude protein, fibre components, *in vitro* gas production, and predicted %OMD, ME and SCFA. The sole moringa fodder produced the highest gas volume (43 ml/200 mg sample) at 96 hours of incubation. Increasing gas volumes were observed with increasing levels of moringa in the fodder combinations. 50% and 75% moringa inclusion levels produced comparable gas volumes to reported values for sole moringa fodder and the leaves of some other commonly used multi-purpose trees. Positive correlation was observed between gas production and crude protein levels while negative correlations were observed between the fibre components and gas production. Species had significant effects on gas production at all incubation periods for all the fodder combinations while significant PEG effect on gas production was observed for only 25% moringa:75% bamboo at 96-hour incubation interval. Species was also observed to have significant effects on the predicted parameters while no significant PEG effects were observed. Moringa and bamboo leaves can be combined at 50% and 75% moringa inclusion levels to supplement ruminant diets, particularly during the dry season when nitrogen is limiting in most tropical pastures.

Growth curve parameters and performance of local Ghanaian chicken and SASSO T44 Chicken

R. Osei-Amponsah[1], B.B. Kayang[1], A. Naazie[1], M. Tixier-Boichard[2], R. Xavier[2], P.F. Arthur[3] and I.M. Barchia[3], [1]Department of Animal Science, P.O. Box 25, University of Ghana, Ghana, [2]INRA, Génétique et Diversité Animales, UMR 1236, 78352, Jous-en Josas, France, [3]Elizabeth Macarthur Agricultural Institute, NSW Department of Primary Industries, PMB 8, Camden NSW 2570, Australia

Body weights of local and SASSO T44 chicken kept under improved management were measured from hatch to 40 weeks of age. Growth curve parameters of body weight were estimated using the Gompertz growth model. Asymptotic mature weights were estimated as 2911.9g and 3628.9g for female and male SASSO T44 chicken; 1232.3g and 1708.3g for female and male local Forest chicken; 1321.3g and 1798.3g for female and male local Savannah chicken. Age at point of inflection was estimated as 73.78 days and 78.19 days for female and male SASSO T44 chicken; 72.4 days and 78.89 days for female and male local Forest chicken; 72.38 days and 77.35 days for female and male local Savannah chicken. The maturing rate parameter was estimated as 0.0228 g/day and 0.0230 g/day for female and male SASSO T44 chicken; 0.0232 g/day and 0.0216 g/day for female and male local Forest chicken; 0.0238 g/day and 0.0229 g/day for female and male local Savannah chicken. Local chicken compared favourably with SASSO T44 chicken in terms of the maturing rate parameter and the age at which they attain maximum age. Selection in local chicken should reduce the time taken by birds to reach the point of maximum growth.

---

no. 15

Age, ecotype and sex effects on growth performance of local chickens and SASSO T44 Chickens under improved management in Ghana

R. Osei-Amponsah[1], B.B. Kayang[1], A. Nazie[1], M. Tixier-Boichard[2] and R. Xavier[2], [1]University of Ghana, Animal Science Department, P.O. Box 25, University of Ghana Legon, Ghana, [2]INRA/AgroParis Tech, Génétique et Diversité Animales, UMR 1236, 78352, Jous-en Josas, France

Body weights of local chicken from the Forest and Savannah ecological zones of Ghana were compared to a control population. Data on 213 Forest chicken, 160 Savannah chicken and 183 SASSO T44 chicken kept under improved management were measured for both sexes from hatch to 40 weeks of age. At all ages the control population had significantly ($P<0.05$) higher weights than the local ecotypes. At hatch, chicken from the forest zone were heavier ($P<0.05$) than chickens from the Savannah zone. This advantage became insignificant at 2 weeks after which period the savannah chicken were significantly heavier ($P< 0.05$) than the forest chicken at all ages. Male birds were heavier ($P<0.05$) than the females from 10 weeks onwards. Actual and scaled growth rates were significantly higher ($P<0.05$) in the control population than the local population. However, size scaling of the growth rates narrowed the gap in actual growth performance between SASSO T44 and the local ecotypes but could not eliminate the effect of ecotype indicating a strong genetic effect of ecotype on growth rate. The significant effect of ecozone on the growth potential of local chicken is an indication that their productive potential could be improved through interventions in the environment.

Economic evaluation of natural suckling in a dairy and meat production system

L.J. Asheim[1], A.M. Grøndahl[2], Ø. Havrevoll[3] and E. Øvren[1], [1]Norwegian Agricultural Economics Research Institute, Research, P.O. Box 8024., Dep., 0030 Oslo, Norway, [2]Norwegian School of Veterinary Science, Companion Animal Clinical Sciences, P.O. Box 8146 Dep., N-0033 Oslo, Norway, [3]Nortura, Pb 75, 2360 Rudshøgda, Norway

The paper deals with economical effects of different systems of start feeding for calves using a Linear Programming (LP) model adapted to a dairy/meat farm in the grain areas of Eastern Norway. The common breed Norwegian Red Cattle (NRF) is bred for both milk and meat and most dairy farmers raise the calves for meat or replacement. Calves are usually separated from their mother shortly after birth and raised in separate pens with milk feeding from teat or bucket. Approximately ten to fifteen percent of the calves are sold for fattening at 3-4 months of age. In a case study natural suckling for 6-8 weeks resulted in weight gain of 1.2 kg/day until 13 weeks for male calves. The calves were slaughtered at 15 months at 301 kg carcass weigh which compares to 296 kg and 18.3 months on an average for NRF bulls. The economics of natural suckling is compared with that of standard artificial rearing by restricted or non restricted milk feeding. Improved start feeding enhances the live weight and health condition of calves and their dams, but farmers need more cows to fill the milk quota.

---

no. 17

Study on comparative advantage of small holder dairying

K. Kothalawala[1], S. Kumar[2] and H. Kothalawala[3], [1]Division of Livestock Planning and Economics, Department of Animal Production and Health, Gatambe, 20400,Peradeniya, Sri Lanka, [2]Division of Livestock Economics and Statistics, Indian Veterinary Reserch Institute, Bareilly, 243122,U.P., India, [3]Veterinary Research Institute, Department of Animal Production and Health, Gannoruwa, 20400,Peradeniya, Sri Lanka

The study was planned and carried out with an objective of investigating income generation capacity by smallholder dairying in Kandy district in Sri Lanka in the last quarter of 2007. A total of 120 number of households were selected using random stratified sampling method. Data were analysed to estimate the hours of labour utilisation (male/female), cost and profit of production of milk, income, etc. The results revealed that the average cost of production was 30.55 SLR (Sri Lankan Rupees) and the profit was negative (-7.14 SLR) per milk litre, when labour was valued at market rate. It was 9.46 SLR when the labour was excluded .Male labour utilization was significatly higher (2.16 hrs/d) to the female labour(38 minutes/d). On an average a farm family can earn 111.44 SLR per day by dairying . It can be concluded that rural families can earn part of their income by dairying and more female labour can be utilizated to enhance the socio economic impact of it.

Local livestock breeds, Livestock Keepers' Rights and the sustainable use of marginal lands

J.B. Wanyama[1], I. Köhler-Rollefson[2] and E. Mathias[2], [1]VETAID, Mozambique Programme, Av. Eduardo Mondlane, 1270, C.P. 1707, Mozambique, [2]League for Pastoral Peoples, Pragelatorasse 20, 64372 Ober-Ramstadt, Germany

Indigenous breeds account for more than 80% of the world's livestock population. These breeds have been created and maintained by local breeders in various environments, often under extreme climatic conditions. The hardy, well adapted animals are relatively independent from expensive inputs, enabling their keepers to make use of areas not suitable for crops. With increasing need for food, energy and transport, indigenous livestock breeds are destined to gain value in providing solutions to these global needs. Worldwide, 640 million subsistence farmers keep livestock. They act as custodians of the world's rich and diverse local breeds. But scientists and policy makers little recognize this role. Unfavorable development policies and programmes such as the introduction of exotic breeds and the promotion of irrigated food and biofuel crops in marginal areas favor large-scale and industrial producers. Under such conditions mobile and small-scale livestock keepers cannot compete and increasing numbers are giving up. If this goes on unchecked, their breeds will disappear one after the other. This paper argues that indigenous livestock breeds and their custodians play a vital role in sustainable use of marginal areas and biodiversity conservation. To continue this role, livestock keepers need a bundle of rights recognising their contribution to breed development and conservation and supporting them to continual to do so.

---

no. 19

Ethnoveterinary medicine: an alternative in managing gastrointestinal parasites in goats

V. Maphosa, J.P. Masika and S. Dube, University of Fort Hare, Livestock and Pasture Sciences, P.B. X1314, 5700 Alice, South Africa

Goats are important to resource-poor farmers, being kept for meat, financial security and socio-cultural activities. High mortalities, especially due to gastro-intestinal parasites limit their productivity. Anthelmintics are used in the control of parasites, but their high costs, development of drug resistant helminths, and demand for organic products worldwide has lead research to focus on alternative control methods. Medicinal plants have been used since time immemorial and *Aloe ferox*, *Elephantorrhiza elephantina* and *Leonotis leonurus* are some plants used to control helminths in goats. There is however, dearth of information on their use, this study therefre validates their efficacy *in vivo*. Forty-eight naturally infected young goats were randomly assigned to eight treatments. Treatments A-F received plant extracts at 50 and 100mg/ml dose concentrations per os, while G received a commercial drug (Albendazole), and group H served as control. Faecal samples were taken on days 1, 3, 6 and 9 for egg counts, body-weights recorded on days 1 and 9, and worms recovered from gastrointestinal tracts of animals on day 9. All plant extracts reduced total faecal egg counts, but *A. ferox* at 100 mg/ml significantly ($P<0.05$) reduced FEC on days 3, 6 and 9, even out-performing Albendazole. Total worms recovered were lowest in animals given *A. ferox* and *E. elephantina* extracts. We conclude that these plants posses anthelmintic properties, however, work is underway for further validation using other methods.

Enhancing food security in rural areas through management and conservation of farm animal genetic resources in Tanzania

S.M. Das[1], D.S. Sendalo[2] and H.A. Mruttu[3], [1]Ministry of Livestock Development and Fisheries, Central Veterinary Laboratory (CVL), P.O. Box 9254, Dar es Salaam, None, Tanzania, [2]Ministry of Livestock Development and Fisheries, Research, Training and Extension, P.O. Box 9152, Dar es Salaam, None, Tanzania, [3]Ministry of Livestock Development and Fisheries, Animal Identification, Registration and Traceability, P.O. Box 9152, Dar es Salaam, None, Tanzania

The dominant issues that would contribute to overall economic development in Tanzania, are reducing under nutrition, enhancing food security, combating rural poverty and achieving rates and patterns of agricultural growth. The existing high Animal Genetic Resource (AnGR) Diversity in the country, the contribution of the livestock sector to the Gross Domestic Product (GDP) would be higher than the current one which is 30% (MoAC, 1998) if the management and conservation of indigenous AnGR Diversity could be improved. There is a strong indication that the Tanzania Zebu cattle possess unique characteristics relating to innate adaptation, climatic tolerance, ability to use poor quality feed and to survive with reduced supplies of feed and water. This paper recommends to have sustainable Animal breeding Programme and revision of all studies previously implemented and formulation of the National breeding policy as well as involving livestock communities in breeding programmes such as, Open Nucleus Breeding Scheme.

---

no. 21

Cattle loading evaluation in farms of southern Chile

N. Sepulveda, F. Artigas, B. Schnettler, S. Bravo and R. Allende, Universidad de La Frontera, Prod. Agropecuaria, Av. Fco. Salazar 01145, Temuco, Chile

A diagnosis was made through an observational study of the management and facilities pre-morten in the farm during loading of cattle during for transport to a slaughterhouse. The results will be used for the design and implementation of an improved operating system. Were evaluates the loading process in 30 farms in the south of Chile. Farmers were between 80 and 12,000 animals each, which form a total of 45,420 head of cattle. Were evaluated the infrastructure such as pens, chutes, loading chutes, and management before and during the loading to a truck. The data obtained subsequent to the questionnaire were tabulated and a descriptive analysis was conducted. The main findings were that a total of 30 sites surveyed, 100% of them have pens, chutes and loading ramps. The 63.4% of respondents believed that the state of the loading ramps are good or very good and 76.7% of those said that chutes and the floor of the chutes is in good or very good condition. A 100% of the respondents presented a bat design of the chutes and a 96.7% of the loading ramps are not covered, as well as a 93.3% of the loading ramps have openings greater than 20 cm. We found that access to water (13.3%) and food (3.3%) is very low by preventing a prolonged stay in confinement pens. A 96.7% of farmers surveyed used instruments unfit for conducted the animals and only 3.3% of the farms use of banners for loading animals.

Economic assessment of trypanotolerant Orma/Zebu cross cattle in a pastoral production system in Kenya

M.W. Maichomo[1], W.O. Kosura[2], J.M. Gathuma[2], G.K. Gitau[2], J.M. Ndung'u[1] and S.O. Nyamwaro[1], [1]Kenya Agricultural Research Institute, P.O. Box 362, 00902 Kikuyu, Kenya, [2]University of Nairobi, College of Agriculture and Veterinary Sciences, P.O Box 29053, Nairobi, Kenya

Infectious and parasitic diseases remain a major constraint to improved cattle productivity among the pastoralists in Sub-saharan Africa. Use of animal health economics to support decision making on cost-effective disease control options is increasingly becoming important in the developing world. Trypanotolerant indigenous Orma/zebu cattle breed in a trypanosomosis-endemic area of Kenya was evaluated for economic performance using gross-margin analysis and partial-farm budgeting. Orma/zebu and Sahiwal/zebu cross breed cattle were exposed to similar husbandry practices and monitored for growth rate, incidence and treatment cost of common infections. Questionnaires were also used to assess the preference rating of the two breeds. Results indicated that incidence of infection was trypanosomosis 3%, anaplasmosis 58%, babesiosis 11%, East Coast Fever 22% and helminthosis 28%, with no significant difference between breeds. The Orma/zebu and Sahiwal/zebu breeds had comparable economic benefits, hence a pastoralist in Magadi division is likely to get similar returns from both breeds. This study therefore recommends adoption of not only the S/Z but also the O/Z breed for cattle improvement and conservation of indigenous genetic resources.

---

no. 23

Development of animal genetic resource information management system

S.S. Lee[1], H.Y. Kim[2], B.J. Park[2], S.H. Yeon[1], C.D. Kim[1], C.Y. Cho[1], S.H. Na[1] and D.S. Son[1], [1]National Institute of Animal Science, Korea, AnGR Station, Yongsan-li, Unbong-eup, Namwon-si, Jeonbuk, 593-830, Korea, South, [2]Insilicogen, Inc., #909, Venture Valley, Gosaek-dong, Gwonseon-gu, Suwon, Gyeonggi, 441-813, Korea, South

In Korea, the conservation of own native animal species and breed is a crucial subject for sustainability and international competitiveness in animal industry. The mission of the system is to coordinate the gathering, accumulation and utilization of animal genetic resource information by standardization and integration of live animal, germplasm and DNA profile data. This system was developed as web-based database(www.angr.go.kr) by AGRML(animal genetic resources markup language) which is specifically designed for animal genetic resource with XML(extensible markup language). The emphases of this system are standardization, internalization, user-friendliness for exact and convenient data sharing. The data can be input, output and converted with various formats such as Excel or XML. The Korean nationalwide product traceability identification system was adopted as the identifier of each animal in this database. The primary species in this system are Korean native beef cattle, pig, chicken and goat at the moment, and more species and breeds will be added to continuously. This system deal with external features, production traits, DNA polymorphism and germplasm storage information with 2D image. The system gives a analytical tools for statistics and phylogenetics, and linked with other databases.

Effect of concentrate supplementation on performance of Red Chittagong cows

S.S. Islam[1], M.J. Khan[1], A.K.F.H. Bhuiyan[2] and M.N. Islam[3], [1]Bnagladesh Agricultural University, Animal Nutrition, Mymensingh, 2202, Bangladesh, [2]Bnagladesh Agricultural University, Animal Breeding and Genetics, Mymensingh, 2202, Bangladesh, [3]Bnagladesh Agricultural University, Dairy Science, Mymensingh, 2202, Bangladesh

Red Chittagong Cattle (RCC) is a potential indigenous genetic resource (*Bos indicus*) of Bangladesh which is under threat. Efforts have been made to develop an appropriate feeding system for this variety and as a part of that activity a feeding trial was conducted to study the effect of levels of concentrate supplementation to RCC cows fed urea molasses treated straw (UMS) based diet on their milk yield and composition. Sixteen RCC cows were assigned into four dietary treatments and supplemented a concentrate mixture containing 10.5 MJ ME and 200 g CP per kg DM at the rate of 0, 10, 15 and 20% of total DM requirement. Rate of supplementation increased total feed, CP and ME intake with a tendency of decreasing voluntary UMS intake. The highest digestibility coefficients of proximate components were observed in 20% concentrate group. Milk yield (kg/d) was highest ($P \geq 0.09$) in 20% concentrate group (2.52 kg/d) and lowest in unsupplemented group (1.57 kg/d). Protein content of milk increased ($P<0.05$) and fat content decreased ($P<0.05$) with increasing concentrate supplementation. Feed cost per kg milk yield was lowest ($P<0.05$) in 10% supplemented group. Ten percent concentrate supplementation was found to be suitable for economic milk yield.

---

no. 25

A survey of beef cattle marketing in Taupye extension area of Mahalapye in Botswana

M. Monkhei, Botswana College of Agriculture, Dept of Agric Economics, Education and Extension, P/Bag 0027, Gaborone, Botswana

The aim of this study was to find out the marketing channels for beef cattle in Taupye extension area, identify marketing problems as well as suggest strategies to minimize marketing problem. A questionnaire was used to collect primary data from beef cattle farmers and buyers. Stratified random sampling was used to select 55 farmers There were ten Beef cattle buyers and all were included as respondents Data collected was analyzed using Statistical Package for Social Sciences (SPSS 12.0.1) It was found that Butchers, BMC, individuals, cooperatives, middlemen, and auctions were the marketing channels used in the area. Major problems faced by beef cattle farmers were lack of transport, low prices paid by buyers. Buyers' problems included lack of cheap and reliable transport, poor roads, farmers offering beef cattle at higher prices, and farmers attempting to sell cattle which did not have bolus (cattle identification mechanism). Farmers are encouraged to use weight when selling their cattle and to observe peak demand period when BMC prices are higher. Strategies to minimize beef cattle marketing problems include forming beef cattle farmers' association and improving beef cattle breeds and head health management. The Government should improve road infrastructure.

The role of Taq-polymerase enzymes for PCR amplification of damaged or degraded bovine forensic samples

A.A. Nemakonde[1], B.J. Greyling[1] and E. Van Marle Koster[2], [1]Agricultural Research Council, Animal Genetics, P/Bag X2, Irene, 0062, Pretoria, South Africa, [2]University of Pretoria, Animal and Wildlife sciences, Pretoria, 0002, South Africa

DNA profiling of exhibits that originate from forensic stock theft cases is routinely used as a tool to link suspects to either a crime scene or the crime itself. DNA derived from aged or degraded samples however is often highly fragmented which severely compromises the efficiency with which a complete profile can be obtained using PCR. In reality forensic samples are often of poor quality and degraded to the extent that results are inconclusive. Conventional polymerases such as Taq, lack certain repair mechanisms and thus can not alleviate the problem of fragmented and degraded DNA templates, typically encountered in tissue samples that are in an advanced state of decomposition. New generation polymerases however show much promise to address this challenge, since they are engineered to be highly processive and have high fidelity characteristics. The aim of this study is to determine the efficiency of different polymerases such as Restorase, Herculase and Faststart High Fidelity PCR System for their ability to enhance the profiles obtained from degraded forensic meat samples that have been subjected to different degrees of degradation/decomposition

---

no. 27

Successful pro-poor goat development under heifer Uganda

W.S. Ssendagire, Heifer Internatonal Uganda, Programme, 28941 Kampala, 256, Uganda

Goats are among livestock used by Heifer International to improve livelihoods of resource-limited households worldwide. Heifer International is a non-profit, Non-Government Organization founded by Dan West in 1944 in USA with a mission of ending hunger, poverty and caring for the earth. Twelve values based, holistic demand driven approaches and zero grazing farming system are used to implement activities. Beneficiaries own 1-5 acre pieces of land. In Uganda, over 20,000 families have directly received training and livestock and over 1,500,000 persons have indirectly benefited. Over 5,000 families have benefited from imported dairy and meat (Boer) goats plus crossing breeding with indigenous nannies. Benefits include: rapid increases in output of goat milk and meat goat off-take, goat sales, manure for improved food crop, fruit and vegetable production in addition to management of the environment. This has translated into improved family nutrition, income, home hygiene and sanitation. Government goat development strategy has been strengthened, gender and HIV/AIDS awareness mainstreamed amongst families assisted by Heifer International. Challenges include: low beneficiaries' attitude and cultural change towards integrated modern farming, insufficient quality breeding stock, ban on livestock importation, and prolonged local quarantine due to outbreaks of notifiable livestock diseases.

The effects of replacing, partially or totally, cane molasses by a concentrated taste enhancer (Molasweet)

L. Mazuranok[1], C. Moynat[1] and R. Van Der Veen[2], [1]Pancosma, Voie des Traz 6, 1218 Le Grand-Saconnex (Geneva), Switzerland, [2]allied Nutrition (Pty) Ltd, 101 Kiaat steet, 0157 Dorningkloof, South Africa

Molasses is a valuable feed ingredient. Besides its nutritional value, it can enhance palatability and the physical quality of the feed products. Due to increasing raw material prices, on occasions a better cost alternative to molasses is sought by the nutritionist. A recent trial in South Africa studied the effects of replacing molasses with a taste enhancer Molasweet (MS). 120 intensively reared lambs were allocated into 3 treatments containing 4 replicates of 10 lambs each: negative control (C), MS 1 (240 g/t) substituting 60% of the molasses, MS 2 (400 g/t) substituting 100% of the molasses. Performance and economical benefits were evaluated. Concerning individual performance, M1 and M2 improved statistically the slaughter weight vs. negative control (respectively 59.96 and 58.48 vs. 55.59 kg, $P<0.01$). The average daily gain and the cold carcass weight followed the same way (respectively 0.34 and 0.31 vs. 0.26 kg/d, $P<0.01$; 30.26 and 30.13 vs. 28.17 kg, $P<0.01$). Compiling pen data, the average daily feed intake was statistically increased in M1 and M2 groups compared to the negative control (respectively 2.08 kg/d for both M1 and M2 vs. 1.85 kg/d, $P<0.01$). The feed conversion ratio was also improved in both M groups (respectively 5.15 and 5.45 vs. 5.72, $P<0.1$).

---

no. 29

Effects of bacterial inoculant and surfactant on fermentation of high DM alfalfa silage

J. Baah and T.A. Mcallister, Agriculture & Agri-Food Canada, Research Centre, Lethbridge AB, T1H 4S6, Canada

Surfactant (sodium dodecyl sulfate, SDS) and a bacterial inoculant (Sil-All; Alltech Canada, Guelph, ON) containing *Lactobacillus plantarum*, *Enterococcus faecium* and *Pediococcus acidilactici* (LEP) were used in a 2 × 3 factorial experiment to study their potential to improve fermentation characteristics of high DM alfalfa silage. Chopped, wilted whole plant alfalfa (50% DM; chop length 0.95 cm) was treated with LEP at 0 or 4×10^5 CFU/g fresh weight and SDS at 0, 0.125% or 0.250% (w/w). Each treated forage was packed into12 mini-silos (2.5 kg capacity), capped, weighed and stored for 1, 3, 7 or 42 d (n=3) before analysis. Decline in pH was more rapid, and terminal pH lower (by an average of 0.7 units) with LEP than without ($P<0.001$). The inoculant also increased d-42 lactic acid concentrations ($P<0.001$). In the 0, 0.125 and 0.25% SDS treatments without LEP, these were 22.7, 23.0 and 21.9 g/kg DM vs 7.5, 10.4 and 8.3 g/kg DM, respectively. Total N, DM loss, and populations of total culturable and lactic acid-producing bacterial co similar among treatments ($P>0.05$). At 0.250%, SDS decreased ($P<0.05$) NH_3-N concentration (as % total N), indicating reduced proteolysis. No yeasts or molds were recovered in silage with LEP only, but were present at 10^2 to 10^3 CFU/g in all other treatments ($P=0.15$). These findings indicate that LEP (± SDS) has potential to enhance fermentation of alfalfa silage, and that SDS at 0.250% (± LEP) could reduce proteolysis. These additives could therefore improve ensiling of alfalfa when the DM content is higher than optimal.

Effect of processing method on nutritive value of palm kernel cake in non-ruminant diets

M. Boateng[1], D.B. Okai[1], A. Donkoh[1] and J. Baah[2], [1]KNUST, Dept. Anim. Sci., Kumasi, Ghana, [2]Agriculture & Agri-Food Canada, Lethbridge, AB, T1J 4B1, Canada

The feed value of palm kernel cake (PKC) for pigs depends on the method used to extract the oil from the kernels. The nutritive value of PKC from two expeller sites and two hydrothermal production facilities were assessed using laboratory rats as a model. Following chemical analyses, the PKC were incorporated into maize-based isonitrogenous (2.9% N) diets at 0% (CON) or 35% (w/w) and fed to 30 individually caged albino rats for 28 d (n=6). All PKC diets included 0.5% (w/w) Alzyme Vegpro (Alltech Canada, Guelph, ON). PKC from expellers (E1, E2) contained more fatty acids (FA) and less crude protein (CP) than did PKC from hydrothermal production (H1, H2), averaging 15.8% vs. 7.7% FA and 13.3% vs. 19.7% CP (DM basis; $P<0.05$), respectively. Lauric, oleic, myristic and palmitic acids accounted for 84% of total FA in all PKC. Essential amino acid (AA) concentrations were similar (47-48% of total AA; $P>0.05$), but lysine content in E1 and E2 (6%) was higher ($P<0.05$) than in H1 and H2 (3%). Weight gains were similar ($P>0.05$) in rats fed E1, E2 or CON (average 2.2 g/d) but higher ($P<0.01$) than in rats fed H1 (1.3 g/d) or H2 (1.5 g/d). Feed efficiency (g DM:g gain) in rats fed E2 was better than in rats fed CON, H1or H2 (4.3 vs. 5.3, 7.0 and 7.1, respectively; $P<0.05$) but similar ($P>0.05$) to that of rats fed E1 (4.7). This study indicated that 35% expeller-produced PKC could potentially be included in maize-based diets for starter pigs with no adverse effects on growth.

---

no. 31

Influence of phosphorus supplementation on growth and reproductive characteristics of beef cows in the semi-arid bushveld of South Africa

M. Orsmond[1], E.C. Webb[1] and I. Du Plessis[2], [1]University of Pretoria, Dept. Animal & Wildlife Sciences, Hilcrest, South Africa, 0002, South Africa, [2]Mara Agricultural Development Institute, Dept Agriculture, Mara Agric Dev Inst, 0002, South Africa

This study evaluated the effect of different levels of phosphorus supplementation on the growth and reproductive characteristics of extensive beef cows in the semi-arid bushveld of South Africa. Trial 1 studied a control, phosphorus supplementation all year round, phosphorus supplementation in the summer months and summer phosphorus supplementation with a winter lick. Trial 2 included two dietary supplementation groups, notably a control group and a Kimtrafos 12 P supplementation all year round. Phosphorus supplementation did not influence cow weight in Trial 1, but was significant in Trial 2 ($P<0.001$). Treatment influenced weaning weights and weight gain, but no improvement in calving interval or conception rates were observed. Phosphorus supplementation does not seem to be essential in grazing beef cattle in the semi-arid bushveld of South Africa.

Use of body condition scores (BCS) versus phenotype score (PTS) as tools to estimate milk yield and composition in different goat breeds raised in small scale production systems

R.G. Pambu[1], E.C. Webb[2], L. Kruger[1], L. Mohale[1] and M. Grobler[1], [1]Agricultural Research Council (ARC), Irene, Animal Production Institute, Private Bag X2, Irene, 0062, South Africa, [2]University of Pretoria, Animal & Wildlife Sciences, Hilcrest, 0002, South Africa

Small scale farmers are generally not well equipped to manage a dairy goat business. Most of their decision making depends on visual appraisal of their goats. The aim of this study was to examine the validity of body condition scoring (BCS) and phenotype scoring (PTS; which is a new concept which includes BCS, age, breed and udder characteristics) of goats raised in small scale farming systems. Thirty-two goats (8 indigenous, 8 British Alpine, 8 Saanen and 8 Toggenburg) were raised in a free ranging system at the ARC-Irene experimental farm. Milk samples were collected from parturition onwards on a weekly basis for a period of two months for chemical analyses. Additional data recorded included breed, age, BCS, udder size (USz) and udder attachment (UAt). Statistical analyses indicate that breed influenced fat and protein content of milk (especially in the Saanen and Toggenburg goats). BCS influence fat content, lactose, milk proteins, milk urea nitrogen and somatic cell count but not milk yield. Both USz and udder attachment influenced milk fat, lactose, milk proteins and milk yield. Results indicate that BCS is an acceptable tool to evaluating general body condition, but phenotype score is a better tool to predict milk yield of goats raised in small scale systems.

---

no. 33

Degradation of slowly degradable N from manure in storage and manure in soil on small farming systems in Kenya

H.A. Markewich[1], A.N. Pell[1], D.M. Mbugua[1], D.J.R. Cherney[1], H.M. Van Es[2], C.J. Lehmann[2] and J.B. Robertson[1], [1]Cornell University, Department of Animal Science, Morrison Hall, Ithaca, New York, 14853, USA, [2]Cornell University, Department of Crop and Soil Sciences, Bradfield Hall, Ithaca, New York, 14853, USA

Livestock manure is a soil amendment used frequently in sub-Saharan Africa. Slowly degradable plant materials in manure add to soil organic matter. Nitrogenous compounds bound in slowly degrading fractions may be mineralized to plant-available N (PAN) during manure storage and when manure is applied to soil. Two storage experiments were conducted; each used manure from two different Kenyan farms. Farm A and C cattle had higher quality diets and produced manure with less bound N than B and D. The neutral detergent insoluble N (NDIN) content of manure was measured to assess the degradation of bound N in manure. Effects of storage duration on manure quality were measured. Decomposition in soil of manure stored for 30d, including the effects of soil insect activity, was assessed by litterbag experiments. Manure A degraded faster than B in storage during the dry season ($P<0.01$) losing 38.4% of NDIN vs 30.6%. Degradation of manure C in soil was nearly 4 times greater than that of D ($P<0.01$). Soil insects increased manure disappearance ($P<0.01$). Results suggest better-fed cattle produce manure with a more immediate benefit to plants in terms of PAN release. NDIN from lower quality manure will persist in soil while much of the NDIN from higher quality manure will be degraded to PAN in one growing season.

Innovative use of agroforestry technologies in semi-arid smallholder farming areas of Zimbabwe to improve livestock production and address rangeland degradation

L. Mukandiwa, S. Ncube, P.H. Mugabe and H. Hamudikuwanda, University of Zimbabwe, Animal Science, P.O. Box MP167, Mt. Pleasant, Harare, Zimbabwe

This paper is a synthesis of different research findings on the relevance of agroforestry technologies in semi-arid smallholder farming communities in Zimbabwe in improving livestock production as well as alleviating poverty. *Acacia angustissima*, a legume shrub with high nutritive value as a livestock feed, was utilized in participatory on-farm research in two smallholder-farming communities. In Bubi smallholder farming area, this forage shrub species mixed with pearl millet was used as feed for goats. *A. angustissima*, at 45% inclusion in the goat diet produced significantly higher goat growth rates than natural rangeland. The farmers in this study demonstrated a keenness not only to use this feeding technology, but to grow their own *A. angustissima*. The second study was done to address the problems of feeding broiler chickens in Chikwaka smallholder farming area. It was concluded that 10% *A. angustissima* leaf meal could be used in broiler diets. The relevance of these interventions in improving livestock production is by providing feeding resources that can be home grown and are cheaply available. These research findings also demonstrated that rangeland degradation could be addressed by finding technologies that are economically worthwhile for farmers, unlike previous approaches that have expected farmers to address environmental degradation problems without any demonstrable economic benefits for the farmers.

---

no. 35

Feeding value of hazelnut leaves in organic farming system of goat rearing

M.R. Alam[1], G. Rahmann[2] and R. Koopman[2], [1]Bangladesh Agricultural University, Dept. of Animal Science, Mymensingh, 2202, Bangladesh, [2]Institute of Organic Farming, Trenthorst, 23847, Germany

The benefits of integration of feeding shrub to goats in organic farming system were investigated. Two groups of German Ziege goats consisted 16 goats/group of 23.9 kg live weight were housed indoor and during 35 days of experiment offered grass hay, mineral lick and water *ad libitum*, 0.4 kg flaked wheat/goat and one of the groups was offered freshly cut hazelnut leaves (*Corylus avellana*) *ad libitum* in the morning. All goats were weighed and grab fecal sampled every week. Hazelnut leaves contained 37.6, 20.0, 17.0, 14.6 and 2.8% dry matter (DM), crude protein (CP), crude fiber, total tannins and condensed tannins (CT), respectively. On average, goats consumed 0.579 and 0.116 kg of hazelnut leaf DM and CP/d, respectively and gained live weight of 0.206 kg/d compared to 0.059 kg/d by controlled goats ($P<0.001$). Goats supplemented with hazelnut leaves were shown 2% reduction and maintained low level of gastrointestinal nematodes (GIN) infestation as compared to 35% gain of infestation in unsupplemented group ($P<0.01$). Both the feeding groups were able to reduce coccidian oocysts counts by 91%. Higher live weight gain and reduction of GIN were probably mediated by enhancing nutrient supply, better utilization of basal feed and CT in leaves on protein metabolism and its antiparasitic properties on nematode viability. Hazelnut leaves may be used as an alternate feed supplement and to conventional control of GIN in organic farming system of goat rearing.

Community evaluation of the importance of different plants used as food, medicine and cosmetic and their population trends

E.C. Kirwa, E.N. Muthiani, A.J.N. Ndathi and W. Ego, Kenya Agricultural Research Institute, Range Research, P.O. Box 12 Makindu 90138, +254, Kenya

A study was carried out in 2006 in 3 sub-locations in Makindu Division of Makueni District in Kenya to list and rank important locally available plants with nutritional medicinal, cosmetic and other uses (basketry, ropes, dyes), their marketing systems and plant species population trends and reasons for the trends. The participants stratified into four groups of men and women under 45 years of age and those above 45 ranked *Adansonia digitata* as the most important fruit tree in all the sites though its population is declining due to clearing for cultivation, destruction by elephants, overexploitation as livestock feed, poor germination and droughts. *Tamarindus indica* and Berchemia discolor were ranked second and third important fruit trees respectively. There was significant variation in the highly valued type of vegetables between sites due to variation in dryness. *Kedrostis pseudogijef* and *Amaranthus hybridus* were ranked first and second, respectively as important vegetable plants. *Ocimum suave* Willd, currently used as mosquito repellant in most homesteads was the most important plant used as cosmetic. Plants with medicinal value varied across sites and between groups and were valued based on the severity of the disease or on the variety of diseases they treat. *Zanthoxylum chalybea*, *Solanum incanum*, *Terminalia brownii* and *Aloe* sp. (Kiluma) featured as important plants at different levels of ranking in the 3 sites.

---

no. 37

Challenges for rare breeds in the market for regional food products

J.K. Oldenbroek[1] and H.S. Van Der Meulen[2], [1]Animal Production Systems Group, Animal Sciences, Wageningen University, P.O. Box 338, 6700 AH Wageningen, Netherlands, [2]Rural Sociology Group, Social Sciences, Wageningen University, P.O. Box 8130, 6700 EW Wageningen, Netherlands

Worldwide the interest for regional and local food products is growing, which results in a large number of niche markets. For rare breeds it creates opportunities to increase their profitability and thus the chances of preservation. Rare breeds are often linked to specific regions and have cultural and historic aspects, which makes them interesting for consumers. Several groups of breeders are setting up projects to become less dependent on public support programs, because they expect the financial contributions of national and provincial administrations to decrease in the near future. Three Dutch cases are described (a chicken, sheep, and cattle breed) to illustrate the market potential of food products from rare breeds. The roles of the stakeholders are elaborated and factors determining the success of regional products from rare breeds are discussed. Finally, some recommendations for the organization of the supply chain for regional products of rare breeds and marketing are made.

Delivering systematic information on indigenous farm animal genetic resources of developing countries

T. Dessie, B. Asrat, Y. Mamo, J.E.O. Rege and O. Hanotte, International Livestock Research Institute (ILRI), Animal Genetic Resources Group, Addis Ababa, Ethiopia, P.O. Box 5689, Ethiopia

This paper describes the rationale, objectives, historical development, structure, functionality, content, utility and future prospects of the Domestic Animal Genetic Resources Information System (DAGRIS) and its status of development. DAGRIS aims at delivering systematic information on indigenous farm animal genetic resources of developing countries. It is a public-domain information resource designed to cater for the needs of different stockholders. It has been developed and managed by ILRI since 1999. At the start DAGRIS covers three ruminant livestock species (cattle, goat and sheep) and countries in Africa, and it is being expanded to cover more livestock species (poultry, pigs, pigs, yak and buffalo) of Africa and selected countries in Asia. The database now includes 176, 170, 82, 124, 165, 30 and 141 breeds of cattle, sheep, goat, chicken, pigs, Yak and buffalo breeds from Africa and Asia, respectively, and about 24,536 trait records with a total data size of 41.13MB. DAGRIS is available free of charge both on the web (http://DAGRIS.ILRI.CGIAR.ORG/) and on CD-ROM.

---

no. 39

Water development and utilization in arid and semi-arid areas of Kenya: a review

W.B. Muhuyi and S.M. Mbuku, Kenya Agricultural Research Institute, Livestock, Box. 3840, 20100 Nakuru, Kenya

Today's African pastoral systems have their origins in the prehistoric Sahara, where they emerged as a means of securing food resources in a drying, variable and unpredictable climate. It is widely recognised that developing countries stand to suffer disproportionately from the effects of climate change, they are in the weakest position to mitigate the adverse effects and stand to lose some of the current development gains. In arid and semi-arid areas (ASAL) of Kenya, provision of adequate and clean drinking water for human and livestock is cited as a major constraint. In these areas, which are occupied by pastoralists, shortage of drinking water occurs in every dry season and the situation is aggravated by drought. Droughts have to be considered as a normal part of the climate cycle and they must be planned for as an integral component of property management based on weather forecasting information. However, with the current marginalisation of pastoralists, their adaptive capacities may have been eroded and they may be more susceptible to climate change than other communities. Conversely, climate change could conceivably lead to the creation of more dryland resources that are suited to pastoralism, thus creating new opportunities for pastoralists to exploit. However, the likelihood and the implications of such changes are very uncertain.

Adapting livelihoods to climate change in livestock-based systems: is there anything new under the sun?

M. Herrero and P.K. Thornton, International Livestock Research Institute, P.O. Box 30709, Nairobi, Kenya

Climate change will have significant impacts on livestock keepers in the developing world in the coming decades. These changes will pose threats to some communities but will also provide opportunities for others. There is not always much clarity as to what precisely is meant by "adaptation" and how it differs from what has gone before. In this paper, we argue that livestock keepers in rain-fed areas have been continually adapting to climatic conditions for centuries, with different degrees of success. The commercial sector has been able to deal with climate variability through technology, information, innovative market schemes, and other practices. Unfortunately the poor and vulnerable have been less able to adapt, and they will require public and private support to be able to reduce the increased risks associated with, as well as to make the most of the opportunities that may arise because of, climate change. This will require a research and development portfolio including a mixture of the old and the new. Significant changes will be required in institutions and policies in both the developed and developing world to support pro-poor adaptation options. A revitalised agricultural research agenda will also be essential, that provides both traditional (i.e crops germplasm development,) and novel research outputs (incentives for dealing with risk, insurance schemes, landscape genomics, etc). Research has a key role to play in preserving or enhancing food security and helping to alleviate poverty as Africa adapts to climate change.

---

no. 2

Baseline of greenhouse gas (GHG) and reduction on poultry and swine industries of ASEAN 8 countries

K. Kaku,

Baseline on GHG from poultry and swine industries were estimated in accordance with 2006 IPCC guidelines, National Reports under UNFCC and FAO bulletin of statistics during 1997-2002 in ASEAN 8 countries (Vietnam, Philippines, Indonesia, Thailand, Malaysia, Myanmar, Cambodia, Laos). GHG (CO_2-equivalent) from poultry and swine industries in the 8 countries increased by ±8% from 12.8 million to 13.8 million tons/year during 1997-2002. Around 10% of GHG (CO_2-equivalent) from poultry and swine farming in the 8 countries was emitted from enteric fermentation and the residual 90% was emitted from manure management during 1997-2002. The reduction on GHG emission, indicating the adoption of manure management technology by poultry farms through the thermal drying of poultry manure and conversion into dry lot systems from liquid/slurry systems on manure, could have brought about 10% reduction of GHG from poultry and swine industries in the 8 countries. The Dec 08 contract at the European Climate Exchange (ECX), a nearby price and benchmark for GHG, was 23.7 euro/ton and prices of forward contracts (24.1 for Dec 09, 24.8 for Dec 10 and 25.7 for Dec11) were higher than the nearby contract at ECX on April 3, 2008. GHG from poultry and swine industries in the 8 countries could be converted in accordance with Dec 08 contracts at ECX. The size of financial economic benefit that developing countries could expect depends on the project size of small-scale Clean Developing Mechanism (CDM) to each farming system in this region.

no. 3

Effect of sodic monensin in the enteric methane production in bovines fed with hay of *Brachiaria brizantha* (Hochst. ex. A. Rich.) *Stapf* cv. Marandu

G. Balieiro Neto, A. Berndt, J.R. Nogueira and J.J.A.A. Demarchi, APTA, www.apta.sp.gov.br, Av. Bandeirantes, 2419, 14030670, Brazil

One of the options to mitigate the seasonality of forage production in Brazil is the postponed pasture at the end of the period of growth, aiming to accumulate forage to use during the dry season. The supplementation of the diet of poor quality with sodic monensin can result in lower methane production and consequently, increase the efficiency of their digestion, because the methane expelled represents loss of energy from food. The forage from postponed pasture is usually deficient in protein and requires protein supplementation, seeking the full development of the ruminal microorganisms. This study aimed to evaluate the effect of sodic monensin in the methane production from enteric fermentation of cattle fed with marandu-grass hay simulating postponed pasture systems. There were used four fistulated cows, and experimental design in 4 x 4 Latin square. The treatments were mineralized salt, mineralized salt with protein, mineralized salt with protein and sodic monensin and mineralized salt with sodic monensin. The SF_6 method was used to measure methane production. The methane production were 17.50, 15.20, 13.95 and 8.12 g CH_4/kg DMI when given mineralized salt, mineralized salt with protein, mineralized salt with protein and sodic monensin, and mineralized salt with sodic monensin, respectively. The sodic monensin was effective in reducing the methane production, when added in mineralized salt or mineralized salt with protein diets.

---

no. 4

Global climate change and animal production

D. Furstenburg and M.M. Scholtz, Agricultural Research Council, Livestock Business Division, Private Bag X2, Irene, 0062, South Africa

The world's climate follows a sequence of cold ice-age, and warm sun-age periods, probably caused by a change in the tilting gradient of Earth's axis and its spinning rate, which affects its atmospheric CO_2 levels and temperature. Median CO_2 levels of between 200-280 ppm maintain an average temperature of 9-21 °C, which is suitable for sustaining life on Earth. Artificial emissions of excessive carbon and greenhouse gasses by the modern developing world have increased CO_2 levels to 315 ppm in 1950 and 396 ppm in 2008, resulting in a rise in temperature of 2.2 °C. A total meltdown of Earth's ice-caps is expected at 450 ppm, with a temperature rise of 4 °C at the equator, and 10-40 °C at the poles. If the current trend of carbon emission continues, CO_2 levels are likely to exceed 800 ppm by 2050. The anticipated global warming will change Africa's rangelands and forests into dry-woodland; arid grassland into Karoo dwarf scrub and desert; and montane sour veldt into low vigour, arid sweet veldt. South African grazing is expected to decline by more than 30% and animal numbers will have to be decreased. Animals will be exposed to other parasites and diseases, and grazers will need to adapt to mixed feeding and browsing. All these factors will certainly impact on animal production. As part of the international combat against artificial greenhouse gasses, animal feedlots excreting millions of tons of CH_4, will also have to give way to more greenhouse friendly animal production systems.

Livestock and climate change: is it possible to combine adaptation, mitigation and sustainability?

K.E. Van'thooft[1], W.B. Bayer[1], J. Wanyama[2] and G. Gebru[3], [1]ETC Foundation, P.O. Box 64, 3830AB Leusden, Netherlands, [2]VETAID-Mozambique, CP 44, Chokwe, Mozambique, [3]GL-CRSP Pastoral Risk Management Project (PARIMA), P.O. Box 5689, Addis Abeba, Ethiopia

In the climate change discussion about livestock the focus is often on ruminants, and specialised ways to reduce its methane production. This paper, which is based on field experiences in several parts of the world, proposes a radically different strategy: the optimisation of livestock-related systems as a whole, rather than maximising individual animal productivity. This is done through farmer-researcher learning and experimentation, in which locally available expertise and resources are optimised. This approach has proven valid in both low-input and high input agricultural systems. Three experiences are presented from different livestock keeping systems: (1) optimising low-input mixed agricultural system in Mozambique, through improved animal traction practices; (2) improved rangeland management and controlled fire with Borana pastoralists in southern Ethiopia; (3) optimising intensive dairy system through the 'cycle approach'- with lower use of concentrates and nitrogen fertiliser – in the Netherlands. In each of these cases the efforts have resulted in better functioning of the systems as a whole, including soil quality, plant production, and life-span of livestock. This has had a positive impact on adaptive capacity and resilience of the farmer communities, as well as mitigation and sustainability.

---

no. 6

Effects of antibacterial substances derived from lactic acid bacteria on rumen methanogenesis

J. Takahashi[1], R. Asa[1], A. Tanaka[1], A. Uehara[2], I. Shinzato[2], Y. Toride[2] and N. Usui[2], [1]Obihiro University of Agriculture & Veterinary Medicine, Graduate School of Animal Science, Obihiro, Hokkaido 080-8555, Japan, [2]Amino Acids Company, Ajinomoto Co., Inc., Fermentation & Biotechnology Laboratory, 1-1, Suzuki-cho, Kawasaki-ku, Kawasaki-shi, 210-8681, Japan

Effects of antibacterial substances (PRA) produced derived from *Lactobacillus plantarum* or *Leuconostoc citreum* on rumen methanogenesis were examined using the *in vitro* continuous methane quantification system. Four different strains of lactic acid bacteria, (1) *Lactococcus lactis* ATCC19435 (Control, non-bacterial substances), (2) *Lactococcus lactis* NCIMB702054 (nisin-Z), (3) *Lactobacillus plantarum* TUA1490L (PRA-1), and (4) *Leuconostoc citreum* JCM9698 (PRA-2) were individually cultured in GYEKP medium. The 80 ml of each supernatant was inoculated in phosphate-buffered rumen fluid. PRA-1 remarkably decreasedcumulative methane production. For PRA-2, there were no effects on CH_4 and CO_2 production and fermentation characteristics in mixed rumen cultures. The results suggested that PRA-1 reduced the number of the methanogens or inhibits utilization of hydrogen in rumen fermentation.

no. 1

The effect of polyethylene glycol and polyvinylpyrrolidone on fermentation dynamics of indigenous varieties of sorghum grain

L.R. Ndlovu[1] and S. Ncube[2], [1]National University of Science and Technolgy, P.O. Box AC939, Ascot, 0000 Bulawayo, Zimbabwe, [2]Makoholi research Station, Private Bag 9182, 0000 Masvingo, Zimbabwe

The effect of polyethylene glycol (PEG) or polyvinylpyrrolidone (PVP) on *in vitro* fermentation dynamics ofsorghum grain from 12 indigenous varieties was examined. The chemical analysis showed that sorghum grains with a brown colour and a testa had the highest soluble proanthocyanidins (PAs) when compared to white or red sorghum grain. The presence of a testa and the colour of the grain had no significant ($P>0.05$) effect on insoluble and fibre bound PAs. PEG had a significant ($P<0.05$) effect on cumulative gas production, except at 6 and 48 hours post incubation but increased ($P<0.05$) the rate of gas production at 6 and 12 hours post incubation. Treatment with PVP increased ($P<0.05$) cumulative gas production at 6 and 12 hours and rate of gas production at 6, 12 and 24 hours post incubation. Treatment with PEG or PVP increased ($P<0.05$) the 72 hour IVDMD of the grain, especially the high PA sorghum grain. The asymptote value of the gas pool size (A) was increased by PEG supplementation, especially for high PA sorghum grain whilst PVP treatment decreased the asymptote value in all the sorghum varieties. Treatment with PEG or PVP reduced time to half asymptote (T/2). The fractional rate of gas production (m%/h) at 12 hours post incubation was increased by PEG or PVP treatment. The study showed that PAs in sorghum grain can be inactivated by PEG or PVP.

---

no. 2

Management of bovine mastitis without antibiotics

R. Mukherjee, Indian Veterinary Research Institute, Medicine, IVRI, Izatnagar, 243 122 (UP), India

Mastitis remains a worldwide problem and major economic threat to the dairy farmers, in spite of improved managemental practices and dry cow therapy. The success in control of mastitis is very difficult due to multitude of microbial involvements and immunosuppression of the mammary gland. Antibiotics are generally used for treatment of mastitis; however, antibiotic therapy is only moderately efficacious. Furthermore, antibiotic therapy reduces the function of the immune cells and cannot prevent the inflammatory reactions. Moreover, the biggest challenge facing the modern dairy industry is the pressure to reduce the use of antibiotics in food producing animals due to detrimental effects of antibiotics on human, animal and environmental health. Thus, in recent years, researchers are investigating the role of non conventional methods of treatment for the management of mastitis. In this regards, botanical therapeutics that include plant derived pharmaceuticals, multicomponent botanical drugs, dietary supplement and plant produced recombinant protein are considered as best bioactive ingredients for the amelioration of various ailments. Thus, this paper reports the successful use of medicinal herbs like *Ocimum sanctum*, *Tinospora cordifolia* and *Azadirachta indica*, which possess antibacterial, anti-inflammatory, antioxidant and immunomodulatory properties along with vitamin E and trace minerals in the treatment of mastitis leading to reducing the incidence of mastitis in dairy farms.

no. 3

Utilization of exogenous lactate by lactate dehydrogenase-C in the midpiece

H. Yamashiro[1], M. Toyomizu[2], N. Aono[1], M. Sakurai[1], F. Nakazato[1], A. Kadowaki[1], M. Yokoo[3], Y. Hoshino[1] and E. Sato[1], [1]Laboratory of Animal Reproduction, Tohoku University, Sendai, 981-8555, Japan, [2]Laboratory of Animal Nutrition, Tohoku University, Sendai, 981-8555, Japan, [3]INBEC, Tohoku University, Sendai, 980-8574, Japan

Recently, we found that raffinose-modified Krebs-Ringer Bicarbonate (mKRB) egg yolk extender with 32.37 mM lactate enhances oxygen consumption and rat sperm motility and protects against freezing injury. This study used the glycolytic inhibitor 2-deoxy-d-glucose (2DG) to examine if the exogenous lactate in mKRB medium containing glucose and pyruvate is involved in glycolysis pathway, thereby the energy supplies for sperm motility. Localization of lactate dehydrogenase-C (LDH-C) in the sperm was also investigated. Sperm motility was significantly higher in the presence of lactate than its absence. When 2DG was added to lactate-free medium, oxygen consumption by sperm decreased during incubation at 37 °C for 3h. 2DG did not significantly affect oxygen consumption by sperm treated with lactate-containing medium. The presence of lactate resulted in a higher fluorescence signal of LDH-C that was concentrated in the midpiece mitochondria. Thus, exogenous lactate increases the utilization of lactate by LDH-C which leads to increased midpiece mitochondrial oxidation, causing increased motility. Hence, lactate may play an important role in providing the energy for motility in rat epididymal sperm.

---

no. 4

Effects of meat consumption of clone cattle meat on reproductive parameters in rabbits

B.C. Yang, S.S. Hwang, Y.G. Ko, G.S. Im, D.H. Kim, S.H. Bae, B.H. Cha, H.N. Kim, J.S. Kim, M.J. Kim and H.H. Seong, National Institute of Animal Science, Division of Animal Biotechnology, #77 Chuksan-gil, Suwon, 441-706, Korea, South

Recent experiments revealed that the compositions of milk and meat produced from clones were not different from those of normal controls. However, the effect of the diets containing cloned animal meat upon reproductive and developmental parameter was rarely studied. Here, we provide the results associated with the developmental toxicity in rabbits fed with the feed containing clone cattle meat powder. Normal Korean Native Cattle's (Hanwoo) meat(control) purchased in market and the SCNT clone meat (treatment) were freeze-dried and then grinded into powder. The feed for rat was mixed with 5 to 10% of the meat powder (meat-based diets), respectively. The rabbits artificially inseminated (gestation day 0) were fed with the diet during gestation periods. The developmental parameter (changes of body weight, body part or organ weight, and skeletal findings of fetuses) of the clone groups were not statistically different from their controls. It is concluded that there are no obvious differences in the developmental parameters in the rabbits fed with clone cattle meat and their progenies compared to the controls.

Gender empowerment through practical training in dairy production

S.H. Raza, University of Agriculture, Faisalabad, Livestock Management, W-76, Umar Plaza, 1st Floor, Jinnah Avn., Blue Area, Islamabad, 44000, Pakistan

The dairy animals are integral part of Pakistan's agriculture based national economy and it is sole responsibility of women in small and landless farmers' social and cultural set up. Almost 48% women are engaged in auricular related activities and Livestock production is the main activity. A study has shown that women spent significantly more time in animal management related activities than men. The income generated by milk sale is mainly utilized by women for house hold activities. The issues faced by women involved in dairying were identified through participatory workshop under a DFID funded project. The women farmers were trained in different aspects of improved dairy management techniques to boost animal productivity. This paper gives details of experience gained during this project and guidelines for researchers who want to work in rural gender community in developing countries. The sharing of experience will help in opening the new windows of research in this unique area of work.

---

no. 6

Mapping quantitative trait loci for fatty acid composition in beef cattle

N.O. Mapholi-Tshipuliso[1] and M.D. Macneil[2], [1]Agricultural Research Council, Animal Genetics, P/Bag X2, Irene, 0062, Pretoria, South Africa, [2]USDA Agricultural Research Service, USA, Livestock and Range Research Laboratory, Miles City, Montana, 59301, USA

Fatty acid composition in beef has received considerable interest recently in view of its implications in human health and meat quality characteristics. The objective of the study was to search for quantitative trait loci (QTL) that affect relative amounts of saturated (SFA), mono-unsaturated (MUFA) and poly-unsaturated fatty acids (PUFA) in beef using data from F_2 families descending from Wagyu and Limousin grandparents. Phenotypic data came from 328 progeny of the F_1 parents. The search was conducted using 217 markers covering the 29 bovine autosomes. A total of six QTL were found which are located on five different chromosomes; on a genome-wide basis two were statistically significant and four were suggestive. On BTA2, a QTL was found that had additive effects on SFA, MUFA and PUFA. Two QTL with dominant effects on MUFA were observed on BTA9. Three additional QTL suggestive of dominant effects on the relative amounts of fatty acids were also detected. Results of this study indicate the relative amounts of SFA, MUFA, and PUFA are under some degree of genetic selection. In conclusion, the fatty acid profile of beef can be improved by locating the relevant QTL through genetic markers and subsequently introgressing the loci through appropriate crossbreeding systems.

Phenolic compounds attenuation evaluation in *Acacia polyphylla* leaves in active grow and fructification, employ calcium oxide

G.E. Nouel Borges, M. Calderon, E. Rodríguez, M.A. Espejo Díaz, E. Molina and J.B. Rojas Castellanos, Universidad Centroccidental Lisandro Alvarado, Unidad de Investigación en Producción Animal, Tarabana, Cabudare, Estado Lara 3023, Venezuela

The attenuation of total polyphenol (TP), simple phenols (SF), total tannins (TT) and condensed tannins (CT) in *Acacia polyphylla* fresh leaves on phonologic stages of active grow and fructification, employ calcium oxide, possible media to reduce the toxic effect of this compounds were evaluated . A factorial arrangement (2x4), two phonologic stages and four levels of CaO (0, 2500, 5000 y 10000 ppt) was used to determine the effect on attenuation of presence of PT, FS, TT and TC, replicated five times for treatment in all randomized design. For the evaluated compounds the concentration were higher in active grow than in fructification. The CaO was the best attenuator for the presence of TP, SF, TT and CT with variation on the magnitude by the type of evaluated compound for two phonologic stages. The level of 2500 ppt was the best to attenuate for both phonologic stages, except for CT where 10000 ppt was the best. This trial, *in vitro*, indicated the potential of the application of CaO to reduced the presence of the biological active form of these substances, that will inhibit the bio-availability of nutrients in this plant as a food source for ruminants and other herbivores.

International aspects of livestock and livestock production no. 1

Boundary management: systems expertise and application for sustainable development of animal production systems

A.J. Van Der Zijpp, Wageningen University, Animal Production Systems Group, P.O. Box 338, 6700 AH Wageningen, Netherlands

Boundary management: systems expertise and application for sustainable development of livestock farming systems The scope of action in animal agriculture is dependent on the function of the actor in animal production. Roles and responsibilities of scientists, farmers, food chain operators, citizens, consumers, policy makers and retailers are different. System science provides insight in their fields of activity and their interdependence. Together they determine the issues effecting the sustainability of livestock farming systems. Livestock farming systems link with subsystems like livestock, land, crops and household and higher system levels like region, nation and world representing the diverse context. And they are part of value chains. Sustainable development of livestock farming systems is complex, because ecological, social and economic issues have to be addressed jointly. Farm management has diverse goals: continuity through provision of financial security, asset building, food production and income. The farming system adapts to changes in context and household resources which vary between and within Africa and Europe. Several issues of boundary management for ecology, social and economic sustainability will be presented for both continents. Knowledge and governance are conditional for sustainable development. Interdisciplinary science has become indispensable to integrate the diversity of issues to achieve optimal solutions, maintain resilience of farming systems and support policy making.

The application of DNA technologies to combat stock theft and improve food security in South Africa

C.M. Pilane, B. Greyling, O. Mapholi and S. Nemakonde, Agricultural Reserach Council (ARC), Animal Genetics and Forensics, P/Bag X2, Irene, 0062, South Africa

In Southern Africa, cross-border and in-land stock theft has become a crisis with farmers losing their genetically favorable livestock and food industries unable to maintain good quality meats and meat products. Cross-border stock theft along the South Africa and Lesotho border has particularly being on the increase in the recent past. Accompanying these increases, are the increase in the production of meat products of unknown animal sources due to illegal poaching. Statistically, stock theft and poaching has costs farmers and meat industries millions of rand, however, most of these cases remain unresolved. The Agricultural Research Institute and the South African Stock Theft Unit, has become instrumental in resolving stock theft, livestock ownership disputes and uncertainties in the animal origins of meats and meat products. Through the use of DNA micro-satellite markers and proper controls, we have resolved the identity of lost or sacrificed livestock, livestock paternity in ownership dispute cases, and the identity of the types of meats and meat products from various meat markets and industries. While some of these cases remain unresolved, those reported and investigated, have led to the increase in the prosecution rate of stock thieves, penalty for illegal meat trading, decreased poaching and a decline in the annual percentages of the reported stock theft cases.

---

no. 3

Development of beef production in Brazil during the last 10 years (1997-2006)

P.M. Meyer[1] and P.H.M. Rodrigues[2], [1]Brazilian Institute of Geography and Statistics/IBGE, R. Duque de Caxias, 1332, 13630-000, Brazil, [2]University of Sao Paulo/USP, College of Veterinary Medicine and Animal Science, R. Duque de Caxias Norte, 225, 13630-000, Brazil

The objectives of this study were to evaluate the development of Brazilian beef production during the last 10 years (from 1997 up to 2006). Bovine herds (number of heads, including beef and dairy), number of slaughtered animals and total carcass weight were collected (per year and per regions) from the IBGE. To calculate the annual average growth rate (%/year), regression analysis was done and the slope was taken between the natural logarithm of each variable and the correspondent year. In 1997, Brazil had approximately 161.4 million bovines and slaughtered 14.8 million heads (9.2% of the total bovine herd), totaling 3.3 billion kilograms of carcass weight. In 2006, Brazilian bovine herds went up to approximately 205.9 million head and doubled the amount slaughtered up to 30.4 million heads (14.8% of the total herd), producing 6.9 billion kilograms of total carcass weight. From 1997 up to 2006, Brazilian bovine herds increased 3.34%/year, while growth rate of slaughter increased at 8.72%/year and of total carcass weight at 8.79%/year. When analyzing by regions, the annual average growth rate of bovine herds was 9.83, 2.47, 0.83, 0.65 and 3.47%/year, respectively for regions North, Northeast, Southeast, South and Midwest. The growth rate of slaughter increased 20.93, 9.86, 8.42, 4.35, 7.98 and of total carcass weight 21.14, 10.74, 8.92, 4.20 and 7.70%/year, respectively for the same regions.

Dairy development in Brazil during the last 10 years (1997-2006)

P.M. Meyer[1] and P.H.M. Rodrigues[2], [1]Brazilian Institute of Geography and Statistics/IBGE, R. Duque de Caxias, 1332, 13630-000, Brazil, [2]University of Sao Paulo/USP, College of Veterinary Medicine and Animal Science, R. Duque de Caxias Norte, 225, 13630-000, Brazil

The objectives of this study were to evaluate the Brazilian dairy development during the last 10 years (from 1997 up to 2006). Milk yield, number of milked cows and formal milk collection were collected (per year and per regions) from the IBGE and productivity (L/cow/year) was calculated. To calculate the annual average growth rate (%/year), regression analysis was done and the slope was taken between the natural logarithm of each variable and the correspondent year. In 2006, Brazil produced 25.4 billion liters of milk and the formal collection by the industry was 16.7 billion liters. Approximately 20.9 million cows were milked with an annual average productivity of 1,212.7 liters. From 1997 up to 2006, milk yield has increased 3.80%/year, while its formal collection has increased at a rate of 5.24%/year. The number of milked cows has increased 2.48%/year, resulting in an increase of 1.28%/year in the productivity (14.55 L/year). When analyzing by regions, the annual average growth rate of milk yield was 9.51, 4.37, 1.65, 5.75, 3.98%/year and of formal collection was 11.72, 6.16, 3.86, 7.10, 4.55, respectively for regions North, Northeast, Southeast, South and Midwest. The growth rate in the number of milked cows was 9.12, 2.18, 0.37, 2.75, 3.08%/year and in milk productivity was 0.36, 2.14, 1.28, 2.93 and 0.87%/year (absolute growth of 2.13, 14.59, 16.29, 52.55 and 9.35 L of milk/cow/year) respectively for the same regions.

Performance of sheep fed on different diets and relationships with in vitro gas production measurements

J.O. Ouda[1,2], I.V. Nsahlai[2] and A.T. Modi[2], [1]Kenya Agricultural Research Institute, Animal Production, P.O. Box 14912, Nakuru, Kiamunyi, Kenya, [2]University of Kwa-Zulu Natal, Animal Science, Private Bag X01, Pietermaritzburg, South Africa, 033, South Africa

In this study, ten diets comprised of varied roughages (RG) and protein supplements (PS) were fed to sheep to investigate the influence of the diets on sheep performance, and the relationships among measurements obtained from *in vivo* and *in vitro* gas production technique (IVGPT) evaluations. The feeding trial lasted for four months. The RG were maize stover (MS) and grass hay (GH). The PS were Lucerne hay (LH), Sericea lespedeza hay (LPZ) and Sunflower oil cake (SFC). Diets affected dry matter (DM) intake (DMI), DM digestibility (DMD) and weight gain (WtGain). Diets also influenced apparent (ApDeg) and True (TruDeg) degradability of DM, gas production (GP) and time taken to produce half of maximum gas volume (T_{hg}). Models with different combinations of ApDeg or TruDeg, fibre content (NDF), T_{hg} and roughage physical form (RFrom) as variables accounted for 75 to 85% of variation in DMI. Models with different combinations of NDF, ApDeg or TruDeg, GP, T_{hg} and degradation efficiency factor-DEF (TruDeg/GP×T_{hg}) accounted for 70 to 89% of the variation in DMD. About 77 to 82% of the variation in WtGain was accounted for by models having various combinations of NDF, ApDeg or TruDeg and RForm. The results demonstrated that IVGPT and RForm measurements can be used to reliably predict sheep performance.

no. 2

Capability of yeast derivates to bind pathogenic bacteria

A. Ganner and G. Schatzmayr, BIOMIN Research Center, Technopark 1, 3430 Tulln, Austria

Adherence of bacteria via their surface lectins to host intestinal epithelial cells is considered an important initial event in bacterial pathogenesis. Mannose-specific (type 1) fimbriae are among the most commonly found lectins in enterobacteria such as *E. coli* and *Salmonella* spp. which are known as causative agents of animal diseases and food-related infections in humans. Yeast cell wall components containing mannanoligosaccharides have been described to display alternative adhesion sites for pathogenic bacteria. By immobilising those bacteria in the gastrointestinal tract they are transported through the gut without colonizing. The present study investigated yeast cell wall products for their ability to adhere *Salmonella-*, *E. coli-*, *Campylobacter-* and *Clostridium* strains. The products were examined with a quantitative microplate-based assay by measuring the optical density as growth parameter of adhering bacteria. Eight out of 10 *S. typhimurium* and *S. enteritidis* strains adhered to defined yeast cell wall product A with up to 2×10^6 CFU/mg. Additionally, 5 different pathogenic *E. coli* strains were tested for their ability to bind to yeast cell wall A. Four of these strains had an average binding capability of 2×10^3 CFU/mg whereas 2×10^6 *E. coli* F4 cells were bound per mg cell wall. *C. jejuni* and *C. perfringens* did not bind to yeast cell wall A. Our results demonstrate that yeast cell wall is able to bind *E. coli* and *Salmonella* spp. up to 10^6 CFU/mg.Thus enteric diseases can be prevented with yeast derivates.

---

no. 3

Robustness of cutin, chromic oxide and acid detergent lignin as a markers to determining apparent digestibility of diets in horses

R.F. Siqueira, R.C. Gomes, P.H.M. Rodrigues, R.S. Fukushima and A.A.O. Gobesso, University of São Paulo, USP, VNP, Av. Duque de Caxias Norte, 225 Pirassununga, S.P., 13635-900, Brazil

It was aimed to evaluate the robustness of fecal output markers cutin, acid detergent lignin (ADL) and cromic oxide (CRO) for estimating organic matter digestibility (OMD) of diets for horses. Four horses (197 kg BW, 10-month old) were randomly assigned to one of four diets differing at contents of soybean oil and alfalfa processing, following a 4x4 Latin Square design of treatments. In each period of the Latin Square, animals were allowed to adapt to the local for eight days and then total feces collection was performed for 3 days. Robustness was evaluate by assessing the linear regression slope between bias (actual OMD – predicted OMD) and actual OMD, diet content of ether extract (EE) and digestible energy (DE), and horse BW. Regardless of the marker, slopes of regressions between bias and EE, DE and BW did not differed from zero ($P>0.05$), therefore, biases were not influenced by varying those factors. Slopes of regressions between bias and actual OMD were not significant ($P>0.05$) for CRO but were different from zero for cutin and ADL ($P<0.05$). All markers showed to be robust to diet and animal body weight variations. Estimative errors of cutin and ADL are influenced by actual organic matter digestibility.

no. 4

Accuracy and precision of cutin as an internal marker for determining apparent digestibility of diets in horses

R.F. Siqueira, R.C. Gomes, P.H.M. Rodrigues, R.S. Fukushima and A.A.O. Gobesso, University of São Paulo, USP, VNP, Av. Duque de Caxias Norte, 225 Pirassununga, S.P., 13635-900, Brazil

This research aimed to compare organic matter digestibility coefficient (OMDC) of diets composed of alfalfa cubes, alfalfa cubes with soybean oil, alfalfa hay and alfalfa hay with soybean oil by total feces collection and the internal markers cutin and acid detergent lignin (ADL) and the external marker chromic oxide. Four male weanling horses (approximately10-months old, 197 kg BW average) were used in a 4 x4 Latin Square design. Evaluation of OMDC estimated by markers was performed by a model considering the bias, i.e., the difference between OMDC estimated by marker and by total feces collection. It was evaluated accuracy and precision. The ADL recovery did not differ from 100% and it was the most accurate (-0.56 bias), therefore the most appropriated marker for this kind of diet. The chromic oxide recovery was 88%, it underestimated the OMDC but it was the most precise (3.14). The cutin recovery was 151% and it overestimated the OMDC showing to be the less accurate and precise among the evaluated markers.

---

no. 5

Efficacy of phytogenic feed additives in farm animals

T. Steiner and M. Rouault, BIOMIN GmbH, Industriestrasse 21, 3130 Herzogenburg, Austria

The significance of gut health for optimized feed efficiency has become more and more evident in present times of rising prices for feed ingredients. Several feeding strategies may be implemented in order to secure gut health and performance of farm animals. Among potential alternatives, phytogenics represent a relatively new and promising group of performance enhancers. Originating from plant raw materials, phytogenics have flavoring properties as well as biological activities. Antimicrobial, antiviral, antioxidant and other biological activities have been found in various phytogenic compounds. However, the mode of action of phytogenics is versatile and needs further scientific evaluation in many cases. A beneficial impact on gut microflora, level of microbial toxins in the gut and nutrient digestibility was reported in recent studies with pigs and poultry. Moreover, it has been speculated that phytogenics may stimulate the secretion of saliva and digestive enzymes. Current data indicates that phytogenics have a pronounced impact on performance and health status of poultry, swine and calves. Considerable improvements in daily weight gain, feed conversion ratio and feed intake were obtained when phytogenics were included in the feed of different species. Moreover, phytogenics may be included into diets specifically in order to overcome stressful periods in the production cycle. Moreover and in comparison to Antibiotic Growth Promoters, phytogenics do usually not bear the risk of cross-resistances and residues in animal products.

no. 6

Effect of full fat flaxseed and antioxidant supplementation on production performance and egg quality of layers

Z. Hayat[1,2], T.N. Pasha[2], F.M. Khattak[2] and G. Cherian[3], [1]University College of Agriculture, Department of Animal Sciences, University of Sargodha, Sargodha, 40100, Pakistan, [2]University of Veterinary & Animal Sciences, Out Fall Road, Lahore, 54000, Pakistan, [3]Oregon State University, Department of Animal Sciences, 122, Withycombe Hall, Corvallis, OR, 97331, USA

The effects of incorporating flax seed and two types of antioxidants (α-tocopherols, butylated hydroxy toluene, BHT) at three levels (50, 100, 150 IU or mg/kg) on production performance and egg quality were investigated. Hens (n=96, 32 weeks old, ISA Brown, 12 birds/treatment) were fed corn-soy diet (no flax, no antioxidant), flax (10% with no antioxidant) or flax 10% +antioxidants for 42 days. Feeding flax has no significant effect on feed consumption, egg production and egg weight of layers. However fatty acid profile was altered significantly ($P<0.05$) with the addition of flaxseed in the diet of layers showing marked increase in the α-linolenic, docosapentaenoic, docosahexaenoic acid and total n-3 fatty acids with a concomitant reduction arachidonic acid in the egg . Egg quality was independent of the incorporation of flaxseed and/or antioxidant supplementation.These data indicated that eggs with increased n-3 fatty acids and tocopherols can be generated by minor diet modifications without affecting production and egg quality parameters.

---

no. 7

Factors influencing energy demand in dairy farming

R. Brunsch, S. Kraatz, W. Berg and C. Rus, Leibniz-Institute for Agricultural Engineering, Max-Eyth-Alle 100, D-14469 Potsdam, Germany

The efficiency of energy use is one of the key indicators for developing more sustainable agricultural practices. The energy inputs in livestock farming are assessed on the basis of direct and indirect energy consumption. The investigations relate on a defined standard technology and show that an efficient production and utilisation of the fodder is an important possibility to reduce the cumulative energy demand. In addition the cumulative energy demand is strongly affected by the composition of the diet. An increasing portion in pasture in the diet causes a decrease in the cumulative energy demand, and an increasing portion of concentrate in the diet causes an increase. An increasing service life of the dairy cows reduces the energy demand. The feed energy requirement per kg of produced milk is decreasing with rising individual performance of the animals. Nevertheless, this effect diminishes gradually with milk yields higher than 8,000 kg FCM per cow and year. Additionally energy demand is increasing with higher replacement rates. Milk yields higher than 8,000 kg FCM per cow and year can not compensate the increase of the cumulative energy demand caused by higher replacement rates. A further improvement of the milk performance is not useful from the energetic point of view in dependence on the cumulative energy demand for the feed-supply.

no. 8

Response of Danish Holstein, Red and Jersey cows to supplementation with saturated or unsaturated fat

M. Thorhauge, M.R. Weisbjerg and J.B. Andersen, University of Aarhus, Faculty of Agricultural Sciences, P.O. Box 50, 8830 Tjele, Denmark

Dairy cows were fed mixed rations based on maize and grass/clover silage, barley, soybean meal and dried sugar beet pulp either unsupplemented (C) or supplemented (substituted barley on weight basis) with saturated C16 rich fat (C16), or with unsaturated rapeseed/linseed in 0.74/0.26 ratio (U). Crude fat concentration were 3.03 (C), 5.65 (C16) and 5.81 (U) % in total ration dry matter (DM). 35 Danish Holstein (DH), 39 Danish Red (DR) and 31 Danish Jersey (DJ) were fed the experimental rations from parturition until 210 days in milk. Statistical analyses were performed within breed, as breeds responded differently to treatments. Treatment responses over lactation were tested using random regression. For the three breeds, the following responses were obtained compared to C. DR: U tended to increase energy intake, but energy corrected milk (ECM, kg/d) (30.9 (U) vs. 31.7 (C), ns) was reduced. C16 increased milk fat (4.42% (C16) vs. 3.87% (C)). DH: C16 decreased DM intake (DMI), but ECM (34.5 (C16) vs. 34.0 (C), ns) was slightly increased. Both C16 and U decreased milk protein (3.30% (C16 & U) vs. 3.49% (C)). DJ: C16 and U tended to decrease DMI and ECM (28.6 (C16) and 28.0 (U) vs. 31.6 (C)), and tended to increase milk fat. In conclusion, C16 reduced protein/fat ratio in milk from all breeds, but U only for DJ. The minor positive (DH), and tendency to negative (DR, DJ) milk yield response to C16 supplementation was unexpected, and seems to be due to a negative effect in early lactation.

---

no. 9

Performance of Ethiopian indigenous goat breeds stall-fed with grain-less diet

A. Sebsibe[1], N.H. Casey[2], W.A. Van Niekerk[3] and A. Tegegne[4], [1]Ethiopian Meat & Dairy Technology Institute, P.O. Box 1573, DebreZeit, Ethiopia, [2]University of Pretoria, Department of Animal & Wildlife Sciences, Pretoria, 0002, South Africa, [3]University of Pretoria, Pretoria, 0002, South Africa, [4]ILRI, P.O. Box, 5689, Addis Abeba, Ethiopia

Performance of the Afar, Central Highland (CHG) and Long-eared Somali (LES) goat breeds were evaluated using three grain-less diets varying in concentrate: roughage ratios (diet 1 was 50:50, diet 2, 65:35 and diet 3, 80:20). The roughage was native grass hay and the concentrate consisted of wheat bran and noug cake (Guizotia abyssinica). Seventy-two eight-month old intact male goats were used and slaughtered after feeding for 126 days. The LES had higher average daily gain (ADG), heavier slaughter, empty body (EBW) and carcass weights than the other breeds. Diet significantly affected ADG, but was similar on most carcass traits. The DP on an EBW basis was the highest on diet 1. Breed affected the DP, which ranged from 42.5-44.6% and 54.3-55.80% on slaughter weight and EBW basis, respectively. The chilling losses were between 2.5 and 3.1%. Breed significantly influenced the carcass fat and crude protein (CP) content and the values ranged from 10.3 to 14.0% and 19.3 to 21.1% respectively. The effect of diet was significant on CP%. Breed and diet significantly influenced the composition of most muscle fatty acids. The findings suggest that potential exist in Ethiopian goat breeds fed grain less diet for the production of meat with specific quality characteristics.

The effect of Lalsil Fresh LB on the fermentation and aerobic stability of ensiled TMR potato hash

B.D. Nkosi[1], R. Meeske[2], I.B. Groenewald[3], D. Palic[1] and T. Langa[1], [1]ARC:LBD-Animal Production Institute, Animal Nutrition, P/Bag x2, Irene, 027 0062, South Africa, [2]Outeniqua Research Farm, Animal Nutrition, P.O. Box 249, George, 027 6530, South Africa, [3]University of the Free State, Centre for Sustainable Agriculture, P.O. Box 339, Bloemfontein, 027 9300, South Africa

A dietary inclusion of 80% potato hash (150 g/kg DM) was used to formulate total mixed rations (TMR) and ensiled with or without a heterolactic inoculant, Lalsil Fresh LB (LFLB) for 90 days. The silages were produced in 1.5l jars under laboratory conditions. Sampling was done on day 0, 4, 10, 20, 40, 60 and 90 and analysed for pH, water-soluble carbohydrates (WSC), volatile fatty acid (VFA), lactic acid, ammonia-N, dry matter (DM), crude protein (CP), crude fibre (CF), fat and minerals. The aerobic stability of silage was determined on day 90. The silages were well preserved as indicated by higher high lactic acid and rapid drop in pH. The LFLB silages had lower ($P<0.05$) pH, CF, fat, butyric acid and ammonia concentration, while causing higher ($P<0.05$) lactic acid, acetic acid, propionic acid, DM, CP, phosphorus and ash than the control silage. The LFLB silage was aerobically stable than the control silage as indicated by lower CO2 production (0.16 g/kgDM vs 2.52 g/kgDM) and higher acetic acid (46 vs 24 g/kgDM). It was concluded that LFLB improved the aerobic stability of TMR potato hash silage through its higher production of acetic acid.

---

no. 11

Evaluating rations for high producing dairy cows using three metabolic models

N. Swanepoel[1], P.H. Robinson[2] and L.J. Erasmus[1], [1]University of Pretoria, Pretoria, 0002, South Africa, [2]University of California, Davis, 95616, USA

Total mixed rations (TMR) and commodity feeds from 16 dairy farms in California (USA) were sampled and chemically analyzed to evaluate their nutrient profiles using the metabolic models Amino Cow, CPM Dairy and Shield Dairy. Objectives were to identify potentially limiting amino acids (AA), determine the impact of the level of maize protein on animal productivity and to determine if there is enough consistency in the nutrient profiles of these rations to produce a ruminally protected AA package to supplement similar rations in the region. Inclusion of maize products in these rations ranged from 31 to 55% of DM, but higher inclusion levels had no detectable impact on performance. The modeled ratio of lys to met in metabolizable protein (MP) delivered to the small intestine did, however, decrease as TMR maize crude protein inclusion levels increased, but it did not impact the final modeled AA profile of MP, or milk component levels. The calculated optimal AA packages varied sharply by model. Amino Cow focused on including met, lys, leu and his, CPM Dairy on ile, leu, lys and met, and Shield Dairy on lys, ile, his and val. However, there was a high consistency within model in the predicted limiting AA sequence among TMR's. This suggests that there may be sufficient consistency in the nutrient profiles among TMR's in this region to support production of a common ruminally protected AA complex, however differences among metabolic models suggest that research to measure animal responses will be required.

no. 12

Administration of citric acid in drinking water of broiler

K.M.S. Islam, M.A. Hossain and M.A. Akbar, Bangladesh Agricultural University, Department of Animal Nutrition, BAU Campus, 2202, Bangladesh

A number of 162 newly hatched straight run broiler chicks (Hubbard Classic) were allocated randomly to 6 treatments each with 3 replicate cages having 9 birds in each. All birds were offered corn-soy based diet on *ad libitum* basis but six different levels of citric acid (CA) such as 0, 0.25, 0.50, 0.75, 1.00 and 1.25% in drinking water for 6 different treatment groups for 4 weeks and CA was withdrawn during week 5. Live weight gain of broilers decreased due to addition of CA during week 4 ($P<0.05$). Although CA was withdrawn from water at week 5, the weight gain was not improved considerably indicating the after-effects CA. There were significant ($P<0.05$) decrease in feed intake during administration of CA. However, at week 5, CA was absent in water, the feed intake was not decreased. Feed conversion efficiency of birds was not affected up to 1.00% level of CA, however, it was significantly ($P<0.05$) reduced at the highest level (1.25%). Water consumption was reduced ($P<0.05$) due to increasing level of CA. On the contrary, at 5th week when acid was not added to water, intake of water was similar in all the treatment groups. Acidity of water was reduced due to addition of CA, similarly pH of different parts of gastrointestinal tract were also reduced in upper part of the tract, but become similar in proventiculas. Carcass weight of broilers of different groups were not significantly affected. So, administration of CA in broilers through drinking water is not suitable for profitable broiler production but lower doses may be tested in further study.

---

no. 13

Effect of dietary citric acid, acetic acid and their combination on the performance of broilers

K.M.S. Islam, J. Islam, Z.H. Khandaker and S.D. Chowdhury, Bangladesh Agricultural University, Department of Animal Nutrition, BAU Campus, 2202, Bangladesh

A total number of 108 day old straight run broiler chicks (Hubbard Classic) were divided into four groups (three cages per group, 8 birds in each) to investigate the effects of feeding without organic acid (control), 0.5% citric acid in feed, 0.5% acetic acid in water and their combination (0.5% citric acid in feed and 0.5% acetic acid in water) on live weight gain, feed consumption, feed conversion efficiency and carcass characteristics. Performance data showed significant differences in body weight gain ($P<0.05$) at (0-5 weeks) of age and feed consumption ($P<0.05$) at week 2 and 3 weeks of age. Analysis of performance data also showed significant differences in FCR ($P<0.05$) at (0-5 weeks) of age. Carcass characteristics did not show significant difference from the control after 4 weeks of supplementation. Highest mortality was found in control and acetic acid group due to excessive hot weather. Results demonstrated that the use of 0.5% citric acid in the diet of commercial broilers may show effects on performance better than the control group with positive significant effects on live weight.

Dose titration and safety margin of citric acid in broiler diet

K.M.S. Islam, S. Islam and M.A. Akbar, Bangladesh Agricultural University, Department of Animal Nutrition, BAU-Campus, Mymensingh-2202, Bangladesh

An experiment was conducted with 108 day old straight run Hubbard Classic broiler chicks for a period of 35 days to know the safety margin of citric acid. The experimental birds were allocated randomly to six dietary treatments each with 3 replication having 6 broilers in each. The citric acid levels were 0, 1.5, 3.0, 4.5, 6 and 7.5% in treatment 1, 2, 3, 4, 5 and 6 respectively. Diet and fresh drinking water were supplied *ad libitum* to the birds. The body weight gain was increased up to 4.5% level of citric acid. Final body weight of birds was 1423, 1430, 1420, 1450, 1420 and 1343 g in respective groups. The addition of citric acid was withdrawn during 5th week. During this week the weight gain were 390, 327, 284, 297, 381, 377 g respectively. At 4th week FCE (g weight gain/kg feed intake) of bird were 366, 435, 477, 495, 427, 361 g and at 5th week they were 256, 264, 230, 232, 314 and 304 g. The best feed conversion efficiency value was observed up to 4.5% level of citric acid. After this level, FCE began to decrease as like weight gain. So, it may be concluded that citric acid up to level of 4.5% in broiler diet increased productivity and feed conversion efficiency and further level may not be show more productivity in broilers.

---

no. 15

Effect of citric acid as alternate source of antibiotic growth promoter flavomycin on the performance and health status of broiler

K.M.S. Islam, M.N. Haque and M.A. Akbar, Bangladesh Agricultural University, Department of Animal Nutrition, BAU-campus, Mymensingh-2202, Bangladesh

An experiment was conducted to determine the influence of citric acid as substitute of antibiotic Flavomycin on the performance and gut health of broiler chicks. A number of 160 broiler chicks (Hubbard Classic) were randomly distributed into four groups, with four replicate cages of 10 birds in each. Standard basal starter diet was given to control group (Group 1). Group 2 offered diet containing citric acid (5 g/kg). Diet of group 3 contained 0.01g Flavomycin per kilogram and group 4 is a combination of citric acid and Flavomycin as mention doses. Addition of citric acid in broiler diet enhanced weight gain than control, Flavomycin as well as its combination with citric acid ($P<0.05$). However, cumulative feed intake was also higher in acidifier supplemented group compared to the antibiotic supplemented group and the combination group. Supplementation of citric acid significantly improved feed conversion ratio (FCR-kg live weight gain/kg feed intake) compared to Flavomycin and its combination with citric acid. Overall 2 birds were died during 35 days feeding trial is lower than usual mortality. Addition of citric acid altered the pH of formulated feed but alteration not exists while feces were tested. Supplementation of 0.5% citric acid in the diet had significantly positive effect on live weight gain, feed intake and feed conversion efficiency. So, citric acid might be an alternate source of antibiotic growth promoter Flavomycin considering performance and health status.

Use of citric acid as alternate source of antibiotic growth promoter availamycin in broiler diet

K.M.S. Islam, R. Chowdhury and M.J. Khan, Bangladesh Agricultural University, Department of Animal Nutrition, BAU-campus, Mymensingh-2202, Bangladesh

An experiment was conducted to determine the influence of citric acid as substitute for antibiotic availamycin on the performance and gut health of broiler chicks. A number of 160 broiler chicks (Hubbard Classic) were randomly distributed into four groups, with four replicate cages of 10 birds in each. Standard basal starter diet was given to control group (Group 1). Group 2 offered diet containing citric acid (5 g/kg). Diet of group 3 contained 0.01g availamycin per kilogram and group 4 is a combination of citric acid and availamycin as mention doses. Both antibiotic and acidifier showed better weight gain than control as well as their combination ($P<0.05$). However, cumulative feed intake was also higher in acidifier supplemented group compared with the antibiotic supplemented group. Supplementation of citric acid significantly improved feed conversion ratio (FCR-kg live weight gain/kg feed intake) compared to availamycin and its combination with citric acid. Overall 1% birds were died during 35 days feeding trial, but any dose related adverse effect was not detected. Addition of citric acid altered the pH of formulated feed but alteration not exists while feces were tested. Supplementation of 0.5% citric acid in the diet had significantly positive effect on live weight gain, feed intake and feed conversion efficiency. So, citric acid might be an alternate source of antibiotic growth promoter availamycin considering performance and health status.

---

no. 17

Use of citric acid in feed of growing rabbit

K.M.S. Islam, M.J. Uddin and A. Reza, Bangladesh Agricultural University, Department of Animal Nutrition, BAU-campus, Mymensingh-2202, Bangladesh

Twelve growing rabbit with (4 to 6wks) age and weight (590 to595g) were allocated at random to 4 dietary treatments having 3 replication with each for a period of 56 days at the Animal Nutrition field laboratory,BAU, Mymensingh .All the dietary treatments named A B C D were iso-energetics and iso- nitrogenous but inclusion level of citric acid was 0,0.5,1.0 and1.5% respectively. The live weight gain among the dietary groups D and C are found higher compare to B&A. Statistically they are not significant but numerically increased due to different level of increasing supplementation citric acid. Dry matter intake and digestibility of Dm CP CF And were not significantly ($P<0.05$) different among the dietary treatments but numerically decreased for the increased amount of citric acid. Feed conversion efficiency (FCR) of different of dietary groups did not differ significantly (($P<0.05$) but improved 6 16 and 8% in 0.5 1.0 and 1.5 CA offering group. This study has shown that CA up to the level of 1.5% in rabbit diet increased the productivity and feed conversion ratio and further level may show more productivity in growing rabbit.

Comparison of the chemical analysis and fermentation characteristics of carob pods and carob pods residue from three regions in Libya

I. Milad and M. Dahoka, Omar Al-Mukhtar Univ., Animal Production, Al-Baida, 119, Libyan Arab Jamahiriya

The study evaluated chemical composition and fermentation pattern of carob pods and carob pods residue from three regions in Libya, namely Al-Jabal, Tripoli, and Alkufra, using proximate analysis and *in vitro* gas production respectively. The crude protein contents of carob pods (g/kg) ranged from 23 to 42. Carob pods residue contains more crude protein, ranged from 34 to 54 compared with the pods. Crude fiber contents (g/kg) were doubled in carob pods residue in comparison with carob pods (59 vs 181), (62 vs 102), and (57 vs 114) for the three regions. Same trend was observed for ether extract (83 vs 181), (134 vs 250), and (151 vs 335). The cumulative *in vitro* gas production (ml/g DM) at 72 hr was significantly different ($P<0.001$). The ranking order for gas production was Alkufra > Tripoli > Al-Jabal. Estimated metabolizable energy (MJ/kg DM) was 3.7, 4.2, and 3.8 for carob pods from Al-Jabal, Tripoli, and Alkufra respectively. The corresponding values for pods residues were 2.7, 3.2, and 3.4 respectively. It was suggested that carob pods and carob pods residue may contain anti-nutritional factors as indicated by poor ME contents and can be used in ruminant diets, only after improvement of their nutritive value.

---

no. 19

Effect of limestone particle size on egg production and eggshell quality during late production

F.H. De Witt, N.P. Kuleile, H.J. Van Der Merwe and M.D. Fair, University of the Free State, Animal, Wildlife and Grassland Sciences, P.O. Box 339, 9300 Bloemfontein, South Africa

A study was conducted to determine the influence of different particle sizes limestone in layer diets on egg production and eggshell quality during the later stages of egg production (>54 weeks of age). Calcitic limestone (360 g Ca/kg), consisting of small (<1.0 mm), medium (1.0-2.0 mm) and large (2.0-3.8 mm) particles were obtained from a specific South African source that is extensively used in poultry diets and included into isocaloric (12.9 MJ AME/kg) and isonitrogenous (155 g CP/kg) diets to ensure a dietary Ca content of 36 g Ca/kg. Sixty-nine, individual caged Lohmann-Silver pullets, 17 weeks of age, were randomly allocated to the three treatments (n=23) for the determination of various egg production and eggshell quality characteristics. Egg production and eggshell quality data recorded on individual basis at 54, 58, 64 and 70 weeks of age was pooled to calculate and statistical analyse parameter means for the late production period. Different limestone particle sizes had no effect ($P>0.05$) on any of the egg production and eggshell quality parameters. These results suggested that larger particles limestone are not necessarily essential to provide sufficient Ca^{2+} to laying hens for egg production and eggshell quality at end-of-lay, if the dietary Ca content satisfies the hen's requirements.

Effect of limestone particle size on bone quality characteristics at end-of-lay

F.H. De Witt, N.P. Kuleile, H.J. Van Der Merwe and M.D. Fair, University of the Free State, Animal, Wildlife and Grassland Sciences, P.O. Box 339, 9300 Bloemfontein, South Africa

A study was conducted to determine the effect of different limestone particle sizes in layer diets on bone quality characteristics at end-of-lay. Calcitic limestone (360 g Ca/kg DM) were obtained from a specific South African source that is extensively used in commercial poultry diets. Limestone particles were graded as small (<1.0 mm), medium (1.0-2.0 mm) and large (2.0-3.8 mm), representing the three treatments, and included into isocaloric (12.9 MJ AME/kg) and isonitrogenous (155 g CP/kg) diets to ensure a dietary Ca content of 36 g Ca/kg DM. Sixty-nine, individual caged Lohmann-Silver pullets, 17 weeks of age, were randomly allocated to the three treatments (n=23) for the determination of various bone dimensional and mechanical properties at end-of-lay. At 70 weeks of age, ten birds per treatment (n=10) were randomly selected and sacrificed for the removal of tibia and humerus bones. Different limestone particle sizes had no effect (P>0.05) on bone weight, length or width at 70 weeks of age. However, an increase in limestone particle size resulted in a significant increase in tibia breaking strength (P=0.0107) and -stress (P=0.0391). These results suggested that larger particles limestone (>1.0 mm) have a beneficial effect on improving bone mechanical properties of older laying hens.

---

no. 21

The effect of dietary ionophores on carcass characteristics of lambs

M.M. Price, O.B. Einkamerer, F.H. De Witt, J.P.C. Greyling and M.D. Fair, University of the Free State, Animal, Wildlife and Grassland Sciences, P.O. Box 339, 9300 Bloemfontein, South Africa

This study was conducted to evaluate the effect of different rumen fermentation modifiers (ionophores) in feedlot finisher diets on various carcass characteristics of S.A. Mutton Merino wethers. A commercial high-protein (330 g CP/kg) concentrate was formulated, incorporating Monensin (16.4 mg/kg), Lasalocid (33 mg/kg), or Salinomycin (17.5 mg/kg) and a Control (no ionophores) treatment. Maize meal (650 g/kg) and lucerne (150 g/kg) were included in the different protein concentrates (200 g/kg) to supply the isonitrogenous (160 g CP/kg) and isocaloric (16.8 MJ GE/kg) total mixed diets used during the trial (63 days). Sixty lambs (BW 29±2.5 kg) were randomly allocated to the treatment groups (n=15/treatment) and each treatment was further subdivided into 5 replicates (n=3/replicate). All animals were slaughtered on Day 63. The warm and cold (4 °C) carcass weights were used to determine weight loss (%). Backfat thickness (mm), and carcass traits such as body length, shoulder- and buttock circumference (cm) were recorded on the cold carcasses. Ionophore treatment had no effect on weight loss (Lasalocid 1.0% vs. Control 1.9%), backfat thickness (Monensin 3.3 mm vs. Lasalocid 4.5 mm) body length (Control 56.7 cm vs. Salinomycin 58.1 cm) or shouler- (Monensin 77.7 cm vs. Lasalocid 79.4 cm) and buttock circumference (Salinomycin 66.5 cm vs. Lasalocid 67.8 cm). Results suggest carcass characteristics are not influenced by rumen fermentation modifiers.

no. 22

Comparison of in vitro rumen fluid and multi-enzyme methods for predicting digestibility of compound feeds for ruminants

D. Palic, K.J. Leeuw, F.K. Siebrits and H. Muller, ARC-Animal Production Institute, Private Bag X2, Irene, 0062, South Africa

The accepted laboratory procedure for determining the organic matter digestibility (OMD) of ruminant feeds is the two-stage *in vitro* Tilley and Terry method (Tilley& Terry). The aim of this study was to compare this method with the *in vitro* Pepsin/Multi-enzyme (PME) incubation procedure and to develop equations for predicting the *in vivo* OMD of compound feeds for ruminants using *in vitro* Tilley& Terry and PME techniques. Six complete diets, with pre-determined *in vivo* OMD obtained in trials with sheep, were analysed by both *in vitro* procedures. The mean OMD values obtained by *in vivo*, Tilley& Terry and PME procedures were 736, 713 and 745 g OM/kgDM respectively and did not differ significantly ($P>0.05$). In a follow up study, 24 new compound feeds, with determined *in vivo* OMD, were used. New samples were analysed for the OMD by both *in vitro* procedures. These values, for total of 30 compound feed samples, were regressed against determined *in vivo* OMD values and equations for predicting the *in vivo* OMD were obtained. Using the Tilley&Terry technique, the prediction equation: OMD(in_vivo)=8.80+0.97 x OMD(in_vitro_Tilley&Terry), (R^2=0.82; RMSE=29.9) has been derived. The equation OMD(in_vivo)=224.30+0.699 x OMD(in_vitro_PME), (R^2=0.90; RMSE=22.2) was obtained using the PME method. The results of this study showed that the OMD of compound feeds for ruminants can be successfully predicted using multi-enzymatic incubation procedure.

---

no. 23

Effect of formaldehyde treatment of cotton seed cake and molasses supplementation on nitrogen utilization by steers fed low quality hay

R.A. Kombe[1], A.E. Kimambo[1], A.O. Aboud[1], G.H. Laswai[1], L.A. Mtenga[1], M. Weisbjerg[2], T. Hvelplund[2] and D.M. Mgheni[1], [1]Sokoine University of Agriculture, Animal Science and Production, P.O. Box 3004 Chuo Kikuu, Morogoro, Tanzania, [2]University of Aarhus, Animal Health, Welfare and Nutrition, Blichers Alle 20, 8830 Tjele, DK, Denmark

Nitrogen utilization was measured using ten steers (2yrs old; 224 kg LWT) allotted to 2 x 5 diets in two trials of 5x5 Latin square each period lasting for 18 days. Control diet was low quality hay sprayed with urea (LQHU) to contain 7% CP. The CP of LQHU was increased to 10 and 13% by supplementing it with two levels of untreated (UCSC) or formaldehyde treated (FTCSC) cotton seed cake, respectively. The same protein treatments were repeated in the second trial where molasses was added to all diets at 300g/steer/d. Rumen and intestinal digestion of protein were measured using *in sacco* and mobile nylon bag techniques. The degradation rate for FTCSC was lower (0.031 vs 0.052; $P<0.001$) than UCSC. The Total tract digestibility of FTCSC and UCSC was similar (91%). Supplementation with molasses reduced both faecal (18 vs. 23 g/d) and urinary (36 vs. 41g/d) nitrogen excretion. Animals fed UCSC excreted more (23.8 vs. 20.2) faecal and urinary (42.7 vs. 38.8) nitrogen than those fed FTCSC. At 13% CP, excretion for both faecal and urinary nitrogen were higher than at 10% and 7% CP. It is concluded that nitrogen retention by steers can be improved by reducing protein degradation in the rumen and addition of molasses in the diet.

no. 24

Nutrients digestibility of browse foliages with differnts levels tannins

R. Rojo, D. López, J.F. Vázquez, S. Rebollar and B. Albarrán, Universidad Autónoma del Estado de México, Centro Universitario UAEM Temascaltepec, km 65.5 Carr. Federal Toluca-Tejupilco, 51300, Mexico

The aim of this study was to evaluate the *in situ* digestibility of dry matter (DDM), neutral detergent fiber (NDF) and crude protein (CP) of browse foliages (*Lysiloma acapulcencis*, *Quercus laeta* and *Pithecellobium dulce*) with different levels of condensed tannins during rain (RS) and drought season (DS) using three inoculums: cow (CO), adapted goat (AG) and unadapted goat (UAG). These animals were equipped with ruminal cannula. Animals were fed with a basal diet content forage:concentrate (80:20) ratio. AG received a high diet in tannins (71.34 g/kg of DM). Incubation period in order to estimate digestibility was 48 h. Forages species, season and inoculums were arranged in a 3x2x3 factorial design. DDM of *P. dulce* was higher ($P<0.01$) in DS, and goats were more efficient in DS (DS: CO:59.69[b], UAG:66.89[a], AG:65.28[a] vs. RS: CO:54.43[c], UAG:54.78[c], AG:51.46[d]), low values were obtained for *L. acapulcencis* in DS, however, AG was better (DS: CO:29.67[ij], UAG:31.07[ih], AG:34.47[gf] vs. RS: CO:24.16[k], UAG:27.31[j], AG:28.19[j]). Season did not affect digestibility of NDF, but, UAG and AG degraded more efficient this fraction, *P. dulce* was the highest (DS: CO:21.47[d], UAG:36.62[a], AG:33.62[a] vs. RS: CO:30.86[bc], UAG:33.24[ab], AG:27.38[c]). Digestibility of CP increased in *P. dulce* (DS: CO: 62.54[b], UAG: 73.09[a], AG: 73.85[a] vs. RS: CO: 58.46[bc], UAG: 58.19[bc], AG: 56.31[c]). In conclusion, goats are more efficient degrading tested browse foliages than the cow.

---

no. 25

Feeding frequency does not affect dorsal and ventral muscle composition of olive flounder, *Paralichthys olivaceus*

J.-D. Kim, Kangwon National University, Animal Life System, Kangwon-Do, 200-701, Chuncheon, Korea, South

The effect of feeding frequency on whole body and muscle composition of juvenile flounder (*Paralichthys olivaceus*) was examined. Fish were kept in each (20 fish/tank) of 18 circular plastic tanks (3 tanks/treatment) for 9 weeks during which they were fed an extruded diet either once (1/D), twice (2/D) and three times (3/D) a day or once (1/EOD), twice (2/EOD) and three times (3/EOD) every other day by hand. Fat content in whole body of fish showed a tendency to increase with an increase in feeding frequency, although fish groups 1/EOD only was significantly ($P<0.05$) lower among treatments. Protein, fat, ash, calcium and phosphorus contents in dorsal and ventral muscle were not affected by feeding frequency suggesting that an excess of dietary lipid would be used for energy purpose or accumulated in other tissue organs.

The *in sacco* and *in vitro* digestibility of sorghum grains from indigenous varieties

S. Ncube[1] and L.R. Ndlovu[2], [1]Makoholi Research Station, Private bag 9182, 0000 Masvingo, Zimbabwe, [2]National University of Science and Technology, P O Box AC939, Ascot, 0000 Bulawayo, Zimbabwe

Two experiments were conducted to examine the feeding value of 12 sorghum grain varieties grown in Zimbabwe. The first examined the *in sacco* degradability of the sorghum grain varieties. The *in sacco* dry matter loss was higher in low PA sorghum grain than higher PA grain. The effective degradability of the grain was highest with a white sorghum grain with a testa. The second examined the *in vitro* dry matter digestibility (IVDMD) of the sorghum varieties. The effect of source of rumen (diet) and polyethylene glycol (PEG) on the digestibility of the sorghum grain was also examined. The diet fed to donor animals had a significant effect on IVDMD of the grain. In the control group (Diet 1), brown sorghum grain gave lower IVDMD than red or white grain (666 vs 838 g/kg DM). The increase in IVDMD due to PEG supplementation was highest with DC 75 (19.5%), a sorghum grain with the highest soluble proanthocyanidins (PAs). In general, brown (high PA) sorghum grain gave lower *in sacco* and *in vitro* digestibility. In the *in vitro* study PEG increased the digestibility of the grain

---

no. 27

Post harvest dry down affects *in vitro* gas production and DM intake of rice straw

M.B.L. Santos[1], G. Nader[1], U. Krishnamoorthy[2], P.H. Robinson[1], S.O. Juchem[1] and D. Kiran[2], [1]UCDavis, Davis, CA, USA, [2]Karnataka Vet., Anim. & Fisheries Sci. Univ, Bangalore, KA, India

Rice straw (RS) has low nutritive value, but it may be higher in fresh (F) vs. dry (D) rice plants. High Si content of RS has been assumed to cause its low digestibility, although there is little supporting evidence. Objectives were to determine *in vitro* gas production (GP; Exp. 1) and voluntary DM intake (VDMI; Exp. 2) of F, during dry down (DD) and D rice plants. In Exp. 1, 2 rice varieties in 2 fields each were used in a split plot design. One sample was collected from each subplot on days -14, -10, -6, -2 (F), 1, 2, 3, 4 (DD) and 6, 8, 12, 19, 33 (D) relative to harvest. Samples were hand harvested and chopped to 2-4 mm lengths prior to *in vitro* GP. In Exp. 2, 8 heifers were housed in a barn near Bangalore and, after 10 d of adaptation, were fed grain-free RS *ad libitum* that was harvested daily from 11 d pre-harvest (F) to 10 d post-harvest (DD: d 1-4; D: d 5-10), with 0.75 kg of concentrate/heifer/d. VDMI was recorded by heifer daily. Stage F GP was highest ($P<0.05$) at 4 h (26.5 vs. 24.2 (DD) and 24.7 (D) ml/g DM), 24 h (132 vs. 115 and 112 ml/g DM) and 72 h (190 vs. 176 and 173 ml/g DM) of incubation. Rate of GP was unaffected by stage, but Si in ADF was higher in D (80.8%) vs. F (74.6%) RS. VDMI by heifers was higher for F (5.1 vs. 4.1 (DD) and 3.7 (D) kg of DM), and this decrease supports similarly lower GP in Exp. 1. Rice plants undergo a substantial decline in nutritive value during dry down, perhaps related to the location of Si in the plant.

The effect of crude glycerol from biodiesel production as an ingredient in broiler feed

O.U. Sebitloane, S.E. Coetzee, A.T. Kanengoni and D. Palic, Agricultural Research Council – Animal Production Institute, Nutrition and Food Science, P/Bag X2, Irene., 0062, Pretoria, South Africa

The rapid increase in production of biodiesel from various oilseeds has resulted in growing quantities of by products such as crude glycerol. The low cost of crude glycerol and its value as an energy source has prompted many researchers to evaluate its nutritive value and use in animal feed. Crude glycerol was used as an energy source in broiler diets formulated to meet typical commercial standards. The aim of this study was to evaluate the performance of broilers fed diets containing either crude glycerol or its combination with sunflower oil as energy sources. Three treatments; (1) Glycerol at 10% and sunflower oil at 2%, (2) sunflower oil at 2% and (3) glycerol at 11% were evaluated replicated six times. The 114 broilers were randomly allocated to 18 pens containing 8 chickens each. The combination of oil and crude glycerol and crude glycerol alone were heavier throughout the growth period than the treatment with an inclusion of sunflower oil ($P<0.05$). Treatments with crude glycerol had higher ($P<0.05$) feed intake and better ($P<0.05$) feed conversion ratio (FCR) than treatment with sunflower oil. The results from this study showed that crude glycerol can be used effectively in broiler diets at 10% level without sunflower oil.

---

no. 29

Effect of weaning age and weight on response of pigs to variable amounts of starter and link feed

A.V. Riemensperger[1,2], P.B. Lynch[1] and J.V. O'Doherty[2], [1]Pig Production Development Unit, Teagasc, Moorepark, Fermoy, Co.Cork, Ireland, [2]School of Agriculture, Food Science and Veterinary Medicine, University College, Dublin 4, Ireland

Ninety-six single-sex pairs of pigs weaned at 21, 27 and 35 days of age were used in a 3 x 4 factorial designed experiment to evaluate the effect of weaning age and amount of post weaning diet on performance to 10 weeks of age. At weaning pigs were allotted to: (A) very low (VL, 1kg starter diet and 3kg link diet per pig), (B) low (L, 2 and 6 kg per pig), (C) medium (M, 3 and 9 kg per pig), or (D) high (H, 4kg and 12 kg per pig) amount of a commercial starter and link diet followed by a common weaner diet (15.0 MJ DE/kg, 20.0% crude protein, 1.3% total lysine). From Day 0 to Day 14 weaning age had a highly significant ($P<0.001$) influence on weight gain (338, 271 and 219g/d for 5, 4 and 3 week pigs respectively) and daily feed intake (447, 317 and 257g/d). From weaning to 10 weeks, weaning age had a highly significant influence on daily gain (476, 402 and 363g/d; $P<0.001$) and feed intake (680, 620 and 560g/d; $P<0.05$) while differences in FCE (1.43, 1.55, 1.57) were not significant. Final weight at 10 weeks of age (26.7, 24.68 and 24.36 kg for pigs weaned at 5, 4 and 3 weeks) did not differ significantly ($P>0.05$). Dietary treatment had no significant effect on performance to 10 weeks of age. An increase of 1.0kg in weaning weight resulted in an increase of 2.3, 2.2 and 1.8kg at 10 weeks of age when pigs were weaned at three, four and five weeks, respectively.

The nutritive value of South African *Medicago sativa* L. hay

G.D.J. Scholtz[1], H.J. Van Der Merwe[1] and T.P. Tylutki[2], [1]University of the Free State, Animal-, Wildlife- and Grassland Sciences, P.O. Box 339, Bloemfontein, 9300, South Africa, [2]Agricultural Modelling and Training Systems, 418 Davis Rd Cortland, NY, 13045, USA

A study was conducted to evaluate the variation- and expand the existing and limiting nutritive value database of *Medicago sativa* L. hay (168 near infrared reflectance spectroscopy spectrally selected samples) in South Africa. The highest moisture content recorded (13.54%) was safely below the critical moisture level of 16% for effective storage. Coefficient of variation (CV) ranged from 1.2% for dry matter (DM) up to 66.2% for acid detergent fibre-crude protein (ADF-CP). The average ash content was 12.97% (7.3 to 29.5%), indicating soil contamination. Fibre fractions varied as follows: acid detergent fibre (ADF) (21.26 to 47.28%), neutral detergent fibre (NDF) (28.89 to 65.93%), lignin (4.32 to 16.25%), cellulose (16.29 to 36.44%) and hemicellulose (5.26 to 19.86%). The mean IVOMD for both 24 and 48hr (69.26 and 73.19% DM, respectively), was representative (CV=±8%) of the *Medicago sativa* L. hay population. Crude protein (CP) (average=20.7%DM) consists of 76.9% true protein. According to ADF-CP, 6% of the samples were heat damaged. High mean calcium (Ca) (1.35%), potassium (P) (2.53%) and iron (Fe) (874ppm) values were recorded. The results emphasises the need for rapid and accurate analysis of chemical and digestibility parameters of *Medicago sativa* L. hay to be used in diet formulation in practice.

---

no. 31

Prediction of chemical composition of South African *Medicago Sativa* L. hay from a near infrared reflectance spectroscopy spectrally structured sample population

G.D.J. Scholtz[1], H.J. Van Der Merwe[1] and T.P. Tylutki[2], [1]University of the Free State, Animal-, Wildlife- and Grassland Sciences, P.O. Box 339, Bloemfontein, 9300, South Africa, [2]Agricultural Modelling and Training Systems, 418 Davis Rd Cortland, NY, 13045, USA

The NIRS to predict chemical and digestibility parameters was investigated. Samples (n=168) representing the spectral characteristics of the South African *Medicago sativa* L. hay population were chemical analysed for the development of calibration equations. Values for r^2 and ratio of prediction to deviation (RPD) used as estimates of calibration accuracy for these parameters were classified as follows: good for dry matter (DM) (r^2=0.97; RPD=4.84), crude protein (CP) (r^2=0.97; RPD=4.57), acid detergent fibre (ADF) (r^2=0.95; RPD=3.97), neutral detergent fibre (NDF) (r^2=0.95; RPD=3.99), lignin (r^2=0.94; RPD=3.61), ash (r^2=0.93; RPD=3.12) and chloride (Cl) (r^2=0.95; RPD=3.74); intermediate for NDF-crude protein (NDF-CP) (r^2=0.91; RPD=2.96), sugar (r^2=0.91; RPD=2.82), *in vitro* organic matter digestibility at 24 hr (IVOMD24) (r^2=0.90; RPD=2.84) and 48hr (IVOMD48) (r^2=0.89; RPD=2.70); and low (RPD<2.31) for soluble protein (SP), ADF-crude protein (ADF-CP), fat, starch, NDF digestibility (NDFD) and the macro minerals (Ca, P, Mg, P, Na and S). The results recorded in the present study indicated that the NIRS technique is acceptable for DM, CP, ADF, NDF, lignin, ash and Cl analysis and for inclusion in quality models.

Evaluation of models for assessing *Medicago sativa* L. hay quality

G.D.J. Scholtz[1], H.J. Van Der Merwe[1] and T.P. Tylutki[2], [1]University of the Free State, Animal-, Wildlife- and Grassland Sciences, P.O. Box 339, Bloemfontein, 9300, South Africa, [2]Agricultural Modelling and Training Systems, 418 Davis Rd Cortland, NY, 13045, USA

A study was conducted to evaluate current proposed models for assessing *Medicago sativa* L. hay quality, using near infrared reflectance spectroscopy (NIRS) analyses and Cornell Nett Carbohydrate and Protein System (CNCPS) milk production prediction as a criterion of accuracy. Application of the theoretically-based summative total digestible nutrients (TDNlig) model of Weiss et al. (1992), using lignin to determine truly digestible NDF, explained almost all of the variation in milk yield (MY) (r^2=0.98). However, this model involves high analysis costs to develop and maintain NIRS calibrations and several of its components were poorly predicted by NIRS and therefore, not suited for quality assessment. Current available models (forage quality index (FQI), relative forage quality (RFQ); relative feed value (RFV)) for assessing *Medicago sativa* L. hay quality revealed lower accuracies (r^2=0.83, r^2=0.76, r^2=0.61, respectively), especially when protein was included in the model (lucerne quality index (LQI), total forage quality index (TFI); $r^2 < 0.49$). The developed empirical equation named lucerne milk value (LMV), including ADF, ash and lignin (Y=b0 – b1ADF – b2ash – b3lignin) (r^2=0.96), proved to be the most practical, simplistic, economical and accurate quality evaluation model for commercial application.

---

no. 33

A model for assessing *Medicago sativa* L. hay quality

G.D.J. Scholtz[1], H.J. Van Der Merwe[1] and T.P. Tylutki[2], [1]University of the Free State, Animal-, Wildlife- and Grassland Sciences, P.O. Box, 339, Bloemfontein, 9300, South Africa, [2]Agricultural Modelling and Training Systems, 418 Davis Rd Cortland, NY, 13045, USA

A study was conducted to identify chemical parameters and/or models for assessing *Medicago sativa* L. (L) hay quality, using near infrared reflectance spectroscopy (NIRS) analysis and Cornell Net Carbohydrate and Protein System (CNCPS) milk prediction as a criterion of accuracy. Milk yield (MY) derived from the CNCPS model, by replacing the average L hay in a complete diet with 168 representative South African L hay samples, was used as a criterion to evaluate and/or develop models for L hay quality grading. The best single predictor of MY was the acid detergent fibre (ADF) content of L hay, which explained 67% of the measured variation. A multiple linear equation (Y=b0 – b1ADF – b2ash – b3lignin) explains 96% of the measured variation in MY. The relatively poor performance of crude protein (CP) (r^2=0.04) and other protein related parameters ($r^2 < 0.25$; adjusted-crude protein, ADF-CP, neutral detergent fibre-CP and soluble protein) in predicting MY suggest that protein content of L hay is an unreliable indicator of L hay quality. It is clear that MY derived from the CNCPS model by replacing L hay in a basal diet with others in the South African L hay population can be significantly predicted with high accuracy by the developed empirical model named lucerne milk value (LMV) consisting of only ADF, ash and lignin.

no. 34

The effect of *Opuntia*-based diets on feed utilization and growth of Dorper wethers

O.B. Einkamerer, H.O. De Waal, W.J. Combrinck and M.D. Fair, University of the Free State, Animal, Wildlife and Grassland Sciences, P.O. Box 339, Bloemfontein 9300, South Africa

Incremental levels of sun-dried and coarsely ground cactus pear (Opuntia ficus-indica var. Algerian) cladodes were used to substitute part of the lucerne hay in balanced diets fed to Dorper wethers (n=28; 33.90±2.98 kg) during a trial period of 70 days. The three treatment diets (T0, T24 and T36) comprised respectively (air dry basis) 0, 240 and 360 g/kg Opuntia; 660, 410 and 285 g/kg lucerne hay; 300 g/kg maize meal; 0, 10 and 15 g/kg feed grade urea; and 40 g/kg molasses meal. The apparent digestibility increased ($P<0.05$; 0.714, 0.732 and 0.756, respectively) with Opuntia inclusion in diets. The DMI ($P>0.05$; 1368, 1345 and 1317 g/day, respectively) and ADG ($P>0.05$; 117.8, 116.4 and 95.6 g/day, respectively) decreased slightly as Opuntia inclusion increased. It is concluded that with incremental inclusion of dried Opuntia cladodes in sheep diets, the DMI as well as ADG may decrease slightly, but the apparent digestibility increases. This may be ascribed to the higher digestibility of Opuntia in relation to the other feed ingredients, but lower nutrient value resulting in the slightly decreased live body weight gains. From these results it seems that adequate nutrients for maintenance and production of sheep were supplied by the treatment diets.

---

no. 35

The effect of *Opuntia*-based diets on carcass characteristics of Dorper wethers

O.B. Einkamerer[1], H.O. De Waal[1], W.J. Combrinck[1], M.D. Fair[1] and A. Hugo[2], [1]University of the Free State, Animal, Wildlife and Grassland Sciences, P.O. Box 339, Bloenfontein 9300, South Africa, [2]University of the Free State, Microbial, Biochemical and Food Biotechnology, P.O. Box 339, Bloenfontein 9300, South Africa

Incremental levels of sun-dried and coarsely ground cactus pear (*Opuntia ficus-indica* var. Algerian) cladodes were used to substitute part of the lucerne hay in balanced diets fed to Dorper wethers (n=28; 33.90±2.98 kg) during a trial period of 70 days. The three treatment diets (T0, T24 and T36) comprised respectively (air dry basis) 0, 240 and 360 g/kg *Opuntia*; 660, 410 and 285 g/kg lucerne hay; 300 g/kg maize meal; 0, 10 and 15 g/kg feed grade urea; and 40 g/kg molasses meal. Carcass weight decreased slightly ($P>0.05$; 19.57, 19.11 and 17.77 kg, respectively) with *Opuntia* inclusion in diets. Fat thickness measured 35 and 110 mm from mid spinal cord (3.73 & 6.10 mm; 3.50 & 5.09 mm; 3.05 & 5.54 mm; respectively), surface area of musculus longissimus dorsi (1488, 1549 & 1431 mm2), as well as carcass tissue coefficients (fat: 0.24, 0.20 & 0.21; muscle: 0.52, 0.55 & 0.53; bone: 0.23, 0.24 & 0.25; respectively) varied slightly ($P>0.05$) with no visible trend as *Opuntia* inclusion increased between treatments. Incremental inclusion of dried *Opuntia* cladodes in sheep diets had no effect on neither one of the carcass characteristics measured in this study. These results suggest that the overall effect of *Opuntia* on the carcass weight and quality of the Dorper wethers were small.

The effect of age on prussic acid concentration in improved forage sorghum in semi-arid Kenya

K.R.G. Irungu, G.B.T. Ashiono, K.A.N. Muasya and D.J.N. Kariuki, Kenya Agricultural Research Institute (KARI), Box 25, 20117 Naivasha, Kenya

Two varieties of fodder sorghum were studied to ascertain the effect of age on prussic acid (hydrocyanic acid) concentration and establish the threshold at which it would be safe to feed the sorghum to young ruminants. Varieties E6518 and E1291 planted in randomized complete block design layout with three replicates were sampled for prussic acid at 5 and 10 weeks (first and second weeding and thinning) after emergence and thereafter, every 2 weeks up to 14 weeks. The data was subjected to analysis of variance and regression which showed that prussic acid concentration was negatively correlated to sorghum age. Variety E1291 contained less prussic acid compared to E6518 during the study, E6518 and E1291 contained 186.7 and 90.8, 167.5 and 139.8 mg at 5 and 10 weeks respectively which decreased significantly to 81.7 and 70.8 mg at 14 weeks. Prussic acid concentration in E1291 during the whole study did not surpass the toxic threshold of 200 mg. The concentration in E6518 approached this threshold below 10 weeks' growth but thereafter declined to below the toxic threshold. Therefore, E1291 can be fed to young ruminants at 5, 10, 12 and 14 weeks but E6518 can only be fed after 10 weeks growth without toxic effects.

---

no. 37

Influence of species and stage of maturity on intake and partial digestibility by sheep fed two tropical grass silages

W.A. Van Niekerk, A. Hassen and F.M. Bechaz, University of Pretoria, Department of Animal and Wildlife Sceinces, Pretoria, 0002, South Africa

The aim of this study was to compare intake and partial digestibility of organic matter (OM) and nitrogen (N) of two ensiled tropical grass species [*P. maximum* (PM) and *D. eriantha* (DE)] either at the boot or full bloom stage of growth. Intake and digestibility were determined by the double marker technique where Yb and Cr were infused continuously into the rumen with a peristaltic pump. Except for OM disappearance in the gastro intestinal tract, neither species nor stage of harvesting had an effect on intake, digesta flow and OM disappearance within the rumen and small intestine. For PM silage, intake (g/d) was higher for the full bloom than boot stage. Total N flow (g/d), non-ammonia nitrogen (NAN) flow (g/d), NAN flow per N intake, the NAN disappearance (g/d) in the ileum and NAN disappearance as a % of N intake were higher for lambs fed on DE than PM silage at full bloom. In DE silage NAN disappearance in the ileum was higher for silage at full bloom than DE silage at boot stage. The true N-digested (%), however, did not differ significantly between the species or stage of maturity. From the results it is evident that silage made from DE at the full bloom stage is superior to the boot stage in terms of NAN disappearance in the lower digestive tract as well as to silage made from PM.

no. 38

Silage fermentation attributes and certain rumen parameters of lambs fed two grass silages differing in maturity stage

A. Hassen, W.A. Van Niekerk and F.M. Bechaz, University of Pretoria, Department of Animal and Wildlife Sciences, Pretoria, 0002, South Africa

The aim of this study was to compare two tropical grass species [*P. maximum* (PM) and *D. eriantha* (DE)] in terms of silage fermentation attributes and certain rumen fermentation characteristics of silage made either at boot or full bloom stage of growth. A lower silage pH was recorded for DE than PM silage. Neither species nor maturity stage had a significant effect on silage ammonia nitrogen, lactic, acetic, butyric, valeric and total volatile fatty acid concentrations. For PM silage total N was higher at full bloom than boot stage. DE had higher total N than PM silage at the boot stage. Species had no effect on propionic acid (PA) concentrations of the silage, but for DE higher PA concentrations were recorded at full bloom than boot stage. Rumen pH was lower in lambs fed on DE than those on PM silage. In PM fed lambs, full bloom stage had resulted in higher rumen NH_3-N concentration than boot stage silage. In lambs fed on DE silage higher concentrations of acetic, propionic, butyric and total VFA were recorded at full bloom than boot stage silage. Neither species nor stage of maturity had significantly influenced the acetic: propionic acid ratio. The results suggested that full bloom stage had a higher nutritive value and better preservation and nutritive value achieved in DE compared to PM silage.

---

no. 39

Influence of species and season on the quality of Atriplex

W.A. Van Niekerk, A. Hassen, P.J. Vermaak, N.F.G. Rethman and R.J. Coertze, University of Pretoria, Pretoria, 0002, South Africa

Three Atriplex shrubs [Atriplex canescens Santa Rita (ACSR), Atriplex canescens Field Reserve 1(ACFR) and Atriplex nummularia (AN)] were compared in terms of chemical composition and *in vitro* digestibility. The plant material was sampled at different seasons (autumn and winter) and from different sites (Hatfield in the Gauteng province and Mier and Lovedale, both in the Northern Cape Province) and these were analysed for crude protein (CP), neutral detergent fibre (NDF) and *in vitro* digestibility (IVOMD). The CP and IVOMD of the leaves were higher than the stem for all Atriplex species. Leaf to stem ratio of the autumn samples was not affected by species at Hatfield and Mier, but ACFR had lower leaf:stem ratio at Lovedale as well as for winter samples at Hatfield. At Lovedale, however, AN had the highest leaf: stem ratio compared to ACSR and ACFR. AN had a higher CP concentration than ACFR and ACSR at both Mier and Lovedale, but species had no effect on CP concentration at Hatfield. Autumn samples had a higher CP concentration than winter samples at both Mier and Lovedale, but season has no effect on the N concentration at Hatfield. AN had a lower NDF and higher IVOMD concentration compared to ACSR and ACFR. Autumn samples are less fibrous and more digestible than winter samples. AN seems better in terms of its nutritive value compared to ACSR and ACFR with better quality forage in autumn than winter.

Nutritive value of *Cassia sturtii* and *Sutherlandia microphylla* as compared to *Medicago sativa*: 1. Chemical composition and degradability of dry matter and NDF

W.A. Van Niekerk, A. Hassen, J. Els, N.F.G. Rethman and R.J. Coertze, University of Pretoria, Pretoria, 0002, South Africa

The aim of this study was to assess the potential nutritive value of two drought tolerant leguminous shrubs (*Cassia sturtii* and *Sutherlandia microphylla*) in terms of chemical composition and degradation parameters when compared to that of *Medicago sativa*. *S. microphylla* had higher NDF, ADF, ADL and lower CP, Ca, Mg, Cu and Fe concentrations than either *M. sativa* or *C. sturtii*. The CP, ADL, Mg and Fe concentrations of *C. sturtii* didn't differ significantly from that of *M. sativa*. However, *M. sativa* is more fibrous and had a higher P and Cu concentration level than that of *C. sturtii*. *M. sativa* had higher a-values for both DM and NDF degradation compared to the two shrub species at a rate constant of 0.02/hr. Similar trend was observed for the ED values, where the two shrub species had lower ED values than *M. sativa*. *C. sturtii* had a higher b-value for DM degradation compared to *S. microphylla*. For NDF, however, the b-values didn't differ among the species. Species had also no effect on the c-values of both DM and NDF. The two shrub species seems to have a lower nutritive value compared to *M. sativa*. Among the two shrubs, however, *S. microphylla* is more fibrous and lignified compared to *C. sturtii*, which may have a negative effect on intake and digestibility.

---

no. 41

Nutritive value of *Cassia sturtii* and *Sutherlandia microphylla* compared to *Medicago sativa*: 2. Digestibility, certain rumen parameters, intake and nitrogen balance by sheep

W.A. Van Niekerk, A. Hassen, J. Els, N.F.G. Rethman and R.J. Coertze, University of Pretoria, Pretoria, 0002, South Africa

This study compared the potential nutritive value of two drought tolerant leguminous shrubs (*Cassia sturtii* and *Sutherlandia microphylla*) with *Medicago sativa* in terms of digestibility, rumen fermentation parameters, intake, microbial nitrogen synthesis and N-balance when fed to sheep. Sheeps fed the two shrub species had a lower crude protein (CP), neutral detergent fibre (NDF) and organic matter (OM) intake than those fed *M. sativa*. The forages didn't differ in terms of CP digestibility, but the NDF digestibility was higher for *M. sativa* than *C. sturtii* and *S. microphylla*. OM digestibility didn't differ between *C. sturtii* and *M. sativa*, but *S. microphylla* had a lower OM digestibility than *M. sativa*. Sheep fed *C. sturtii* had a lower rumen NH_3-N concentration compared to those receiving *S. microphylla* or *M. sativa*. Sheep fed *C. sturtii* had a lower rumen acetate concentration and acetate: propionate ratio compared to sheep on *M. sativa*, but the species didn't differ in terms of propionic acid concentration. Sheep fed *M. sativa* had a higher butyric acid concentration compared to the two shrub species. Although sheep fed on *M. sativa* had a higher microbial protein synthesis and nitrogen balance compared to those fed the two shrub species, the nitrogen balance for all the three diets were positive.

The effect of Bonsilage mais Flussig and Lalsil Fresh LB on the fermentation and aerobic stability of whole crop maize silage

B.D. Nkosi[1], T. Langa[1], R. Meeske[2], I.B. Groenewald[3] and D. Palic[1], [1]Animal Production Institute-ARC, Animal Nutrition, P/Bag x2, Irene, 0062, South Africa, [2]Outeniqua Research Farm, Animal Nutrition, P.O. Box 249, George, 6530, South Africa, [3]University of the Free State, Centre for Sustainable Agriculture, P.O. Box 339, Bloemfontein, 9300, South Africa

The effects of heterolactic inoculants Bonsilage mais (BM) and Lalsil Fresh LB (LFLB) as additives for whole crop maize (300 g/kgDM) were studied. The silage was produced with BM, LFLB and untreated maize in 1.5l jars under laboratory conditions. Samples were collected on day 0, 4, 10, 21 and 60 of ensiling and analysed for pH, water-soluble carbohydrates (WSC), volatile fatty acid (VFA), lactic acid, ammonia-N, crude protein (CP), crude fibre (CF), dry matter (DM), energy, fat and ash. The aerobic stability of silage was determined on 90 of ensiling. The BM and LFLB caused lower ($P<0.05$) pH, ammonia-N, butyric acid and DM, while increasing WSC, lactic acid, CP and acetic acid than the control. Treatments did not impair ($P>0.05$) CF, fat, ME, ash and propionic acid contents of the silages. The BM and LFLB treated silage were more ($P<0.05$) aerobically stable than the control silages as indicated by lower CO_2 production. It was concluded that BM and LFLB improved the aerobic stability of maize silage due to their higher acetic acid productions.

no. 43

The effect of different ionophores on production performance of feedlot wethers

M.M. Price, F.H. De Witt, O.B. Einkamerer, J.P.C. Greyling and M.D. Fair, University of the Free State, Animal, Wildlife and Grassland Sciences, P.O. Box 339, 9300 Bloemfontein, South Africa

A trial was conducted to evaluate the effect of different rumen fermentation modifiers (ionophores) in feedlot finisher diets on the production performance of S.A. Mutton Merino lambs. A commercial high-protein (330 g CP/kg) concentrate was formulated, incorporating Monensin (16.4 mg/kg), Lasalocid (33 mg/kg) or Salinomycin (17.5 mg/kg) and a Control (no ionophores) treatment. Maize meal (650 g/kg) and lucerne (150 g/kg) were mixed with the different protein concentrates (200 g/kg) to produce the four isonitrogenous (160 g CP/kg) and isocaloric (16.8 MJ GE/kg) total mixed diets used during the experimental period of 63 days. Sixty lambs (BW 29±2.5 kg) were randomly allocated to the 4 treatment groups (n=15/treatment). Each treatment group was further subdivided into 5 replicates (n=3/replicate). Individual body weights and average feed intake (AFI) was recorded weekly and used in the calculation of feed conversion ratios (FCR) and average daily gain (ADG). Type of ionophore had no effect on AFI (1427.5 g/animal/day for Control group vs. 1563.4 g/animal/day for Salinomycin group), FCR (Lasalocid 4.84 vs. Control 5.15), and ADG (Control 278.3 g/day vs. Lasalocid 319.7 g/day) during the experimental period. Results suggest the efficiency of the different rumen fermentation modifiers to be similar and financial implications could influence their usage or preferance in diets.

Effect of roughage to concentrate ratio on ruminal fermentation and protein degradability in dairy cows

L.J. Erasmus, W.A. Van Niekerk and H. Nienaber, University of Pretoria, Department of Animal and Wildlife Sciences, Lynnwood Road, 0001,Pretoria, South Africa

Published research suggest it might be feasible to decrease ruminal protein degradability by lowering rumen pH through dietary manipulation of rumen fermentation. Three ruminally cannulated Holstein cows were used in a 3x3 Latin square design experiment and fed three total mixed rations differing in roughage:concentrate ratio (60:40; 45:55 and 30:70).The roughage portion consisted of equal parts of lucerne hay and Eragrostis curvula hay. Protein degradability of three feedstuffs differing in potential degradability (sunflower oilcake meal, cottonseed oilcake meal, roasted soya) were estimated using the *in situ* technique. Mean ruminal pH over 24h differed and were 6.44, 6.27 and 6.00 respectively for the 3 diets. Time below a rumen pH of 5.8 was 2.5h, only for the high concentrate (70%) diet. Mean ruminal NH_3^-N and VFA concentrations did not differ although the high concentrate diet resulted in a lower acetic:propionic acid ratio. Dietary treatment did not affect protein degradability within feedstuff suggesting that roughage:concentrate ratios ranging from 60:40 to 30:70 and pH ranging from 6.0 to 6.4 are still within the physiological boundaries where rumen ph do not affect ruminal protein degradability. Alternatives such as chemical or heat treatment would be more feasible than dietary manipulation of pH to increase dietary RUP content when formulating practical dairy cattle diets.

---

no. 45

Rumen pH and NH_3-N concentration of wethers fed temperate pastures supplemented or not with sorghum grain

M. Aguerre[1], J.L. Repetto[1], A. Pérez[2], A. Mendoza[1], G. Pinacchio[2] and C. Cajarville[2], [1]Facultad de Veterinaria, UdelaR, Departamento de Bovinos, Lasplaces 1550, Montevideo, 11600, Uruguay, [2]Facultad de Veterinaria, UdelaR, Departamento de Nutrición, Lasplaces 1550, Montevideo, 11600, Uruguay

The aim of this work was to evaluate the effect of sorghum grain supplementation on ruminal pH and NH_3-N concentration of wethers consuming a temperate pasture (*Lotus corniculatus*). Sixteen Corriedale x Milchschaf wethers fed temperate pastures *ad libitum*, were non-supplemented or supplemented with ground sorghum grain in 2 equal meals at 0.5, 1.0 or 1.5% of their BW. Rumen liquor samples were collected through permanent tubes inserted in the rumen at 0, 1, 2, 3, 4, 5 and 6 h after supplementation. Ruminal pH was measured immediately and NH_3-N concentration (mg/100 ml) was determined by direct distillation. Mean daily pH values for non-supplemented wethers and supplemented with 0.5, 1.0 and 1.5% of their BW were 6.45, 6.14, 6.09 and 5.43, respectively (SEM=0.15). Significant differences on pH were found between 1.5% supplemented group, and non-supplemented, 0.5% and 1.0% supplemented groups ($P<0.01$), while a trend was found between non-supplemented and 1.0% supplemented group ($P=0.07$). After 0 h, all pH values for 1.5% supplemented group were under 6.0. Meanwhile no differences were found in NH_3-N concentration among groups (mean=37.15, SEM=3.60). Ground sorghum grain supplementation significantly reduced rumen pH when was provided at 1.5% of BW to wethers fed temperate pastures, but did not affect NH_3-N concentration.

no. 46

Dry matter intake and digestibility of wethers and heifers fed temperate pastures supplemented or not with sorghum grain

M. Aguerre[1], C. Cajarville[2], V. Machado[1], G. Persak[2], S. Bambillasca[2] and J.L. Reprtto[1], [1]Facultad de Veterinaria, UdelaR, Depto. Bovinos, Lasplaces 1550, Montevideo, 11600, Uruguay, [2]Facultad de Veterinaria, UdelaR, Depto. Nutrición, Lasplaces 1550, Montevideo, 11600, Uruguay

The aim of this work was to evaluate the effect of sorghum grain supplementation on dry matter intake (DMI) and dry matter digestibility (DMD) of wethers and heifers consuming a temperate pasture. Twenty four Corriedale x Milchschaf wethers (45.6±6.2 kg BW) and 24 crossbred heifers (210.0±42.5 kg BW) housed in metabolic cages, fed temperate pasture (*Lotus corniculatus*) *ad libitum*, were non-supplemented or supplemented in 2 equal meals with ground sorghum grain at 0.5, 1.0 or 1.5% of their BW. Feed offered and refused was measured for 11 days and feaces voided were recorded daily during 5 days. Samples of feeds and feaces were collected daily and analyzed for dry matter (DM) (60 °C for 48 h). Supplemented wethers consumed less than non-supplemented ones (1.91 vs 1.47 kg DM/day, P<0.01). Meanwhile, supplementation led to higher DMI in heifers (5.98 vs 7.65 kg DM/day, P<0.01). Dry matter digestibility was 69.0% (SEM=2.7) for wethers and 65.1% (SEM=2.6) for heifers, no differences where found for any treatment. Meanwhile when all animals were analyzed together the non-supplemented group had lower DMD than supplemented groups (64.1 vs 68.1, P<0.05). Sorghum grain supplementation affected differently DMI of wethers and heifers. Dry matter digestibility of supplemented diets was higher probably due to grain digestibility.

---

no. 47

Performance of crossbred dairy calves on different feeding strategies under smallholder farms in sub humid eastern part of Tanzania

H.L. Lyimo[1], G.H. Laswai[2], L.A. Mtenga[2], A.E. Kimambo[2], D.M. Mgheni[2], T. Hvelplund[3] and M.R. Weisbjerg[3], [1]Ministry of Livestock Development and Fisheries, Box 9152, Dar es Salaam, Tanzania, [2]Sokoine University of Agriculture, Box 3004, Morogoro, Tanzania, [3]University of Aarhus, Box 50, 8830 Tjele, Denmark

An on-farm study was carried out to assess performance of crossbred (Frisian/Ayrshire x Tanzania Shorthorn Zebu) dairy calves reared under different feeding strategies in the sub humid eastern part of Tanzania. Strategy 1 (S1) involved feeding of calf concentrate formulated using locally available feed resources to contain per kg DM, CP 189 g and ME 13 MJ. Strategy 2 (S2) involved feeding of a dairy cow concentrate (per kg DM, CP 130 g and ME 13 MJ) commonly used by farmers in the study area. Strategy 3 (S3) was a control, where farmers were let to rear their calves with no interference in feeding. Restricted suckling and *ad libitum* feeding of forages and concentrate (up to 1 kg) were used for the calves in S1 and S2. Weaning was at 12 weeks. Feed intake and growth performance were recorded for 28 weeks. Total dry matter intake was not different (P>0.05) between S1 and S2. Growth rates (kg/d) of calves under S1 were higher (P<0.05) both pre (0.40) and post (0.46) weaning than those on S2 (0.35 and 0.32) and S3 (0.35 and 0.27). There is a potential of improving performance of calves under smallholder dairy production systems by feeding balanced calf concentrates formulated using locally available feed resources.

no. 48

Feeding schedule had no effect on ruminal pH of wethers fed *Lotus corniculatus*

A. Pérez Ruchel[1], C. Cajarville[1], F. Sanguinetti[1], C. Iturria[1], W. Saavedra[1] and J.L. Repetto[2], [1]Facultad de Veterinaria, UdelaR, Nutrición Animal, Lasplaces 1550, 11600, Uruguay, [2]Facultad de Veterinaria, UdelaR, Bovinos, Lasplaces 1550, 11600, Uruguay

The effect of the feeding schedule on ruminal pH of animals consuming a high quality pasture was studied. Eighteen lambs (49 kg BW) were fed a pasture (80% *Lotus corniculatus*) and allocated into 3 groups. Groups 1 and 2 were housed in metabolic cages and group 3 was grazing. Group 1 was continuously fed during all the day, group 2 was fed once daily during 6 hours and group 3 grazed once daily, during 6 hours. Ruminal pH was determined hourly during 24h using permanent ruminal tubes. Data were analyzed as repeated measures using the mixed model including treatment (t), hour (h) and t x h interaction effects. pH was strongly influenced by h (P<0.001), ranging mean values from 6.09 to 7.27. There were no differences between treatments (Group 1: 6.65, group 2: 6.66, group 3: 6.62; P>0.05), but the shape of the curves was different between them (t x h was significant, P=0.043). Grazing and housed animals had similar pH values. Data suggest that the pasture offered led to an adequate ruminal pH for fiber digestion instead of the restricted feeding time.

---

no. 49

Fibre management in high grain feedlot diets for beef cattle

V. Beretta, A. Simeone, J.C. Elizalde, M. Collares, M. Maccio and D. Varalla, University of Uruguay, Ruta 3 km 363, Paysandu, 60000, Uruguay

An experiment was conducted to evaluate the effect on animal performance of different strategies of including the roughage source (wheat straw, 350 kg bales, CP: 10%, ADF 42.4%) to the feedlot diet of Hereford steers (n=24, 343 kg) and calves (n=24, 168 kg): 1) roughage (15%, DM basis) offered together with the concentrate (85%) in a total mixed ration (TMR); 2) same quantities as in 1) but roughage and concentrate fed in separate troughs (RS); 3) same quantity of concentrate as in 1) but roughage offered separate and *ad libitum* (RAD), putting the wheat straw bale in the yard. In none of the treatments wheat straw was chopped. Animal weight was recorded every 14 days and feed intake daily. Four animals per treatment were observed for their intake behaviour from 8:00am to 6:00pm within 15 min interval. Daily gains were only affected by animal category (calves: 0.96, steers: 1.22 kg/day, P<0.001). Feed refusals were nil for TMR and RS. Roughage intake was increased for the RAD treatment compared to RS (P<0.01), however, it did not affect concentrate intake (P>0.05), both in steers or calves. Feeding the roughage separately increased eating total time in steers (P<0.05) independent of roughage level (P>0.05). For calves, feeding time was only increase for RAD (P<0.01). These results suggest that it is possible to modify roughage delivery form depending on operative restrictions at the feedlot without affecting animal gains, however feed conversion rate may be increased.

Ingestive behaviour of beef cattle under different grazing managements

V. Beretta, A. Simeone, O. Bentancur, S. Fariello and M. Pérez, University of Uruguay, Ruta 3, km 363, Paysandu., 60000, Uruguay

Grazing animals modify their behaviour to adapt to variations in pasture condition in an attempt to maintain nutrient intake. Understanding this process under specific combination of grazing method and stocking rate would help to decide on the best option. This study evaluated the effect of the grazing method (GM, varying residence time on paddock from 1 day (daily strip), to 3-4 days, or 7 days) and forage allowance (FA, 2.5 or 5.0 kg dry matter/100 kg liveweight), on forage intake (FDMI) and ingestive behaviour of Hereford steers and heifers (n=48; 283±25 kg). Experiment was conducted in Uruguay on a mixed grass/legume pasture (3600 kg DM/ha, PC: 14.7%) during autumn (18/4 to 14/6/2007). FDMI was estimated as the difference between forage on offer and residue. Grazing activity (GA), rumination (R) and idling (I) were recorded from 8:00am to 6:00pm every 15 minutes by visual appraisal, and bite rate (BR, bites/min) every two hours. FDMI was not affected by FA, GM or FAxGM (P>0.05). Differences were observed in ingestive behaviour: daily strip grazing promoted higher GA (P<0.01), less R and I (P<0.01) independent of FA (P>0.05), compared to 3-4 or 7 days GM, which did not differ (P>0.05). GA tended to change between days depending on GM (P=0.06), increasing from day 1 and day 4, for the 7-day GM. No differences were observed in BR (>0.05). Under these grazing conditions, results show there is no benefit of reducing residence time on paddock, as the animal manages to compensate for variations on pasture conditions along the days getting a similar FDMI.

---

no. 51

Effect of removing long fibre from beef cattle feedlot diets

A. Simeone, V. Beretta, J.C. Elizalde, J. Franco and G. Viera, University of Uruguay, Ruta 3 km 363, Paysandu, 60000, Uruguay

An experiment was conducted to evaluate the effect of removing the roughage source from a high grain diet (85% DM basis) fed to Hereford steer. Animals (n=30; 341 kg) were randomly allotted to one of the following treatments: 1) diet with long fibre (LF), animals received a diet 85% concentrate (ground sorghum grain, sunflower meal and urea) and 15% wheat straw (not chopped); 2) diet without long fibre (NLF), wheat straw was substituted for wheat bran, and a total mixed ration formulated for equal levels of NDFe, ME and CP, in both treatments. Trial was conducted during late winter in Uruguay from August 1 to September 27/2007. Animals where introduced to diets during first 14 days, and slaughtered after the feeding period. Feed was offered at 2.45% liveweight in four meals. Cattle were weighed every 14 days and feed intake was measured daily. Carcass weight, dressing percentage, subcutaneous fat depth, and pH 24-hours were measured. Liveweight increased linearly (P<0.01), with no differences en daily gains between treatments (LF: 1.646 kg/d; NLF: 1.745 kg/d; P>0.05). Feed refusals were nil and no differences where observed in feed intake. Feed conversion ratios were 5.9:1 and 5.4:1 for LF and NLF, respectively. No differences were observed in carcass traits (P>0.05), except for pH 24-hours which was lower for steers fed NFL compared to LF (5.50 vs. 5.61; P=0.02). These results suggest that when roughage is included at low levels it is possible to remove this fibre source while maintaining NDFe intake, without affecting animal performance.

no. 52

NO-production of macrophages stimulated with yeast products

A. Ganner, C. Stoiber, A. Klimitsch and G. Schatzmayr, BIOMIN Research Center, Technopark 1, 3430 Tulln, Austria

The ban of antibiotic growth promoters in the European Union 2006 has stimulated research in the field of functional feed additives. Yeast derivates are potential alternatives for antibiotics to improve animal health. Yeast β-glucan and yeast cell walls have been described to have immunmodulatory properties enhancing antimicrobial resistance by activating various points of host defense mechanisms. Activated macrophages secrete O_2 radicals and the bactericidal nitrogen metabolite NO, which are able to attack extracellular pathogens or intracellular pathogens that survive phagolysosome fusion. The present study investigated yeast derivates for their ability to activate macrophages by measuring their NO-production with an *in vitro* assay using a chicken bone marrow-derived macrophage cell line (HD11). NO is determined after 48 hours of incubation in the supernatant using a Griess reagent which provides a colorimetric reaction. Notably, 3 of 6 β-glucans and 8 of 13 yeast cell walls tested stimulated HD11 up to 100% of the positive control; the other 3 β-glucans and 4 yeast cell walls stimulated up to 50%. Three out of 6 yeast nucleotide products stimulated HD11 up to 100% and another nucleotide product showed a stimulation of 50% of the control. This study revealed that yeast derivates are potent NO-inducer´s in macrophages. This might improve animal health by alerting the immune system, preparing it to respond quickly to infections and counteracting the effects of weakened immunity.

---

no. 53

Evaluation of efficacy of a multi-strain probiotic feed additive in comparison to an antibiotic growth promoter on broiler performance

M. Mohnl[1], A. Ganner[2] and R. Nichol[3], [1]BIOMIN GmbH, Industriestrasse 21, 3130 Herzogenburg, Austria, [2]BIOMIN Research Center, Technopark 1, 3430 Tulln, Austria, [3]BIOMIN Laboratory Singapore Pte Ltd, 3791 Jalan Bukit Merah #08-08, 159471 Singapore, Singapore

The present trial was conducted to evaluate the efficacy of a multi-strain probiotic product in comparison to a commonly used AGP (antibiotic growth promoter) on broiler performance under field conditions in Thailand. The trial was conducted with day-old broiler chicks (AA+) with 32,600 birds per group, 4 replicates per group and 8160 birds per replicate. Experimental groups included a negative control group (NC), a group which received a multi-strain probiotic product (Biomin® Poultry5Star, Biomin GmbH) via the drinking water on day 1, 2, 3 and on three consecutive days around feed change and a positive control group (PC) which received Avilamycin (5 ppm) via the feed. The birds were kept under observation for 35 days and performance parameters like live weight, body weight gain, feed intake, feed conversion ratio (FCR) and mortality were determined. Probiotic group and AGP group increased body weight by 4.4% and 2.6% respectively and reduced mortality by 25.7% and 2.8% in comparison to control. Similarly, feed conversion was improved in broilers receiving the probiotic when compared to broilers in the NC group by 5.1%. In the present study the probiotic product had a better potential to improve broiler performance as Avilamycin and might therefore be a promising alternative to the use of AGPs in broiler production.

Effect of lignin type on rate of neutral detergent fiber digestion and potential energy yield

E. Raffrenato, P.J. Van Soest and M.E. Van Amburgh, Cornell University, Animal Science, Morrison Hall, 14853, Ithaca, USA

The objective of this work was to study the relationship between Klason lignin (KL) and Acid Detergent Lignin (ADL) with *in vitro* NDF digestion (IVNDFd) in an effort to assess if acid labile phenolic compounds affect the rate of degradation (k_d). Eighty five forages (alfalfas, corn silages and grasses) were analyzed for NDF, ADL, KL and IVNDFd (6, 12, 24, 30, 36, 48, and 96 hr fermentations were used for k_d estimations). Correlations were estimated among lignin types (KL versus ADL), lignin and extent of IVNDFd, and lignin type and IVNDFd k_d and tested for significance ($P<0.05$). Within and among all forage types, the correlation between ADL and KL was high and positive (0.77 to 0.90). Within and among all forages, only ADL was consistently negatively correlated with IVNDFd at 24, 30, 48 and 96 hours (-0.54 to -0.94). Correlation among forages for IVNDFd k_d and lignin type were not consistent. Among all forages, KL was negatively correlated with IVNDFd and IVNDFd k_d. The correlation between IVNDFd and ADL increased, as fermentation time increased among all forages (0.24 to 0.90), however, the correlation of KL and IVNDFd was greater up to 48 hr of fermentation suggesting that the soluble phenolics affected both the rate and extent of IVNDFd. Compared with ADL, KL disappeared during IVNDF and there was a high negative correlation with the difference in KL and ADL (ΔL) and IVNDFd k_d in those forages. In particular in those forages, a unit increase in ΔL corresponded to an average 18% increase in the k_d.

---

no. 55

Fatty acid profile in the meat of Santa Inês Lambs fed with different roughage:concentrate ration and fat sources

I. Garcia, A. Almeida, T. Costa and I. Leopoldino, Federal University of the Vale of the Jequitinhonha and Mucuri, Zootecnia, Diamantina, Minas Gerais, Brazil

This research was developed in the University of Vale of the Jequitinhonha and Mucuri (UFVJM) – Brazil. Twenty-four lambs were used. Were fed with diets with 30(30F) or 70(70F) % of Tifton hay with fat (control) or without a fat source, bypass fat (BF) or fat soybean (FS). The lambs were slaughtered with the average of live weight of 36.4 kg. After the carcass cooling, the muscle Longíssimus was taken for the analysis of the fatty acid profile. The diets with 70F, the fatty acid contend was higher for C18:0, and lower for C18:1-C9, which was decreased with the inclusion of BF. The C18:2-C9T11, conjugate linolenic acid (CLA) in the diets was higher with the inclusion of BF; however the increase was lower in the diet that contended more roughage. There were not differences in the percentage of CLA between the control and the use of FS in the diets with 30F, but using 70F, the inclusion of FS provided increase in the percentage of CLA. The meat of lambs fed with 70F has the percentage of saturated fatty acid increased and decrease of unsaturated and monounsaturated fatty acids. The polyunsaturated fatty acid (PUFA) was superior for animals that received BF and 70F, there were not differences from control to that ones that received FS. It follows that the use of bypass fat improve the profile of fatty acids in meat, especially in diets with higher proportions of roughage, no changes when the use of soybean.

no. 56

Weights and percentages of carcass of Santa Inês breed lamb fed with diets containing different

I. Garcia, G. Santos, N. Lima, F. Alvarenga and G. Dissimoni, University of the Vales of the Jequitinhonha and Mucuri, Zootecnia, Diamantina, Minas Gerais, Brazil

This experiment was carried out in Moura Experimental Farm, Curvelo-MG, in Sheep Production Sector of Universidade Federal dos Vales do Jequitinhonha e Mucuri. The performance of 24 Santa Inês male lambs was evaluated feed with a diet containing different roughage rations (30 and 70%) with (bypass fat or soybean) or without fat source *ad libitum*. The objective was to determinate the influence of each treatments under carcass weights and performances. Lambs feed with 30% of roughage without fat source, with bypass fat and soybean showed a similar warm carcass weight. The warm carcass performance was better in lambs feed with 30% of roughage and bypass fat compared with that ones receiving 70% of roughage. The add of fat source, in rich concentrated diets, can give a greats carcass performance, more efficient in bypass fat diet than in soybean diet.

---

no. 57

Study on feeding system for increased feed intake in dairy cow pre- and postpartum

H.S. Kim, H.Y. Jeong, H.J. Lee, K.S. Ki, W.S. Lee, S.B. Kim and B.S. Ahn, Dairy Science Division, National Institute of Animal Science, Rural Development Administration, #9, Eoryong-ri, Seonghwan-eup, Cheonan-si, Chungnam, 330-801, Korea, South

The objective of this study was carried out to investigate the effect of RFV and feed additives (cow-ban) on dry matter intake, milk yield, milk composition and metabolic disorders of dairy cow prepartum and postpartum. Dry matter intake was tended to increased feed intake in cows fed high quality forage of RFV 140 compared to RFV 110 for 3 weeks pre- and postpartum. Especially, DMI of dairy cows fed forage of RFV 110 was significantly higher in feed additive (cow-ban) than in control. Cows fed RFV 140 produce more milk than those fed RVF 110. Concentration of glucose was significantly different among treatments, especially higher in plasma of cows fed RFV 140 and feed additives (cow-ban) at 3 weeks pre- and postpartum. On the contrary, NEFA increased non significantly for non-treated cows at 3 weeks pre- and postpartum. Ca Concentration not different among treatment prepartum and postpartum. Displaced placenta was happened only 1 cow in control groups.

no. 58

Thermo-protective coating does not affect bioefficacy of a phytase in broilers fed mash diets

A. Owusu-Asiedu[1], P.H. Simmins[1] and O. Adeola[2], [1]Danisco (UK) Ltd, Box 777, SN8 1XN Marlborough, United Kingdom, [2]Purdue University, Department of Animal Sciences, 1151 Lilly Hall, West Lafayette, IN 47907, USA

The study evaluated bioefficacy of bacteria-derived phytase (EC 3.1.3.26) coated with thermo-protective coating technology (TPT; C-phytase) in broiler chicks fed corn-SBM-based basal mash diets. The trial tested the hypothesis that bioefficacy of same uncoated phytase molecule (U-phytase) is similar to C-phytase in mash diet. Three-day old male Ross 308 chicks (240) were randomly assigned to four dietary treatments with 6 cage replicates each (10 chicks/cage). The diets were: Positive control (PC), Negative control (NC) and NC+500 U/kg U-phytase and NC+500 U/kg C-phytase. Basal PC and NC diets were formulated to be similar in nutrient contents, except for Ca and P. Available P and Ca in NC diet was reduced by 0.20% and 0.15%, respectively. Chromic oxide was added to the diets as a marker and each diet was fed for 21 days. Birds were weighed on days 0 and 21, and were slaughtered on day 21 for P digestibility and tibia ash determination. Birds fed PC diet gained more weight and had higher tibia ash content ($P<0.05$) than birds fed NC diet. Addition of phytase regardless of source to NC diet improved ($P<0.05$) weight gain, P digestibility and tibia ash of birds. In conclusion, both coated and uncoated phytase were equally effective in improving weight gain, P digestibility and tibia ash content in chicks receiving diets lower in P and Ca indicating that TPT coating does not affect phytase release and bioefficacy in mash diet.

---

no. 59

The effect of concentrate feeding strategies on milk production and milk quality of Jersey cows grazing short rotation ryegrass (*Lolium multiflorum*) or Kikuyu (*Pennisetum clandestinum*) pasture

M.D.T. Joubert[1,2], R. Meeske[1,2] and C.W. Cruywagen[1], [1]Stellenbosch University, Department of Animal Sciences, Private bag X1, Matieland, 7602, Stellenbosch, South Africa, [2]Institute for Animal Production, Department of Agriculture Western Cape, Outeniqua Research Farm, P.O. Box 249, 6530, George, South Africa

The aim of the study was to investigate the effect of different concentrate allocation strategies on milk production and milk composition of Jersey cows grazing either ryegrass (*Lolium multiflorum*) or kikuyu (*Pennisetum clandestinum*) pasture. Concentrate feeds were offered twice daily, at 06:00 and 15:30. Treatments were: 3 kg concentrate in the morning and 3 kg in the afternoon (Control); 5 kg in the morning and 1 kg in the afternoon (Treatment 1); or 4 kg in the morning and 2 kg in the afternoon (Treatment 2). Treatment had no significant effect ($P>0.05$) on milk production or milk composition on either of the pasture types. For cows on ryegrass pasture, mean milk production, FCM, milk fat content and milk protein content values were 20.1 kg/cow/day, 20.8 kg/cow/day, 43 g/kg and 35 g/kg, respectively. For cows on kikuyu pasture, the respective values were 18.3 kg/cow/day, 18.7 kg/cow/day, 42 g/kg and 32 g/kg. It was concluded that feeding more concentrate in the morning and less in the afternoon, as compared to equal amounts in the morning and afternoon, did not affect milk yield or milk composition, irrespective of pasture type.

Dilemma of diagnosing selenium status in sheep exposed to high levels of selenium

J.B.J. Van Ryssen, C. Jansen Van Rensburg and R.J. Coertze, University of Pretoria, Animal & Wildlife Sciences, University of Pretoria, 0002, Pretoria, South Africa

Selenium (Se) is considered one of the more toxic essential elements. Overfeeding of Se is quite possible especially when the Se status of livestock has not been established before supplementation. Animals obtain Se mainly in two forms: inorganic Se, e.g. inorganic supplements and Se in drinking water; and organic Se, viz. supplements such as Se-enriched yeasts and Se naturally present in feed. For five months sheep (7/treatment) were fed different levels (all toxic levels depending on authority) of either inorganic (INORG, sodium selenite) or organic (ORG, seleno-yeast) Se. Levels were 1.0, 1.8 and 6.3 mg Se/sheep/day. A control group received the basic diet supplying 0.38 mg presumably organic Se/d. No toxic symptoms were observed. Liver concentrations were (mg Se/kg DM): Control=976, at 1 mg Se/d: INORG=4071; ORG=5432, at 1.8 mg Se/d: INORG=10083; ORG=12050, at 6.3 mg Se/d: INORG=16480; ORG=25180. Whole blood Se concentrations showed high positive correlations with Se in livers, though the sources followed completely different trends. INORG: y (liver Se)=-6717+ 32x (blood Se) R^2=0.41, ORG: y=-127+13x (R^2=0.85). It is often claimed that inorganic Se is more toxic than organic Se sources. However, especially at high intakes, tissue and fluid levels would be poor measures of Se status of animals. Without information on the source of Se in the diet, tissue Se levels would not be very informative on the risk of Se toxicity.

---

no. 61

Fractional rate of degradation (kd) of starch in the rumen and its relation to *in vivo* rumen and total digestibility

T. Hvelplund, M. Larsen, P. Lund and M.R. Weisbjerg, University of Aarhus, Department of Animal Health, Welfare and Nutrition, P.O. Box 50, 8830 Tjele, Denmark

Fractional rate of degradation (k_d) is an important parameter in modern feed evaluation. Estimates of k_d for starch was obtained on twenty starch sources originating from barley, wheat, oat, maize and peas and treated in different ways both chemically and physically. The starch sources were fed in mixed diets together with grass silage and soy bean meal and allocated *ad libitum* to fistulated dairy cows. The starch content varied between 13 and 35% in ration dry matter for the different starch sources. The design was a series of cross over experiments with 2 cows and two periods. Ruminal starch pool was estimated from rumen evacuation and starch flow was estimated by duodenal and faeces sampling. Fractional rate of rumen degradation was estimated from the equation [k_d=rumen degraded/rumen pool] and rumen and total digestibility of starch from flow measurements. Relation between k_d (h^{-1}) and rumen and total digestibility was calculated by the GLM procedure in SAS. The relation between k_d and rumen digestibility (RD) was estimated to [k_d=1.139 -3.580*RD+3.078*RD^2; R^2=0.45; valid for 0.58d=7.888-19.33*TD+11.89*TD^2; R^2=0.57; valid for TD>0.81]. It is concluded that the relations between k_d and starch digestibility can be used to estimate this important parameter (k_d) for feed table use from experiments where starch digestibility is known and within the range valid for the equations estimated.

Detoxification of aflatoxins by yeast sludge in the feed of dairy cattle and its impact on milk production

T.N. Pasha, M. Amjad, M. Aleem and M.A. Jabbar, University of Veterinary & Animal Sciences, Lahore, 54000, Pakistan

To determine the effects of feeding yeast sludge in the diet of dairy cattle on the detoxification of aflatoxins and milk production, twelve Sahiwal cows were randomly divided into three groups having four cows each to receive either no aflatoxins A (control), B (500 ppb aflatoxin) or C (500 ppb aflatoxin and 1% yeast sludge). Feed intake per group and milk production per cow was recorded daily and milk samples were analyzed for aflatoxin M1, milk fat, solid not fat (SNF) and total solids. The feed intake (kg) of cows was lowest (163.37±2.27) in group B and highest in group C (177.00±1.99) than group A (170.14±1.98) and differed significantly among groups. However, the milk production was not different statistically and was 7.58 ±1.32, 7.10±0.79 and 8.37±0.47 kg for groups A, B and C, respectively. The aflatoxins M1 (ppb) residues in milk at the end of treatment were 2.87±0.76, 15.51±2.63 and 12.74±1.02, respectively and differed significantly among the groups. The mean values of milk components for groups A, B and C showed non-significant differences which were 4.09±0.43, 4.06±0.43 and 4.29±0.34, respectively for fat; 8.89±0.49, 8.58±0.15 and 8.74±0.21, respectively for SNF and 12.94±0.47, 12.64±0.46 and 13.03±0.55, respectively for total solids. The results of the study revealed that yeast sludge can detoxify feed aflatoxins and improves feed intake as well as having tendency to improve milk production and milk quality.

---

no. 63

Performance of beef cattle and sheep fed diets based on non-conventional by-products

T.E. Bhila, M.M. Ratsaka, B.D. Nkosi, K.-J. Leeuw, M.J. Chipa and T.M. Langa, ARC-API, Irene, Animal Nutrition, P/Bag x2, Irene, 0062, South Africa

A survey of byproducts from the agro-industry was conducted whereby; cabbage, beetroot and potato hash were found in abundant quantities. The agro-industry is faced with dumping areas for these by-products. This can be both an economical and environmental problem. The use of these waste/by-products is restricted due to the poor understanding of their nutritional and economic value, as well as their proper use in ruminant rations, and the presence of anti-nutritional factors which may have negative impact on the animal performance. The by-products were collected, processed and analysed for nutritional qualities. The by-products where incorporated in different dietary blends and fed to cattle and sheep. Data for feed intake, live weight gain, feed conversion ratio (FCR), and diet digestibility were collected at regular intervals. When formulated to be isonitrogenous and isoenergetic, it was found that the by products replaces some of the ingredients in the different diets partially and sometimes completely. There is enough evidence indicating feeding value for the by-products in this study. These non-conventional by-products have been proven to have good feeding value; however, alternative and cheaper methods of processing the by-products will need to be explored.

The effect of sugar, starch and pectin as microbial energy sources on *in vitro* forage NDF digestibility

M. Malan and C.W. Cruywagen, Stellenbosch University, Department of Animal Sciences, Private bag X1, Matieland, 7602, Stellenbosch, South Africa

The aim of the study was to evaluate the effect of purified energy sources, sugar (sucrose), starch (maizena) and pectin (citrus pectin), on *in vitro* neutral detergent fiber (NDF) digestibility (IVNDFD) of forages. Forage substrates used included wheat straw (*Triticum aestivum*), oat hay (*Avena sativa*), lucerne hay (*Medicago sativa*), ryegrass (*Lolium multiflorum*) and kikuyu grass (*Pennisetum clandestinum*). *In vitro* digestibility determinations were done with the aid of an Ankom Daisy II incubator and forage substrates were incubated, with or without the respective energy sources, for 12h, 24h, 48h and 72h. Rumen fluid was collected from two lactating Holstein cows receiving a diet consisting of oat hay, wheat straw and a concentrate mix. No forage x energy source interactions were observed. Ryegrass and kikuyu had a higher IVNDFD than lucerne, oat hay and wheat straw ($P<0.01$). At 72h, IVNDFD was 35.2% for lucerne, 40.4% for oat hay and 38.5% for wheat straw, compared to 68.4% for kikuyu and 66.4% for rye grass. A possible explanation is that kikuyu and rye grass samples came from freshly cut material, harvested after a 28d regrowth period. Starch and pectin lowered IVNDFD ($P<0.01$). Mean forage IVNDFD values (including all forages across all times) were 42% for the control treatment, 41% for sucrose, 37% for starch and 36.5% for pectin. It is hypothesized that microbes fermented the easily fermentable energy sources first before attacking forage NDF.

---

no. 65

Effect of enzymes on growth performance and intestinal health in pig weaners

A.T. Kanengoni, O.G. Makgothi, S. Monesi, R. Thomas and M.L. Seshoka, Agricultural Research Council, Pig Nutrition, Pvt Bag X2, Pretoria 0062, South Africa

The objective of this experiment was to evaluate the effect of adding enzymes to diets differing in buffering capacity on growth performance traits in weaner pigs and on their intestinal health. Eighty pigs, 40 of each sex, 28 days of age, with an average live weight of 7±1.5 kg were used. Four diets 1) pellets 2) control diet (normal acidic diet without enzymes); 3) control+enzyme, and; 4) alkaline+enzyme were used in the trial. The 80 animals were randomly allocated to the four treatments in a completely randomized design. Animals were weighed at the start of the trial and weekly thereafter. Scouring was recorded per pen on a scale of 1 to 5. Feed and water were supplied *ad libitum* in automatic feeders. The commercial diet outperformed ($P<0.05$) the other three diets in daily gain, average daily feed intake and in feed conversion ratio. There was no difference ($P>0.05$) among the other three diets. There was no difference in scouring index in the piglets in the control, control+enzyme and alkaline+enzyme diet ($P>0.05$). There was however a lower incidence of scouring ($P<0.05$) in the commercial pellets than the other diets. The use of enzymes and the change in pH of the diet had no effect on intestinal health and improving growth of weaners in soyabean based diets.

no. 66

Development of grassland establishment techniques for grazing in abandoned paddy fields of Japan

K. Ikeda[1], M. Higashiyama[1], Y. Higashiyama[1], M. Nashiki[2] and T. Kondo[1], [1]National Agricultural Research Center for Tohoku Region, Forage Production and Utilization Research Team, akahira4 Shimokuriyagawa Morioka Iwate, 020-0198, Japan, [2]National Institute of Livestock and Grassland Science, Nasu Research Station, 768 Senbonmatsu, Nasushiobara,Tochigi, 329-2793, Japan

In Japan, to control the overproduction of rice, rice cultivation areas in paddy fields is restricted by the government. Since in these paddy fields the cultivation for crop production has been stopped in many cases, the utilization of the paddy fields for grazing of breeding cattle is increasing. However, since soil moisture content is high even if a paddy field has been drained, we cannot utilize the techniques of conventional grassland establishment in these areas. Therefore, we investigated the influence of grazing on vegetation of sown grass and soil hardness, and compared the persistency among some grass species in paddy fields used for grazing. By this study, we proved that soil becomes softer as soil moisture content increases and the sown grass trampled by cattle is greatly degraded. Moreover, *Phalaris arundinacea* L. and *Agrostis alba* L. were more persistent in paddy fields than *Lolium perenne* L. and x *Festulolium* spp.

---

no. 67

Energy cost and equivalent liveweight gain cost associated to the ammonia–N detoxification for steer grazing oats

A. Simeone[1], V. Beretta[1], J.C. Elizalde[1], J. Rowe[2] and J. Nolan[2], [1]University of Uruguay, Ruta 3 km 363 Paysandu, 60000, Uruguay, [2]University of New England, Armidale, NSW, 2351, Australia

Beef cattle grazing lush pastures during the autumn in temperate areas commonly experience growth rates lower than expected given apparently high quality of pastures. This response has been attributed to a low efficiency of N use in the rumen. However, restrictions on intake could be playing a major role. The energy cost of removing the excess NH_3-N in the rumen was calculated for Hereford steers (327 kg, sd 20.8) grazing oats pasture (CP=17.8%, DMD=74.6%), at 5% forage allowance from 31 May to 15 July, and compared that value to the energy required for liveweight gain. Procedure was based on observed liveweight gains, dry matter intake records, pasture chemical composition, and *in situ* data from rumen cannulated steers. Energy cost of removing excess NH_3-N was calculated based on the methodology proposed by Cohen (2001). Excess of NH_3-N from the rumen was estimated to be 10.4 g/day (5.98% of total dietary N). The energy cost associated with the removal of excess N was 0.707 MJ ME/day, which represented an increase in ME requirements for maintenance of 1.64%. This, energy would be equivalent to the ME required for the gain of 22g of liveweight per day (AFRC, 1993) a value that is not relevant in terms of animal performance of growing cattle. It is probable that higher forage CP levels than obtained in our study are necessary, to impact negatively on animal performance.

Effect of live yeast and monensin supplementation on the productivity of feedlot cattle

L.J. Erasmus[1], M.N. Leviton[1], R.F. Coertze[1] and R. Venter[2], [1]Dept of Animal and Wildlife Sciences, University of Pretoria, Lynnwood road, 0001,Pretoria, South Africa, [2]Vitam International, Lenchen road, 0046,Centurion, South Africa

The effects of live yeast(Levucell SC 1-1077) and monensin supplementation on the productivity of feedlot cattle fed a hominy chop based diet were examined. The 3 treatments were: monensin (33ppm)(M), yeast (0.4g/head/day)(SC),and a combination (M+SC). Ninety Bonsmara-cross weaner calves were used in a randomised block design and fed one of the three dietary treatments for a period of 99 days. Each treatment consisted of 30 animals subdivided into three treatment pens with 10 animals per pen. The M+SC treatment resulted in a lower (P<0.05) DMI when compared to M or SC. No difference was observed between the feed conversion ratios of the 3 treatments (P>0.05), however the SC treatment and the M+SC treatments respectively, tended (P<0.10) to be more efficient in terms of feed conversion ratio when compared to the M treatment (5.62 & 5.58 vs. 5.88 kg feed/kg gain). These results suggest a potential (additive) interaction between monensin and live yeast, but the mode of action needs further investigation.The ADG varied between 1.98 (M) and 2.06 kg/d (SC) and did not differ (P>0.05) suggesting similar effects on growth in diets containing around 45% NFC. Rumen scoring was done to describe the extent of rumen damage as a result of rumen acidosis; only 18 out of 90 animals showed rumen damage and due to the low numbers no conclusions could be drawn on the effect of treatment on rumen damage and the incidence or severity of acidosis.

---

no. 69

Effects of dietary vitamin A level on growth performance and intramuscular fat of pork meat in growing/finisher pigs

S.Y. Ji, J.C. Park, S.D. Lee, H.K. Moon, H.J. Jung, I.C. Kim and J.B. Lim, National Institute of Animal Science, Swine Science Division, #9 Eoryong-ri Seonghwan-eup Cheoan-si Chungnam, 330-801, Korea, South

This study was performed to validate whether no vitamin A supplementation could be a practical method for improving intramuscular fat content in pork meat. Intramuscular fat (IMF) content has been known to be determined by adipocyte differentiation within muscle fibers, which was inhibited by vitamin A. In this experiment, 54 of Landrace x Yorkshire pigs were fed with 3 levels of supplemental vitamin A (A: none, B: 1300 IU/kg, and C: 4000 IU/kg) for 9 weeks. Their average initial body weights were 68.0±5.0 kg. Average daily gain(kg/d) 0.89±0.13, 0.92±0.10, 0.92±0.12, and gain/feed were 0.25±0.04, 0.29±0.03, and 0.29±0.04 in 3 groups, respectively. And all parameters of performance in pigs did not show any significant differences between groups. On the 63th day, 3 pigs every group were slaughtered for measuring characteristics of their meats. IMF contents were 3.06±0.70, 2.15±0.73, 2.03±0.29 in each group without any significance, but it was inversely proportional to level of vitamin A supplementation in diet. In conclusion, although low level of vitamin A in feed premix did not increase IMF content of pork meat significantly, it seemed to have considerable correlation with the development of IMF.

Effect of 2-bromoethanesulfonic acid on *in vitro* fermentation characteristics and methanogen population

S.Y. Lee[1], S.H. Yang[2], M.A. Khan[2], W.S. Lee[2], H.S. Kim[2] and J.K. Ha[1], [1]Seoul National University, Department of Agricultural Biotechnology, Kwanak-gu, Seoul, 151-742, Korea, South, [2]National Institute of Animal Science, Dairy Science, Seonghwan, Chungnam, 330-801, Korea, South

An *in vitro* incubation study was conducted to investigate effects of 2-bromoethanesulfonic acid (BES) on fermentation characteristics and methanogen population. BES at the final levels of 0, 1 and 5mM with two different substrates having different ratio of timothy and concentrate (100% timothy vs 40% timothy-60% concentrate) were incubated for 0, 24, 48 and 72hr in the 39⊠ incubator. Total DNA extracted from culture fluid was used as a template for real-time PCR to measure the population of methanogenes. The four different primer sets were used for amplification of bacteria, total methanogens, the order Methnobacteriales and the order Methanomicrobiales. BES reduced ($P<0.01$) total gas and methane production in dose-dependent manner regardless of type of substrate. However, hydrogen production was increased by BES treatment ($P<0.01$). Total VFA concentration was not affected, but molar percentage of propionate and butyrate was increased and acetate to propionate ratio was reduced by BES treatment ($P<0.01$). BES also reduced ($P<0.01$) the population of total methanogens, the order Methnobacteriales and the order Methanomicrobiales in dose-dependent manner and the type of substrate did not influence the trend although the magnitude of response was different between all roughage and 40% roughage.

---

no. 71

Macadamia oilcake as an alternative protein source

S.E. Coetzee[1], A. Baoteng[2], O.U. Sebitloane[1], A. Kanengoni[1] and D. Palic[1], [1]Agricultural Research Council Animal Production Institute, Nutrition and Food Science, P. Bag X2, 0062 Pretoria, South Africa, [2]University of Venda, Department of Animal Science, School of Agricultural, Rural Development and Forestry, University of Venda, Thohoyndo, South Africa

The traditional dietary feedstuffs fed to broiler chickens consist of maize and soybean oilcake meal (SOC) as the main energy and protein source. In general, plant protein sources are essential to contribute to the fast protein accretion of the broilers. However in the Southern African context these sources of protein are expensive and relatively scarce especially to rural farms. Therefore the identification of alternative readily available and cheaper sources of protein is of great value to producers. Agro-industrial by-products such as avocado meal and macadamia oilcake meal (MCD) have been identified as potential alternative protein sources in animal feeds. The study was conducted to evaluate MCD meal as an alternative protein source to SOC meal. Treatments were as follows; 1) SM0, 2) SM1,3) MS2, 4) MS1 & 5) MS0 respectively. Performance parameters measured were Average weight gain (AWG), Feed Conversion Ratio (FCR), European Performance Efficiency Factor (EPEF) and dressing percentage (DP). There were no significant differences ($P>0.05$) between different treatments for both EPEF and DP. However, MS2, SM1 and SM0 were significantly lower in weight than MS0 and MS1 ($P<0.05$). A similar pattern was found for weight gain and cumulative feed intake throughout the growth period.

no. 72

Effect of dietary lipid sources on production performances of broilers

S.P. Els[1], F.H. De Witt[1], H.J. Van Der Merwe[1], M.D. Fair[1] and A. Hugo[2], [1]University of the Free State, Animal, Wildlife and Grassland Science, P.O. Box 339, Bloemfontein 9300, South Africa, [2]University of the Free State, Microbiology, Biochemistry and Food Biotechnology, P.O. Box 339, Bloemfontein 9300, South Africa

A study was conducted to determine the influence of different dietary lipid sources and inclusion levels on production performances of male broiler birds. Eight isocaloric (13.57 MJ AME) and isonitrogenous (200 g CP/kg) diets were formulated, using sunflower oil (SO), high oleic sunflower oil (HOSO), fish oil (FO) and tallow (T) at a 30 g/kg and 60 g/kg inclusion level. Eight hundred, day-old Ross 788 broiler males were randomly allocated to the 8 treatments (n=100) and were further subdivided into 4 replicates/treatment (n=25). Birds were reared inside a natural ventilated building on cement floor pens (12 birds/m^2), using sawdust as bedding. All birds receive a standard commercial diet for the first 14 days. Experimental diets were fed for 28 days until termination of the study at 42 days of age. Feed intake and body weights were recorded weekly, while mortalities were recorded and weighed daily. Mortalities were brought into consideration during the calculation of average daily feed intake (g feed/bird days) and feed conversion ratio (g feed/g body weight gain). Body weight (BW), average feed intake (AFI), average daily gain (ADG) and feed conversion ratio (FCR) did not show any differences between dietary lipid sources. However, as expected, AFI and FCR values were reduced ($P<0.05$) by a higher (60 g/kg) dietary inclusion level of lipids.

---

no. 73

Evaluation of the small ruminant nutrition system model using South African Mutton Merino and Dorper growth and body composition data

A. Cannas[1], A. Linsky[2], L.J. Erasmus[2] and W.A. Van Niekerk[2], [1]Dipartimento di Scienze Zootecniche, Universita di Sassari, 07100, Sassari, Italy, [2]Department of Animal and Wildlife Sciences, University of Pretoria, 0001, Pretoria, South Africa

Recently the biological model CNCPS for sheep led to the development of the Small Ruminant Nutrition System (SRNS) model. Our objective was to evaluate predictions of the SRNS model under local conditions using data from the SA Mutton Merino (late maturing) and Dorper breeds (early maturing). The prediction variables evaluated were feed intake, average daily gain (ADG), empty body gain (EBG) and composition of EBG. Two different equations were compared to estimate energy value of gain and five different equations were compared to estimate efficiency of conversion of ME to NE for gain. Overall, the original SRNS model gave the best predictions when compared to any of the modifications tested. Predictions of feed intake were accurate and precise with a low systematic bias (mean bias (MB)=40 g/d; RMSPE=50 g/d; r^2=0.95). Predictions on ADG were also accurate and precise (MB=2.4 g/d; RMSPE=21.4 g/d; r^2=0.76), while the EBG was over predicted (MB=31 g/d; RMSPE=38 g/d; r^2=0.69). The model slightly over predicted the fat and protein content of EBG (by 7.2% and 3.9%, respectively), showing high accuracy but very low precision, probably because the measured range of variation of fat and protein content of gain was narrow.

Growth performance and body composition of South African Mutton Merino and Dorper lambs
L.J. Erasmus[1], A. Linsky[1], W.A. Van Niekerk[1], R.J. Coertze[1] and A. Cannas[2], [1]Department of Animal and Wildlife Sciences, University of Pretoria, 0001, Pretoria, South Africa, [2]Dipartimento di Scienze Zootecniche, Universita di Sassari, 07100, Sassari, Italy

The objective of this study was to generate growth and body composition data of local sheep breeds in order to validate a biological model, the Small Ruminant Nutrition System (SRNS). The SRNS model accounts for energy and protein requirements of sheep and goats under diverse environmental conditions. Thirty two Dorper ((D) early maturing) and thirty two SA Mutton Merino lambs ((SAMM) late maturing); with equal numbers of males and females, were used. Lambs were slaughtered at the start (20 kg LW), middle (30 kg D, 35 kg SAMM) and end (40 kg D,50 kg SAMM) of the growth period and the carcass and intestines analysed at each slaughtering. Growth was therefore measured over two periods. During the first growth period the ADG of male D and SAMM were higher than the female D (P<0.05). During the second growth period feed conversion ratios were similar for all groups, but the male SA Mutton Merino group had the highest average daily gain (P<0.05). On average over the growth period, CP content of the carcasses and intestines decreased from 50.6 and 55.5% to 38.1 and 39.0% respectively and fat content increased from 29.9 and 30.0% to 50.4 and 49.2% respectively. The male SAMM lambs yielded the highest average end carcass weight (30.6 kg) and the female D lambs the lowest (25.5 kg).

---

no. 75

Influence of dietary lipid sources on carcass traits of broilers
S.P. Els[1], A. Hugo[2], F.H. De Witt[1], H.J. Van Der Merwe[1] and M.D. Fair[1], [1]University of the Free State, Animal, Wildlife and Grassland Science, P.O. Box 339, Bloemfontein 9300, South Africa, [2]University of the Free State, Microbiology, Biochemistry and Food Biotechnology, P.O. Box 339, Bloemfontein 9300, South Africa

A study was conducted to determine the influence of different dietary lipid sources on dressing percentage, breast meat yield and breast weights of broilers. Four isocaloric (13.57 MJ AME) and isonitrogenous (200 g CP/kg) diets were formulated, using sunflower oil (SO), high oleic sunflower oil (HOSO), fish oil (FO) and tallow (T) at a 60 g/kg dietary inclusion level. Four hundred, day-old Ross 788 broiler males were randomly allocated to the 4 treatments (n=100) and further subdivided into 4 replicates/treatment (n=25). All birds receive a standard diet for two weeks where-after experimental diets were fed for an additional four weeks. At six weeks of age 3 birds/replicate (n=12/treatment) were randomly selected, weighed and slaughtered at a commercial abattoir. Chilled carcasses (4 °C) were weighed to determined dressing percentage. Breast muscles were removed from the chilled carcasses, de-skinned and weighed for the calculation of breast meat yield. Breast meat yield were expressed as percentage of live body weight (TBBW) as well as carcass weight (TBCARC). This study showed that the broilers fed a diet supplemented with T had better (P<0.05) dressing percentages, breast meat yield and breast weights compared to all other treatments. These results suggested that dietary lipid sources could be used to improve certain carcass characteristics of broilers.

no. 76

Gas production parameters of four browse species without and with Polyethylene glycol (PEG)

A.G. Mahala[1], A. Abdalla[1] and I.V. Nsahlai[2], [1]Faculty of animal production, University of Khartoum, Khartoum North, 13114, Sudan, [2]SASA, UKZN, Animal and Poultry Science, PMB, 3209, South Africa

Nutritive values of leaves and pods of four browse species (*Acacia torrtlis, Ziziphus spinachristi, Acacia mellifera* and *Acacia seyal*) were evaluated by determining chemical composition and *in vitro* gas production (GP) with and without PEG. Crude protein (CP) contents ranged from 16.0 to 24.1%. Acid and neutral detergent fibre and total condensed tannin (TCT) contents ranged from 18.5-43.8%; 36.1-72.3%, and 0.5 to 6.2%, respectively. TCT was negatively correlated with maximum GP whereas ADF was negatively correlated with all GP parameters. The GP parameters varied significantly among the species. There were also significant differences among browse species in the cumulative GP, *in vitro* organic matter digestibility (IOMD) and calculated metabolizable energy (ME). PEG significantly ($P<0.05$) increased the GP, IOMD and ME. The IOMD values of leaves and pods ranged from 39.7 to 64.5% without PEG and from 37.9 to 80.0% with PEG. The ME contents ranged from 5.7 to 13 MJ/kg DM and from 5.9 to 10.6 MJ/kg DM with and without PEG, respectively. The best improvement in gas production, IOMD and ME were observed with 50% PEG w/w. Generally, this improvement indicates the negative effect of the tannin on digestibility.

---

no. 77

Evaluation of relative biological efficiency of additives for sugarcane ensiling

L. Maria Oliveira Borgatti[1], A. Luiz Veiga Conrado[1], J. Pavan Neto[1], P. Marques Meyer[2] and P. Henrique Mazza Rodrigues[1], [1]University of São Paulo – FMVZ/USP, Departament of Animal Nutrition and Production, Duque de Caxias Norte, 225, Pirassununga, SP 13630-000, Brazil, [2]Brazilian Institute of Geography and Statistics –IBGE, Agriculture Supervision, Duque de Caxias, 1332, Pirassununga, SP 13630-000, Brazil

This study aimed at evaluating the effects of adding alkalis to sugarcane silage. A completely randomized design was used, with 6 additives in two different concentrations (1 or 2%), plus a control group, totalizing 13 treatments [(6x2)+1] with four repetitions. The tested additives were: sodium hydroxide, limestone, urea, sodium bicarbonate, quicklime and hydrated lime. Silages were produced in laboratory silos that were opened 60 days after ensiling. The Relative Biological Efficiency (RBE) was calculated by the slope ratio technique, using the data obtained from ratio between desirable and undesirable products of silage, according to the equation: D/U ratio=[lactic/(ethanol+acetic+butyric)]. Limestone showed the best RBE, in relation to sodium hydroxide (89.4%). The RBE of urea was 49.2%, sodium bicarbonate 47.7% and hydrated lime 34.3%. In general, these additives altered fermentative pattern of sugarcane silage, inhibiting alcoholic fermentation and improving lactic acid production. Additives increased pH silage and these increased was according with fermentation data.

no. 78

Effect of dietary lipid sources on lipid oxidation of broiler meat

S.P. Els[1], A. Hugo[2], F.H. De Witt[1] and H.J. Van Der Merwe[1], [1] University of the Free State, Animal, Wildlife and Grassland Science, P.O. Box 339, Bloemfontein 9300, South Africa, [2] University of the Free State, Microbiology, Biochemistry and Food Biotechnology, P.O. Box 339, Bloemfontein 9300, South Africa

The objective of this study was to investigate the effects of different dietary lipid sources and inclusion levels on lipid oxidation of thigh and breast muscle of male broilers. Eight isocaloric (13.57 MJ AME) and isonitrogenous (200 g CP/kg) diets were formulated, using sunflower oil (SO), high oleic sunflower oil (HOSO), fish oil (FO) and tallow (T) at a 30 g/kg and 60 g/kg inclusion level. Eight hundred, day-old Ross 788 broiler males were randomly allocated to the 8 treatments (n=100) and further subdivided into 4 replicates/treatment (n=25). All birds received a standard diet for two weeks where-after experimental diets were fed for an additional four weeks. Birds were slaughtered under commercial abattoir conditions at 6 weeks of age. Carcasses from twelve birds per treatment were trimmed for breast and thigh cuts by removing skin. Twelve breast and thigh samples from each treatment group were stored at 4 °C for 7 days and another twelve breast and thigh samples were stored at -18 °C for 100 days. Meat samples were used for thiobarbituric acid reactive substances (TBARS) analysis. Birds fed FO showed more ($P<0.05$) oxidation in both thigh and breast meat than birds from any of the other treatments during storage. These results indicated that dietary lipid sources do influence the lipid oxidation processes of broiler meat.

---

no. 79

Influence of dietary lipid sources on sensory characteristics of broiler meat

S.P. Els, A. Hugo, C. Bothma, F.H. De Witt and H.J. Van Der Merwe, University of the Free State, Faculty of Natural and Agricultural Sciences, P.O. Box 339, Bloemfontein 9300, South Africa

A study was conducted to determine the influence of different dietary lipid sources and inclusion levels on sensory characteristics of breast meat. Eight isocaloric (13.57 MJ AME) and isonitrogenous (200 g CP/kg) diets were formulated, using sunflower oil (SO), high oleic sunflower oil (HOSO), fish oil (FO) and tallow (T) at a 30 g/kg and 60 g/kg inclusion level. Eight hundred, day-old Ross 788 broiler males were randomly allocated to the 8 treatments (n=100) and further subdivided into 4 replicates/treatment (n=25). All birds receive a standard diet for the first 14 days, where-after experimental diets were fed for 28 days. At 42 days of age, 3 birds/replicate (n=12/treatment) were randomly selected, weighed and slaughtered at a commercial abattoir. Breast muscles were removed from the chilled carcasses (4 °C) and de-skinned. Meat samples were wrapped in aluminium foil and steamed (200 °C) before cutting into smaller pieces (2.5 cm^3) and served to the respondents (n=75) of a consumer panel. Each respondent tasted 8 meat samples before completing a nine-point hedonic scale questionnaire. Meat samples of the HOSO treatment were preferred ($P<0.05$), while FO samples were the least acceptable ($P<0.05$) to the respondents. These results suggested that dietary lipid sources could be used to manipulate sensory characteristics of broiler breast meat according to consumer preferences.

The effect of forage particle size and water addition to barley based TMR on eating and rumination behavior and ruminal pH in early lactation period

R. Valizadeh, A. Hosseinkhani, A.,A. Naserian and S. Sobhanirad, Ferdowsi University Of Mashhad, Animal Science, P.O. Box 91775-1163, Mashhad, Iran

Eight Holstein cows in early lactation were allocated to 4 dietary treatments in a change over design with factorial arrangement of 2×2 including two levels of alfalfa particle size (5 and 20 mm) and two levels of TMR dry matter (without and with adding water up to 50% DM) in order to measure the changes of ruminal pH and eating behavior of dairy cows. Diets were similar in chemical composition. Treatments had no effect on eating time, eating rate and meal numbers, but alfalfa hay particle size reduction, decreased daily ruminating time (P=0.02) and total daily chewing time (P=0.03). TMR particle size had no effect on ruminal pH but interestingly water addition resulted to a significant decline in ruminal pH (P=0.02). Daily patterns of chewing behavior of the experimental cows showed that water addition resulted a 2 hours delay to reach the peak of rumination time. This delay can be a problem in diets with high and rapid fermentation rate such as barley based diets and may be the reason of low pH in the wet diets in this study.

---

no. 81

Effects of inclusion levels of banana (*Musa* spp.) peelings on degradability, rumen environment and passage kinetics of cattle fed basal elephant grass

J. Nambi-Kasozi[1], E.N. Sabiiti[1], F.B. Bareeba[1], E. Sporndly[2] and F. Kabi[1], [1]Makerere University, P.O. Box 7062, Kampala, Uganda, [2]Swedish University of Agricultural Sciences, P.O. Box 7070, Uppsala, Sweden

The effect of different levels of banana peelings (BP) on degradation characteristics, rumen environment and digesta kinetics of cattle fed basal elephant grass (EG) was evaluated using three rumen fistulated steers in four treatments. The steers were fed BP at 0, 20, 40 & 60% levels of the daily ration with EG to constitute the four diets. All animals were supplemented with maize bran, cotton seed cake & Gliricidia to make the diets iso-nitrogenous. The nylon bag technique was used to measure BP & EG degradabilities under each treatment. Rumen fluid samples were collected to determine pH & volatile fatty acids (VFA) concentrations. The animals were later dosed with chromium-mordanted hay & faecal grab samples collected to determine digesta kinetics. Effective DM, CP & NDF degradabilities of BP & EG were lower at higher BP levels. Effective DM, CP & NDF degradabilities of BP ranged between 574.3–806.9; 629.0–801.7 & 526.9–688.9 g/kg respectively. Effective CP degradability of EG was relatively high (547.5–569.1 g/kg) while that of DM & NDF were comparatively low (381.3–402.7 & 335.9–373.4 g/kg respectively). Rumen pH & VFA were lower at higher BP levels. Total tract retention time was lowest at the highest BP level. Higher levels of BP negatively affected degradability & rumen environment.

no. 82

Effects of essential oils on ruminal environment and performance of feedlot calves

J.I. Geraci[1], A.D. Garciarena[1], D. Colombatto[2] and D. Bravo[3], [1]EEA Balcarce INTA, Balcarce, 7620 Buenos Aires, Argentina, [2]University of Buenos Aires, Viamonte 430 st, Buenos Aires City, Argentina, [3]Pancosma Research, Voie-des-Traz 6, 1218 Geneva, Switzerland

Essential oils (EO) have the potential to modify the ruminal environment and to replace the use of antibiotics for specific markets. Twenty four Aberdeen angus calves (135.4 kg initial weight) were blocked by weight in four groups and randomly allocated to 8 pens of 3 animals each. Treatments were EO (800 mg Xtract 7065/animal/d), monensin (430 mg Rumensin® kg/DM) added into a mineral mixture. Diets were fed twice a day and consisted (DM basis) of 70% coarsely ground corn, 28% pelleted sunflower meal and 2% mineral mixture, plus 200 g alfalfa hay/animal/d. The experiment lasted 85 days, and DM intake, average daily gain (ADG), feed conversion ratio (FCR) were determined throughout the study. Also, two ruminally fistulated steers were also used in a cross over design in order to examine ruminal variables (pH and NH_3). Compared to monensin, EO did not alter DM intake (6.50 vs. 6.45 kg, P=0.79), ADG (1.31 vs. 1.23 kg/d, P=0.35), FCR (4.96 vs. 4.99, P=0.91), or RFD (1.80 vs. 1.68 mm/month, P=0.65) for EO and monensin, respectively. Ruminal pH was lower in EO than with monensin (5.58 vs. 6.09, P=0.002), and the same held true for NH_3 (10.78 vs. 20.05 mg/dl, P=0.01). Lower NH_3 concentrations with EO addition can be partially explained by the lower pH, conducive to lower protein degradation, and possibly due to some EO action on specific, deaminative bacteria in the rumen.

---

no. 83

Influence of polyclonal antibody preparation on *in situ* degradability of three energetic sources

W.G. Otero[1], C.T. Marino[2], F.R. Alves[3], F.A. Ferreira[4], M.D. Arrigoni[5] and P.H.M. Rodrigues[6], [1]FMVZ/USP, Pirassununga, 13635900, Brazil, [2]FMVZ/UNESP, Botucatu, 18610307, Brazil, [3]FMVZ/USP, Pirassununga, 13635900, Brazil, [4]FMVZ/USP, Pirassununga, 13635900, Brazil, [5]FMVZ/UNESP, Botucatu, 18610307, Brazil, [6]FMVZ/USP, Pirassununga, 13635900, Brazil

The objective was to evaluate the effect of polyclonal antibody preparation (PAP) against specific ruminal bacteria on *in situ* degradability of dry-grounded corn grain (CG) and high moisture corn silage (HMCS) starch and citrus pulp (CiPu) pectin. Nine cows ruminally cannulated were used in a 3x3 latin square replicate 3 times with a factorial arrangement of treatments 3x3 regarding 2 rumen modifiers represented by monensin and PAP plus a control group and 3 energetic sources supplemented in the diets represented by CG, HMCS and CiPu. Each period had 21 days, 16 for treatments adaptation and 5 for the data collection. There wasn't effect of interaction between modifiers and energetic sources for anyone of the ruminal parameters studied. It was observed effect of energetic source in effective degradability(Ed) of starch of CG and HMCS at 0.02 (P=0.0001); 0.05 (P=0.0001) and 0.08/h (P=0.0001), where the animals supplemented with HMCS had greater Ed than the animals from CiPu and CG group. There wasn't difference between CG and CiPu group. When potential degradability (Pd) was evaluated, effect of energetic source was observed (P=0.0001), the CG group showed greater values than the groups fed with HMCS and CiPu that did not differ between them.

Effect of saponified high fat sunflower oilcake and lipoic acid on fat quality of lambs

F.K. Siebrits[1], A. Makgekgenene[1] and A. Hugo[2], [1]Tshwane University of Technology, Animal Sciences, Private Bag X680, Pretoria 0001, South Africa, [2]University of the Free State, Microbial Biochemical and Food Biotechnology, P.O. Box 339, Bloemfontein 9300, South Africa

Sheep fat contains relatively high levels of saturated fatty acids while poly-unsaturated fatty acids (PUFA) are toxic to cellulolytic bacteria and are saturated in the rumen. Stabilization of residual oil in sunflower oilcake by conversion into calcium salts would be advantageous. Alpha lipoic acid acts as an anti-oxidant to ameliorate the effects of oxidative stress caused by high levels of PUFA. Residual oil (14%) in mechanically extracted sunflower oilcake (MSFOC) was saponified and compared in complete feedlot diets containing 12% crude protein and 4.3% fat fed to 40 SA Mutton Merino weaner lambs (ca. 23 kg) for 9 weeks. Commercial extracted oilcake (LFOC) with 2.4% residual oil was used in a control diet (2.9% fat, 10% crude protein). Both diets were fed with, or without a weekly oral dosing of 500 mg α-lipoic acid. Fatty acid composition was determined on back fat samples while thiobarbituric acid-reactive substances (TBARS) were determined on samples of *M. Longissimus dorsi* stored for 0 and 6 months and displayed for 6 days. The backfat of the lambs on the saponified MSFOC diets contained higher levels ($P \leq 0.05$) of saturated fatty acids and lower levels of mono-unsaturated fatty acids. PUFA were unaffected. High TBARS levels (> 1.0) were found after 6 months storage. Non significant increases in TBARS were observed in the groups that received lipoic acid.

---

no. 85

Effect of polyclonal antibody preparation on *in vivo* digestibility in cattle fed high concentrate diets

C. Tobias Marino[1], W. Guimarães Otero[2], P. Henrique Mazza Rodrigues[2], A. Di Costanzo[3], C. Ludovico Martins[1] and M. De Beni Arrigoni[1], [1]FMVZ/UNESP, Department of Animal Breeding and Nutrition, Fazenda Lageado, Botucatu, SP, Brazil, [2]FMVZ/USP, Departament of Animal Nutrition and Production, Pirassununga, SP, 13635-900, Brazil, [3]University of Minnesota, Departament of Animal Science, St. Paul, MN, 1364, USA

Nine ruminally fistulated cows were used to test an avian-derived polyclonal antibody preparation (PAP) against specific ruminal bacteria. The experimental design was a 3x3 latin square replicated 3 times with a factorial arrangement of treatments 3x3 regarding 2 rumen modifiers (monensin and PAP) plus control group and 3 energetic sources. The energetic sources utilized were dry-grounded corn grain, high moisture corn silage and citrus pulp. Total dry matter apparent digestibility and its fractions were estimated by the external marker chromic oxide (Cr_2O_3). There was no interaction between ruminal modifier and energetic source ($P > 0,05$) for any of the digestibility coefficients analyzed. In relation to rumen modifiers, both PAP and monensin did not alter ($P > 0.05$) coefficients of digestibility of DM, OM, CP, EE and pectin. At PAP group, NDF and ADF digestibility and TDN was lower ($P < 0.05$) in this group compared to control, without difference of these two groups to monensin. Starch digestibility was lower ($P < 0.05$) for PAP when compared to control and monensin.
PhD Scholarship from CNPq. Project financed by FAPESP.

Relationship between calf daily gain and stocking rate on Japanese Shorthorn cow-calf grazing pastures

M. Higashiyama[1], M. Muramoto[2] and T. Kondo[1], [1]National Agricultural Research Center for Tohoku Region, Research Team for Forage Production and Utilization in Tohoku, 4 Akahira Shimokuriyagawa Morioka Iwate, 020-0198, Japan, [2]University of Iwate, Faculty of agriculture, 3-18-8 Ueda Morioka Iwate, 020-8550, Japan

This study sought to assess the stocking rate for maximum gain per calf on Japanese Shorthorn cow-calf grazing pastures. We examined the relationship between daily gain of calves (castrated males and females) and stocking rate using data from nine pastures for seven years in the Kitakamisanchi hills and mountains of northern Japan. The pastures were dominated by *Phleum pratense* and *Poa pratensis*. Calves were born from February to April and grazed in pastures from May to October, with their dams from birth to the end of grazing season every year. Daily gain was estimated from a difference between a constant weight of 35 kg at birth and weight after grazing in October. Daily gains of castrated male and female calves were relatively constant with increasing stocking rate to 1.9 animal unit equivalents (AUE) per hectare and decreased with higher stocking rates. Castrated male (female) calves had a constant daily gain of 1.00 (0.92) kg/head/day. The results indicate that the stocking rate for maximum gain per calf and high performance per unit area is 1.9 AUE/ha. The study also concluded that the pasture area per cow-calf pair for maximum calf daily gain is 0.71 ha or more.

---

no. 87

Dry matter production and nutritive value of *Brachiaria brizantha* cv.MG5 and reproductive performance of beef cows and live weight gain of calves grazing on MG5 pasture in Japan

Y. Nakanishi, K. Hirano, Y. Nakamura, E. Tuneishi and A. Yamada, National Agricultual Research Center for Kyushu Okinawa Region, Animak Grazing, 2421suya, koushi, kumamoto, 861-1102, Japan

Trial was conducted using Japanese Black cows to evaluate nutritive value and productivity of *Brachiaria brizantha* (MG5) which is new tropical grass in Japan. Dry matter production (ton/ha/year) of Mg5 was greater than rhodesgrass which was main tropical grass in southern area of Japan (44.0 vs. 27.1). Digestibility of MG5 was higher than that of rhodesgrass. TDN content of MG5 was higher than rhodesgrass (62.6% vs 56.2%) and crude protein content of MG5 was higher than rhodesgrass (12.7% vs 10.8%). Beef cows grazing on MG5 pasture indicated good reproductive performance and average daily gains of grazing beef calves were over 0.8kg.

no. 88

Effects of the dose of capsicum extract on intake, water consumption and rumen fermentation of beef heifers fed a high-concentrate diet

M. Rodriguez-Prado[1], S. Calsamiglia[1], A. Ferret[1], J. Zwieten[1], L. Gonzales[1] and D. Bravo[2], [1]Universitat Autonoma de Barcelona, Placa Civica, 08193 Bellaterre, Spain, [2]Pancosma Research, Voie-des-Traz 6, 1218 Geneva, Switzerland

Capsicum extract can be used in beef cattle diets to stimulate DM intake and water consumption without reducing ruminal pH. Four beef Holstein heifers (BW=438±71 kg) fitted with a 1-cm i.d. plastic ruminal trocars were used in a 4 × 4 Latin square design to evaluate the effect of 3 doses of capsicum extract on intake, water consumption and ruminal fermentation in heifers fed a high-concentrate diet. Animals were fed (DM basis) 10% barley straw and 90% concentrate. Treatments were: no additive (CTR), 625 mg/d of capsicum extract (CAP625), 1250 mg/d of capsicum extract (CAP1250), and 2500 mg/d of capsicum extract (CAP2500). Each experimental period consisted of 25 d (15 d for adaptation, 5 d of continuous measurement of DM intake, and 3 d for rumen sample collection). Animals had *ad libitum* access to water and feed offered once daily at 0800. Data was analysed using PROC MIXED for repeated measures (SAS), and differences were declared at $P<0.05$. Intake of water (30.3 and 29.4 vs 27.4 L/d) was higher in CAP625 and CAP2500 compared with CTR, respectively. Intake of concentrate was also higher in CAP2500 vs CTR (8.40 vs 7.64, respectively). As a result of the higher intake, total VFA tended ($P<0.07$) to be higher in CAP625 and CAP2500 compared with CTR (144.8 and 142.9 vs 134.1 mM, respectively). In spite of higher intake, pH was not affected by treatments.

---

no. 89

Effects of cinnamaldehyde-eugenol and capsicum on rumen fermentation and feeding behavior in beef heifers fed a high-concentrate diet

D. Bravo[1], M. Rodriguez-Prado[2], S. Calsamiglia[2], A. Ferret[2], J. Zwieten[2] and L. Gonzales[2], [1]Pancosma Research, Voie-des-Traz 6, 1218 Geneva, Switzerland, [2]Universitat Autonoma de Barcelona, Placa Civica, 08193 Bellaterra, Spain

Four beef Holstein heifers (BW=436±17 kg) fitted with a 1-cm i.d. plastic ruminal trocars were used in a 4 × 4 Latin square design on the effect of feeding combinations of essential oils on feeding behavior and ruminal fermentation in heifers fed a high-concentrate diet. Animals were fed (DM) 10% barley straw and 90% concentrate. Treatments were arranged in a factorial design with cinnamaldehyde and eugenol (CINE, 167 and 334 mg/animal/d, respectively) with or without capsicum (CAP, 500 mg/animal/d). Each experimental period consisted of 25 d. Data was analysed using using PROC MIXED for repeated measures (SAS) with differences at $P<0.05$. Consumption of water was higher in CAP compared with CTR (33.7 vs. 37.8 L/d) and animals fed CAP spent more time eating (11.3 vs 8.0% of the time of the day) and the feed intake pattern was more stable during the day compared with CTR. As a result, the pH fell to its lowest level 6 h after feeding but was lower in CTR compared with CAP, suggesting that the modification of the pattern of intake induced by CAP was responsible for controlling the sharp drop of pH after feeding. Total volatile fatty acids tended ($P<0.07$) to be higher in CAP compared with CTR (162.4 vs. 133.6 mM), but the proportion (mol/100mol) of acetate (59.2), propionate (25.2), butyrate (13.4) and lactate (0.31) was not affected by treatments.

Meta analysis of the effect of a mixture of carvacrol, cinnamaldehyde and capsicum oleoresin in broilers

D. Bravo and C. Ionesco, Pancosma Research, Voie-des-Traz 6, 1218 Geneva, Switzerland

This meta-analysis in broiler involves a botanical composition comprised of carvacrol, cinnamaldehyde and capsicum oleoresin (XT). Data from 13 studies with side by side comparisons of XT to a negative control or a positive control diet (AGP) provided 38 treatment groups pooled for evaluation. The effect of XT addition was investigated using treatment means from trial groups with either a fixed effect of XT (present or absent) or AGP (present or absent) and a random effect of the trial. For each performance measure, mean value of XT or positive or negative controls, the number of groups involved, partial r^2, P value were determined. Addition of XT increased average daily gain (51.3 vs. 48.9 g/d, N=38, r^2=0.61, P<0.001), increased the feed intake (84.1 vs. 81.9 g/d, N=27, r^2=0.23, P<0.01) and decreased feed conversion ratio (1.68 vs. 1.73, N=27, r^2=0.42, P<0.001). Addition of AGP also increased the daily gain (50.4 vs.48.9 g/d, N=16, r^2=0.59, P<0.001), increased feed intake (84.0 vs. 81.9 g/d, N=16, r^2=0.26, P=0.04) but did not alter feed conversion ratio (1.71 vs. 1.73, N=16, r^2=0.13, P=0.172). When XT was compared to AGP, the following results (XT vs. AB) were observed: tendency for higher daily gain (51.3 vs. 50.4 g/d, N=23, r^2=0.14, P=0.060), same feed intake (84.1 vs. 84.0 g/d, N=23, r^2=0.0, P=0.952) and tendency for lower feed conversion ratio (1.68 vs. 1.71, N=23, r^2=0.19, P=0.08). Meta analysis using a data consisting of diverse trials avoided emphasis of a particular trial while illustrating the robust nature of XT.

---

no. 91

Proteic supplement with slow release urea for grazing animals during rainy season

G.J. Braga[1], M. Manella[2] and V.B.O. Leite[1], [1]APTA/SAA, PRDTA Centro-Oeste, R. Sebastião Soares s/n, 17380-000, Brazil, [2]Alltech Brasil, Nutrição Animal, R. Said Mohamad El Khatib, 280, 81170-610 Curitiba-PR, Brazil

The objective of the trial was to evaluate the replacement of soybean meal by slow release urea (SRU) on free choice proteic supplements for beef cattle. The trial was conducted at research facilities of APTA, located at Brotas, São Paulo state, Brazil. The performance of 54 steers (LW=330 kg) were evaluated from November 2006 to May 2007. Steers were distributed in 18 paddocks (0.8 ha) with 3 fixed heads per paddock. The supplement was supplied for a daily intake of 150 g/head. The SRU replaced a portion of soybean meal, keeping the supplements iso-proteic. The experimental design was completely random with two proteic supplements (Control or SRU). The co-variable herbage allowance (HA) was included in analysis to correct variations in herbage mass among paddocks. There was a linear increase of average daily gain (ADG) with HA increase. The maximum ADG (1.2 kg) was reached with 3.5 kg DM/kg PV. There was no interaction between HA and supplement; and ADG for SRU was 10% higher (P<0.06) when compared to Control. The results demonstrated that inclusion of SRU in proteic supplements may replace a portion of soybean meal. SRU use in free choice proteic supplements during rainy season may improve animal performance.

no. 92

Herbage yield in Signal grass pastures as affected by grazing management

G.J. Braga[1], J.N. Portela[2], C.G.S. Pedreira[2], V.B.O. Leite[1] and E.A. Oliveira[1], [1]APTA/SAA, PRDTA Centro-Oeste, R. Sebastião Soares s/n, 17380-000 Brotas/SP, Brazil, [2]Esalq/USP, Departamento de Zootecnia, Av. Pádua Dias, 11, 13418-900 Piracicaba/SP, Brazil

Signal grass is largely planted as cultivated pasture in Brazil, but no management targets have been identified. The objective of the study was to evaluate daily herbage accumulation rate (HAR) in Signal grass pastures under a two intensity x two frequency grazing treatments with steers. Intensities corresponded to 5 and 10 cm post-graze sward height; and frequencies were determined by pre-graze sward light interception (LI=95 or 100%). The trial was carried out from January 2007 to April 2008 on an Entisol. The 100%-LI pastures had higher HAR during the entire period (29 and 36 kg/ha for 95 and 100%-LI, respectively). There was no grazing intensity effect on HAR, except for the summer of 2007 (January to March) when pastures grazed to 10 cm had higher HAR (59 kg/ha) than those grazed to 5 cm (45 kg/ha). Total leaf yield increased when regrowth was extended to reach 100%-LI (7930 vs. 8729 kg/ha for 95 and 100%-LI, respectively), but stem and dead material yield increased as well (4877 vs. 7155 kg/ha for 95 and 100%-LI, respectively). The advantages in extending the rest period due to increase HAR was counterbalanced by the great influence of stem and dead material, decreasing pasture leafiness and probably nutritive value.

---

no. 93

Effect of type of dietary fat and forage to concentrate ratio on performance, digestibility and blood parameters of dairy cows fed alfalfa based diets

S. Kargar, G.R.S. Ghorbani and M. Alikhani, Isfahan University of Technology, Animal Sciences, College of Agriculture, 84156 Isfahan, Iran

Eight early lactating Holstein cows were used to investigate following treatments: 1) 0% fat and 34:66 forage to concentrate (F: C) ratio (Control), 2) 2% HPO and 34:66 F: C ratio, 3) 2% YG and 34:66 F: C ratio and 4) 2% YG and 45:55 F: C ratio (YGHF). Fat supplementation had no effect on DMI, but significantly ($P<0.001$) reduced in cows on high forage diet. YG diet increased milk yield relative to other treatments. Feeding fat tend ($P<0.06$) to increase milk fat and significantly ($P<0.04$) enhanced milk fat yield. Feeding YG compared with HPO and YGHF treatments increased milk protein and content. Fat supplementation was not affected total tract digestibility of DM, but it influenced by type of the fat, so that digestibility was higher in YG compared with HPO diet ($P<0.04$), and tend ($P<0.06$) to decline with high forage diet. Blood parameters such as insulin, glucose, blood urea nitrogen were not affected by treatments. Results of this study suggest that the use of yellow grease, which is both very unsaturated and rumen-active fat, in alfalfa hay based diets did not cause milk fat depression and significantly increased production responses compared with other treatments.

no. 94

Effect of supplementation with molasses block containing gliricidia or moringa leaves on *in vitro* gas production and microbial protein synthesis

H. Soetanto[1] and F. Firsoni[2], [1]Brawijaya University, Animal Nutrition, Jl. Veteran, 65145, Malang, Indonesia, [2]National Nuclear Agency, Centre for Isotope and Radiation Technology, Jl. Cinere Pasar Jumat, 12070, Jakarta, Indonesia

This study aimed at evaluating the beneficial effect of supplementation with molasses block (MB) containing moringa (*Moringa oleifera*, Lamm)g or liricidia (*Gliricidia sepium*) leaves on gas production and microbial protein synthesis (MPS) under *in vitro* conditions. Five treatments were assigned: A=maize stover+concentrate; B=A+MB containing moringa; C=B+PEG 6000; D=A+MB containing gliricidia ; and E=D+PEG 6000. The results showed that addition of either moringa or gliricidia leaves in MB, increased significantly the production of gas at 2, 4, 8 and 12 hours incubation. Feed DM and OM digestibility increased from 64.6% to 66.9% and from 63.5% to 66.0%, respectively. Addition of PEG 6000 had no significant influence on gas production and feed digestibility suggesting that the condensed tannin content of both moringa and gliricidia leaves are not a limiting factor during fermentation in the rumen. There was significant increase in microbial protein synthesis as affected by MB supplementation with the highest value was achieved by treatment B. This indicates that the leaf of moringa is better than gliricidia to supply nutrients required for MPS in the rumen. In conclusion, MB containing moringa or gliricidia leaves can be used as a good source of supplement for ruminant animals in the tropical country like Indonesia.

---

no. 95

Potentials of jackfruit wastes (*Artocarpus heterophyllus* L) as ruminant feeds in East Java

K. Kusmartono, Faculty of Animal Husbandry, Brawijaya University, Animal Nutrition, Jl. Veteran, Malang, East Java, 65145, Indonesia

This paper reviewed a series of study on the use of jackfruit wastes (JFW) as a complementary feed to rice straw (RS) for sheep and crossbred Ongole steers. Urea in the form of molasses-urea-block (MUB) or leaves of *Gliricidia sepium*/cassava were used to improve N content of the ration. Results of the first year trial showed that total DMI by the sheep and the steers were not affected by N supplementation. Sheep ate a much greater proportion of JFW (92%) than cattle (47%). The main treatment effect with the steers was associated with feeding of the molasses-urea cake which led to increases in intake of total DM and CP. In the second year study involving 18 growing fat-tailed sheep showed similar results on DM intake as those in the first year trial. CP intake and digestibility were increased significantly ($P<0.05$) by supplementating either gliricidia or cassava leaves hay. A similar value in CP digestibility was observed between sheep given gliricidia (69.7%) and cassava leaves (73.0), but the amount of N retained was significantly higher in sheep fed cassava leaves (22.2 g/day) than gliricidia (19.22 g/day) and MUB (13.6 g/day). As a consequence, daily gain obtained was higher in sheep fed cassava leaves (112.0 g/day) than gliricidia (97.1 g/day) and molasses-urea-block (95.6 g/day). It is concluded that JFW had a high potential as a ruminant feed, especially in the areas when N supplements (eg. legume or tree leaves) are available in the system.

no. 96

Nutrient composition and fiber digestibility of kikuyu (*Pennisetum clandestinum*) pasture relative to re-growth stage in Mediterranean climate

C.W. Cruywagen[1], P. Botha[2] and F.V. Nherera[1], [1]Stellenbosch University, Animal Science, PBag X1 Matieland, 7602, South Africa, [2]Western Cape Department of Agriculture, Livestock Division, P. O Box 249, 6530, South Africa

Changes in yield, nutritional composition and *in vitro* fiber digestibility of kikuyu grass with re-growth period during a Mediterranean summer (November-December 2006) season were investigated. During a 42 day cycle, plots were harvested at 7 (6 cuttings), 14 (3 cuttings), 21 (2 cuttings), 28, 35 and 42 days re-growth (treatments). Five weeks prior to harvesting the pasture was mowed to 50 mm stubble height. One week before cutting the first treatment all plots were cut again to 50 mm and fertilized. There was no significant change in dry matter (DM) content of pasture with increased re-growth period, mean DM content was 150 g/kg DM. Crude protein (CP) declined significantly from about 250 g/kg DM in forage harvested at 7 and 14 days to 202 g/kg DM at 42 days. Within the 7 days re-growth treatment, CP varied significantly, with the sixth cut having the lowest content. Ether extracts were significantly lower in re-growth at 42 days, about 25 g/kg DM. The content of neutral detergent fiber (NDF) was highest at 28 days (595 g/kg DM). *In vitro* NDF disappearance (48 hours) was highest for forage harvested at 21, 28 and 35 days, ranging between 75-80% and declining to about 73% in the 42 day treatment. Understanding changes in nutritional profiles of grazed pasture provides opportunity to estimate nutrient supply from pasture and enhance ration formulation

---

no. 97

The effects adding mimosa bark extract as a source of condensed tannin in the diet on feed intake, digestibility and milk yield of dairy cows

H. Hartutik and K. Kusmartono, Brawijaya University, Faculty of Animal Husbandry, JL. Veteran, Malang, Indonesia

Our previous study showed that crude protein contained in commercial concentrates available for dairy farmers in East Java are highly degradable (average of 82%). This might have been one of the reasons for the low milk yiled obtained by farmers in most areas, in highland (15 litres/day) and lowland (10 litres/day). The present study aimed to invenstigate the efficacy of including mimosa bark extract as a source of condensed tannin (CT) in the diet to reduce the protein degradation in the rumen which may affect milk yield. 15 lactating dairy cows were used and they were given the following treatmants: T1=elephant grass (*ad lib*)+10 kg concentrate; T2=T1+50 gr mimosa bark extract; T3=T2+PEG; T4=T1+100 gr mimosa bark extract; T5=T4+PEG. Three replicates were applied for each treatment and they were arranged in a randomized block design. Results of the study showed that increasing the amount of mimosa bark extract in the diet did not give any effect on feed consumption and digestibility (DM,OM and CP). There were tendencies that adding mimosa bark extract reduced total solid of milk, but increased milk yield, solid not fat, protein and fat contents of milk. It is concluded the use of mimosa bark extract as a source of CT up to 1% of total dry matter intake did not reduce feed consumption, but increased milk yield by almost 1 litre/head/day.

no. 98

Comparative meat production potentials and carcass evaluation of Nili-Ravi buffalo and different breeds of cattle calves

H. Jalil, T.N. Pasha, S. Ahmad and M.A. Jabbar, University of Veterinary and Animal Sciences, Lahore, 54000, Pakistan

Male Nili-Ravi buffalo and cattle calves of Sahiwal, cross-bred, non-descript, Cholistani and Dajal breeds of approximately one year age with an average body weight of 125 kg were procured. Ten calves of each breed were assigned fattening rations at ad-libitum having CP and TDN 11.23 and 68.70%, respectively for a period of 91 days. Daily feed intake and monthly weight gain were recorded. The average daily gain for Nili-Ravi buffalo, Sahiwal, crossbred, non-descript Cholistani and Dajal breeds calves were 0.822, 0.796, 0.856, 0.746, 0.840 and 0.753 kg. On the completion of feeding trial five calves from each group at random were slaughtered for carcass evaluation. The dressing percentage for Nili-Ravi buffalo, Sahiwal, crossbred, non-descript, Cholistani and Dajal breeds was 48.54, 49.37, 49.19, 52.06, 51.83 and 52.27 percent; percentage ratio of meat was 48.38, 52.33, 54.13, 49.36, 51.39 and 53.19, bone was 33.14, 32.87, 32.61, 31.79, 36.11, 31.38 percent, fat was 7.04, 6.05, 8.43, 12.96, 7.22, 9.19 percent and other tissues was 4.83, 5.13, 4.38, 4.01, 4.61 and 3.73 percent, respectively. The chemical composition of lean meat revealed moisture 73.34, 77.78, 76.86, 75.75, 76.53, 75.90 percent, crude protein 21.0, 20.55, 20.12, 20.19, 20.56 and 20.64 percent and fat 9.71, 9.90, 7.76, 8.71, 8.16 and 8.54 percent for groups one to six, respectively. There was no difference (P>0.50) among different breeds for feed intake, growth rate and carcass characteristic.

---

no. 99

Simplification of urea treatment method for wheat straw for its better adoption by the farmers

M.A. Jabbar, H. Muzafar, F.M. Khattak, T.N. Pasha and A. Khalique, University of Veterinary and Animal Sciences, Lahore, 54000, Pakistan

Efforts have been made in the past to improve the digestibility and protein contents through chemical treatment. Efficacy of treating straw with urea for improving its nutritive value is equivalent to anhydrous or aqueous ammonia. But the adoption rate of urea treatment is low, may be due to relatively tedious technology and requirement of labour. There is need that the method of urea treatment be simplified to overcome there problems. In the present study effort has been made to simplify the existing urea treatment method so that farmers could easily adopt this technique. The new method involves weighing the required amount of urea (4% of straw to be treated), mixing it with double the amount of manure or *Acacia* leaves, giving 30% moisture by adding water, putting this mixture in bag, piling the straw on bag, again moistening the straw with water (60% of straw) and incubating this material for a month under the cover of plastic sheet or mud plaster. In this method three steps including preparation of urea solution, sprinkling of solution on straw and pressing the straw during treatment process, have been eliminated which resulted in saving labour by 55% and time. The efficiency of new method is about 65-70% of the conventional method. The method is slightly less efficient than the conventional one but very simple and economical as it involves less labour as compared with conventional one.

The use of polyethylene glycol in the diagnosis of polyphenolics in the foliage of acacia species

H.K. Mokoboki, Notrh West University, Animal Science, Private Bag X2046 Mmabatho 2735, 2735 Mmabatho, South Africa

The aim of this work was to assess the efficacy of polyethylene glycol (PEG) for measuring tannin content in leaves of species, *Acacia karroo*, *Acacia nilotica*, *Acacia tortilis*, *Acacia galpinii*, *Acacia sieberiana*, *Acacia hebeclada*, and *Acacia rhemniana*. In addition, *in vitro* dry matter and organic matter degradabilities of the browse leaves were assessed. Young leaves from the species were collected from five different trees of each species at the University of Limpopo in April 2002, and their capacity to bind PEG was compared using a completely randomized design. PEG (MW 4000) was added at ratios of plant material: PEG solution (w/v) of 1:05 1:10, 1:15, 1:20 and 1:25. The ratio of 1: 20 (w/v) resulted in the best response in terms of tannin precipitated compared to the other ratios. The PEG binding capacity of *Acacia* species was found to be high ranging from 12.73 to 28.82% dry matter, in direct response to the proportion of tannin content in the leaves. Dry matter and organic matter degradabilities varied from 345.4 to 534.2 and 254.4 to 473.9 g/kg DM, respectively, for all species and were negatively correlated to PEG-binding capacity. *A. hebeclada*, *A. sieberiana* and *A.galpinii* had trace levels of total phenolics, *A. tortilis* contained approximately 90 g/kg DM of total phenolics whilst *A. karroo*, *A. nilotica* and *A. rhemniana* had intermediate amounts. These results suggested that polyethylene glycol can be used to determine the content and effects of tannins in tanniferous plants used by ruminants.

---

no. 101

Growth and slaughter characteristics of ankole cattle and its crossbreds with boran and friesian

D. Asizua[1], D. Mpairwe[1], F. Kabi[1], D. Mutetikka[1] and J. Madsen[2], [1]Makerere, Animal Science, P.O. Box 7062, Kampala, 256, Kampala, Uganda, [2]University of Copenhagen, Large animal sciences, Groennegaardsvej 2, DK 1870 Frederiksberg C., Denmark, +45, Copenhagen, Denmark

One hundred and forty four bulls comprising 48 animals of each breed [pure Ankole (ANK) and its crossbreds with Boran (AXB) and Friesian (AXF)] were assigned to three feeding systems (FS) to evaluate their performance for improved beef production. The bulls averaging 191±9.6 kg LWT and between 1 to 2 years were arranged in a completely randomized design in a 3 X 3 factorial treatment arrangement: T1 (Grazing alone), T2 (Grazing + concentrate supplementation) and T3 (finished in feedlot and fed *ad libitum* maize stover + concentrate supplement). The concentrate comprised 70% maize bran; 20% cotton seedcake and 10% molasses. FS significantly ($P<0.001$) affected daily weight gains (ADG) and hot carcass weight; and ($P<0.05$) dressing percentage. ADG was 850, 550, 270 g/day; hot carcass weights were 145, 132, 110 kg while the dressing percentages varied from 52, 51 and 50 for T3, T2 and T1 respectively. Similar trends were observed for non-carcass components although there were no differences between T3 and T2. Breed differences were observed for ADG where AXF and ANK had higher ($P<0.05$) ADG (620 and 560g/day) than the AXB (500g/day). Carcass characteristics and non-carcass components did not differ ($P>0.05$) among the three breeds. The results of this study indicated that all three breeds had a great potential for producing beef when finished in a feedlot.

no. 102

Effects of protected fat supplements on Primiparous and Multiparous Holstein Cows in early lactation

M. Ganjkhanloo, G. Ghorbani, K. Rezayazdi, M. Dehghan and H. Moravag, Tehran University, Animal Science, Karaj, 4111, Iran

3 Latin square design with 21-d periods. Cows were randomly assigned to three groups. Cows in group This study was conducted to evaluate the production response of multiparous and primiparous cows to rumen protected fat in early lactation. Twelve (Nine multiparous and three primiparous) Holstein cows were used in a replicated 3* 1(control) were fed total mixed ration (TMR), contained 40% farage, and 60% concentrate mix on DM basis, cows in group 2 were fed TMR containing 3% prilled fat (Energizer-10) that was replaced with cracked corn, cows in group 3 were fed TMR containing 3.5% calcium salts of fat (Magnapac) that was replaced with cracked corn. All diets were isonitrogenous. Intake of dry matter (DM), organic matter (OM) and NDF were greater when cows were fed control diet than when they were fed rumen protected fat in multparous cows($P<0.05$); intake of net energy for lactation (NEL) in all cows and intake of DM, OM and NDF in primiparous cows were similar for all diets($P>0.05$). Production of milk and 3.5% FCM and milk composition in primiparous and multiparous cows not affected by supplemental fat ($P>0.05$). In multiparous cows production efficiency (3.5% FCM/DMI) was higher for cows fed supplemental fat diets($P<0.05$), and cows fed fat were more efficient than those fed control diet but in prmiparous cows supplemental rumen protected fat did not influence production efficiency($P>0.05$).

---

no. 103

Effects of genotype and dietary treatments on ph, carcass characteristics and composition of mubende and mubende-boer crossbred wethers

K. Kamatara[1], D. Mpairwe[1], M. Christensen[2], D. Mutetikka[1], D. Asizua[1] and J. Madsen[2], [1]Makerere University, Animal Science, P.O. Box 7062, +256 Kampala, Uganda, [2]University Of Copenhagen, Faculty of Life Sciences, Groennegaardsvej2,DK 1870 Frederiksberg C., Denmark, +45 Copenhagen, Denmark

Sixty wethers comprising 30 animals of each genotype Mubende (MDE) and MubendexBoer crossbreds (MXB) were assigned to three feeding systems (FS) to evaluate their performance for improved meat production. The withers averaging 28 ±1.6kg LWT and between 9-12months were arranged in a completely randomized design in a 2x3factorial treatment arrangement. T1 (grazing alone), T2 (grazing + concentrate), T3 (grazing + concentrate + molasses). Values of pH were measured in the *M. Longissimus dorsi* at 45min, 3, 7, 24 and 48 hours postmortem. Genotype significantly ($P<0.05$) affected hot and cold carcass weights, hind limb length and width. Cold carcass weights were 20 and 18kg for MXB and MDE respectively. Hind limb lengths were 43 and 42cm while hind limb widths were 38 and 37cm for MXB and MDE respectively. Finishing system significantly ($P<0.05$) affected carcass weights, hot weights for T1, T2 and T3 were 18, 22 and 21kg respectively. The ultimate pH was not affected by both genotype and feeding system. Carcass grades and compositions were not affected by genotype and feeding system. The study showed that supplementation and crossbreeding Mubende goats would be an option for improving goat meat production in Uganda.

no. 104

A comparison between hominy chop and defatted maize germ meal as the main energy source in diets of feedlot steers

K.-J. Leeuw, H.H. Meissner and D. Palic, Agricultural Research Council, Animal Nutrition, Private Bag X2, 0062, Irene, South Africa

The feed industry, in particular the feedlot industry, estimates a lower economic value for de-fatted maize germ meal (DFG) in comparison to hominy chop (HC). This is based on the assumption that DFG will give a lower performance value, since it contains less metabolisable energy then HC. The economic value of DFG is estimated to be 15–20% lower than HC. This study was therefore done to test the validity of these estimates. 75 Steers with an initial weight of about 190kg were blocked by weight into five treatments with 15 animals each and housed individually. Feedlot steers were fed diets with either DFG or HC as the main energy component and with three substitution combinations of these energy sources. Treatment 1 (T1) contained only HC as main energy source, treatment 2 (T2) 75% HC and 25% DFG, treatment 3 (T3) 50% HC and 50% DFG, treatment 4 (T4) 25% HC and 75% DFG and treatment 5 (T5) only DFG. The performance of the steers during the feeding period was monitored as well as carcass characteristics, leading on to an economic evaluation and relative price for DFG. Feed conversion ratio was highly significantly (P=0,002) different between treatments. DFG containing diets resulted in a 9 to 14% poorer feed conversion ratio in comparison to the sole HC diet. The estimated difference between DFG and HC is 12 – 14%, i.e. better than the anticipated 15 – 20%.

---

no. 105

Variability in the cut-and-carry feeding system in the Kenyan highlands and its implications

H.A. Markewich, A.N. Pell, D.J.R. Cherney and D.M. Mbugua, Cornell University, Department of Animal Science, Morrison Hall, Ithaca, New York, 14853, USA

Cattle fed by the cut-and-carry system on small Kenyan farms consume a highly variable diet. Daily and seasonal variation in diet composition and quantity fed is not considered when animal nutritionists make predictions of production responses to diet. This study aimed to quantify the compositional variations in cut-and-carry diets. Feed offered to and refused by dairy cattle on 12 farms in Embu, Kenya were weighed and sampled for 5 weeks during the two dry and two rainy seasons from June 2006 to June 2007. More feed was offered during rainy seasons than in the dry seasons (P<0.001). Refusals were largest during the long dry season (P<0.001), likely due to the bulkiness of the low quality feeds offered. Coefficients of variation for mass of feeds offered and refused exceeded 0.4 in the short dry season and equaled or exceeded 0.6 in the other 3 seasons. Preliminary data suggest that lactating cows were offered more feed of a higher quality (lower in fiber and lignin, higher in protein) than heifer calves or bulls. The highly variable nature of cattle diets on Kenyan farms documented by this study suggest the need for this variability to be taken into account when evaluating the nutritive quality of tropical feeds. If the daily and seasonal variation is not taken into account, feeding trials may not accurately predict production responses to local feeds.

The influence of solid feeds and age on rumen development in pre-weaned dairy calves

J.W. Kiragu, B.N. Mitaru and K.R.G. Irungu, Kenya Agricultural Research Institute, National Animal Husbandry Research Centre, P.O. Box 25 Naivasha, 20117, Kenya

Insufficient supply of dairy replacements is a major constraint to the development of smallholder dairy production in the tropics. The objective of this study was to evaluate manipulation of rumen ecosystem in calves using high quality forages and solid feeds to save milk for human consumption. Twenty- four Friesian calves were used in a completely randomized design to evaluate three treatments: milk only, milk plus sweet potato vines and milk, sweet potato vines plus concentrates on intake, growth and rumen development. Calves entered the study at 7 days and all calves were fed whole milk at 10% body weight in addition to their respective dietary treatments. Two male calves from each treatment were slaughtered at 4, 6 and 8 weeks to study rumen development. Samples of rumen fluid were taken from the reticulo-rumen immediately after slaughter to evaluate rumen fermentation. Dry matter intake, growth, rumen tissue weights and papillae development increased significantly with increased solid feeds as the calves matured. Similarly, rumen pH and total VFA increased with intake of solid feeds at 4 weeks of age, but rumen ammonia concentration decreased with age due to increased microbial activities. In conclusion, forages and solid feeds were shown to stimulate early development of rumen function hence enabling calves to efficiently utilize cheap roughages.

---

no. 107

Sample preparation of *Medicago sative* L. hay for chemical analysis

G.D.J. Scholtz[1], H.J. Van Der Merwe[1] and T.P. Tylutki[2], [1]University of the Free State, Animal-, Wildlife- and Grasslandland Sciences, P.O. Box 339, Bloemfontein, 9300, South Africa, [2]Agricultural Modelling and Training Systems, 418 Davis Rd Cortland, NY, 13045, USA

A study was conducted to evaluate the variation- and expand the existing and limiting nutritive value database of *Medicago sativa* L. hay (168 near infrared reflectance spectroscopy spectrally selected samples) in South Africa. The highest moisture content recorded (13.54%) was safely below the critical moisture level of 16% for effective storage. Coefficient of variation (CV) ranged from 1.2% for dry matter (DM) up to 66.2% for acid detergent fibre-crude protein (ADF-CP). The average ash content was 12.97% (7.3 to 29.5%), indicating soil contamination. Fibre fractions varied as follows: acid detergent fibre (ADF) (21.26 to 47.28%), neutral detergent fibre (NDF) (28.89 to 65.93%), lignin (4.32 to 16.25%), cellulose (16.29 to 36.44%) and hemicellulose (5.26 to 19.86%). The mean IVOMD for both 24 and 48hr (69.26 and 73.19% DM, respectively), was representative (CV=±8%) of the *Medicago sativa* L. hay population. Crude protein (CP) (average=20.7%DM) consists of 76.9% true protein. According to ADF-CP, 6% of the samples were heat damaged. High mean calcium (Ca) (1.35%), potassium (P) (2.53%) and iron (Fe) (874ppm) values were recorded. The results emphasises the need for rapid and accurate analysis of chemical and digestibility parameters of *Medicago sativa* L. hay to be used in diet formulation in practice.

no. 108

Feeding characteristics and body developments of growing buffaloes raised by small-scale farms in Tarai, Nepal

Y. Hayashi[1], M.K. Shah[2], Y. Tabata[3], H. Kumagai[3] and S.K. Shah[2], [1]Meijo University, Faculty of Agriculture, Kasugai, 486-0804, Japan, [2]Tribhuvan University, Institute of Agriculture and Animal Science, Rampur, Chitwan, Nepal, [3]Kyoto University, Graduate School of Agriculture, Kyoto, 606-8502, Japan

Thirty farms that raise growing buffaloes in Tarai were selected for the survey of animal census, feeding traits and body dimensions in the pasture-sufficient-period (PSP), the pasture-decreasing-period (PDP) and the fodder-shortage-period (FSP). The average number of the growing buffaloes per farm was 1.3. Seventy five percent of the buffaloes were born in PSP. The mean supply of dry matter (DM) per metabolic bodyweight ($BW^{0.75}$) was higher in PDP than in PSP and FSP (145.1 vs. 88.9 and 57.0 g/kg $BW^{0.75}$, respectively, $P<0.01$). The average provisions of crude protein (CP) and total digestible nutrients (TDN) per $BW^{0.75}$ were lower in FSP than in PSP and PDP (4.2 vs. 10.5 and 8.5 g/kg $BW^{0.75}$ for CP, and 25.5 vs. 48.1 and 61.9 g/kg $BW^{0.75}$ for TDN, respectively, $P<0.01$). The mean supplies of calcium and phosphorus per $BW^{0.75}$ were also lower in FSP than in the other periods ($P<0.01$). The maximum values of bodyweight (BW), body length, wither height, heart girth (HG) and hip-born width (HW) of the buffaloes less than 24 months old were 257.0 kg, 113.6 cm, 113.2 cm, 151.0 cm and 40.5 cm, respectively. The formulae to estimate BW with HG and HW in the buffaloes were established by the multiple-regression analyses. The environments of pasture might have caused the variance in nutrient supplies to the growing buffaloes.

---

no. 109

Rate of intake in wethers fed a temperate pasture with different feeding schedule and supplemented or not with additives

A. Pérez-Ruchel[1], J.L. Repetto[2], M. Michelini[1], L. Pérez[1], G. Soldini[1] and C. Cajarville[1], [1]Facultad de Veterinaria, UdelaR, Nutrición Animal, Lasplaces 1550, 11600, Uruguay, [2]Facultad de Veterinaria, UdelaR, Bovinos, Lasplaces 1550, 11600, Uruguay

The effect of the feeding schedule and the use of additives on the intake and its rate were studied on animals consuming a temperate pasture. Twenty four wethers (47 kg BW), housed in metabolic cages, were fed a fresh pasture (80% *Lotus corniculatus*) and allocated into 4 groups. Group AD had forage available all day, group 1D was fed for 6 h/day, group 1D+B was fed 6 h/day plus 2% DM intake level of buffer (75% $NaHCO_3$-25% MgO) and group 1D+S was fed 6 h/day plus 6.2 x 10^9 UFC/day of Saccharomyces cerevisiae. Daily intake and its rate was measured weighting the amount of offers and orts every 1 hour for 6 hours. Orthogonal contrasts were performed on data to study the effect of feeding schedule, the use of additives and the type of additive used. There were no differences in g of DM ingested/kg $BW^{0.75}$/day (mean value: 52.8; ESM=5.14, P=0.435). Groups fed 6 h/day showed a higher rate of intake for every hour studied (i.e. hour 2: AD: 5.9, 1D:7.4, 1D+B: 8.3, 1D+S: 5.9; $P<0.05$). Among groups fed 6h/day there were no differences. The exception was at hour 2, when buffer supplementation led to a higher rate of intake (3.2 vs 2.3 g of DM ingested/kg BW for 1D+B vs 1D+S; P=0.013). The effect of buffer at hour 2 could be due to an effect on the forage fermentation.

no. 1

Sheep systems to increase profit, manage climate risk and improve environmental outcomes in high-rainfall zones of southern Australia

M.A. Friend and S.M. Robertson, Charles Sturt University, School of Animal and Veterinary Sciences, P.O. Box 588 Wagga Wagga, NSW 2650, Australia

Australian agriculture is faced with environmental issues including increasing land and river salinity and reduced biodiversity. Overlaying this, climate change threatens to challenge the economic sustainability of current farming systems. 'EverGraze' is a national research and extension project developing livestock (principally sheep) production systems based on perennial pastures which will deliver improved profit and environmental outcomes. Integrated bioeconomic and biophysical modelling was used to identify systems which have the potential to deliver improved profit and environmental outcomes, as well as identifying constraints to achieving these outcomes. Modelling indicated that perennial-plant based livestock systems can be more profitable than livestock systems based on annual plants, and that profit can be maximised in the high-rainfall zone of southern NSW by utilising a meat-merino system which lambs in September. Whole farm system research is now underway to test the potential of various livestock systems. Results to date suggest that a system which is more flexible in relation to climatic variability will be more profitable in poor years and may produce a similar gross margin in better years to systems with higher stocking rates. This paper presents modelling results and how they were used to design field experiments, and preliminary experimental results from a large-scale grazing systems experiment.

---

no. 2

Small-scale milk production in South Africa: why is it not working?

S.M. Grobler and M.M. Scholtz, ARC, Animal Production Institute, Private Bag X05, Lynn East 0039, South Africa

In South Africa it appears that small-milk producion systems in the emerging and communal sectors are not thriving, whereas similar systems are thriving in other developing countries. Kenya, for example, dominates milk production and marketing in eastern Africa, with over 85% of the dairy cattle population, most of them kept by smallholders in areas of high and medium cropping potential. Other developing countries where small-scale milk producers contribute significantly to the country's milk demand include Zambia, Sri Lanka and Latin American countries. In South Africa, the average herd size for milk production in the communal sector is 6 dairy cattle or 11 dual purpose cattle and in the emerging sector it is 39 or 42 cattle respectively. The Nguni (34% in communal), Drakensberger (19% in emerging), Brahman (22% in communal and 16% in emerging), Afrikaner (20% in communal) and Bonsmara (13% in emerging) breeds or crosses dominate the mixed dairy and dual purpose systems used for small-scale dairying in the communal and emerging sectors. The most common dairy breeds like the Holstein/Friesian (6%) and Jersey (3%) are rarely used in these sectors. The challenge therefore seems to be, to encourage the use fo appropriate genotypes to enhance small-scale or dual purpose milk production, even if it implies the introduction of new genotypes (e.g. the Gir) into South Africa. Other challenges in these sectors range from inadequate infrastructure, poor understanding of small-scale milk production systems and inadequate support from Government.

Reproduction and production potential of communal sheep in the Eastern cape

C.B. Nowers and J. Welgemoed, Dohne Agricultural Development Institute, Agriculture, Private Bag X15, Stutterheim 4930, South Africa

The objective of this trial was to determine the production and reproduction potential of communal sheep that are found in communal sourveld areas by managing them under favourable commercial conditions. In this study woolled sheep from the communal sourveld farming areas at Wartburg community (similar to that of the sourveld at Dohne A.D.I.) were randomly divided into three treatment groups. One group (Communal treatment) continued to be managed under the normal communal farming practices and remained in the communal area. A second group was transferred to the Dohne where they were subjected to sound sheep management practises of commercial farming but still mated with local communal rams (Commercial + communal ram). A third group was managed identical to the second group but mated with rams from Dohne (Commercial + stud ram). The conception rate of communal ewes (95.5% & 90.4%) reared under commercial conditions at Dohne differed significantly from the conception rate of communal ewes at Wartburg (65.1%). The pre-weaning mortality rate of lambs reared in the communal area of Wartburg is significantly higher than those reared under commercial conditions (21.6% vs 6.6%). The lambing percentage (lambs born/ewes mated) of communal ewes under optimal conditions is about 40% higher than ewes kept under communal conditions (107% vs 65%). Wool production of communal ewes under commercial conditions differed significantly from those managed under communal conditions but is still±1.5 kg less (35%) than those of dual purpose commercial sheep.

---

no. 4

The impact of dairy crossbreds on herd productivity in smallholder farms in western Kenya

W.B. Muhuyi, F.B. Lukibisi and S.M. Mbuku, KARI, National Beef Research Centre, Lanet, P.O. Box. 3840, 20100 Nakuru, Kenya

The dairy crossbreds would serve as a model for determining the potential role of indigenous cattle in contributing to the economic and, efficient milk and meat supply under the prevailing tropical conditions. In this study, milk production traits considered were lactation milk yield (kg), lactation length, (days) while reproductive traits included calving interval (days) of Zebu cattle and dairy crossbreds in different production systems in western Kenya. The different cattle genotypes were kept in extensive, semi-zero grazing and zero grazing systems. The mean lactation yield was 1,763.90±648.67kg (M±SD), the mean lactation length was 357.43±95.77days and the mean calving interval was 553.70±129.42days. The mean lactation yield of Zebu crossbreds was 1500kg and higher than the lactation yield of Zebu cows which was 416.11 ±158.05kg. Productivity indices of yield per day of calving interval (kg/dci) for semi-zero grazed herds were (3.43kg/dci) and zero-grazed (6.28kg/dci) herds and were higher than for the extensively grazed herds (2.60kg/dci). Crossbreeding between the highly productive temperate breeds and the adapted indigenous cattle has allowed for the use of improved germplasm within the prevailing tropical constraints which has resulted in increased superior performance of the crossbreds in various production systems thus increasing the livelihoods of the human populace.

Working to fatten calves in Sweden: working time, working environment and managerial style

E. Bostad[1], S. Pinzke[2] and C. Swensson[1], [1]Swedish University of Agricultural Sciences, Rural Buildings and Animal Husbandry, P.O. Box 59, 230 53 Alnarp, Sweden, [2]Swedish University of Agricultural Sciences, Work Science, Business Economics and Environmental Psychology, p.o box 88, 230 53 Alnarp, Sweden

With increasing national and global demand for meat, the study had the aim to enhance the sustainability of the Swedish beef production. Short-term aims were to improve labour utilization and farm work logistics. A long-term aim was to increase farm efficiency and reduce work strain and hazard risks. An additional aim was to design a cost estimation guide for use at investments or renewals of cattle farms in terms of safe, labour saving organization and implements. Through semi-structured questionnaires 50 out of 83 farms producing 50-1500 calves per year were analysed for working time, working routines, physical and psychosocial factors. Possible relations between labour expense and managerial styles, motivations and attitudes of the individual farmer were investigated. Ten of the 30 largest farms were chosen for field studies with in-depth recordings during daily and seasonal work tasks. Physical work environmental factors like awkward postures and movements related to the work task and its repetitiveness were video recorded and analyzed in association to the risk of developing musculoskeletal disorders. Preliminary results show no clear correlations between herd size and efficiency in working time and work strain or hazard risks. The results are expected to be presented during 2008.

---

no. 6

A systems approach to the South African dairy industry

M.M. Scholtz and S.M. Grobler, ARC, Animal Production Institute, Private Bag X05, Lynn East 0039, South Africa

South Africa experiences periods of milk shortages, as a result of market forces, but can the cow effectively respond to the market forces? Periodically milk production comes under pressure, mainly due to the high price of commercial dairy feed and low milk prices. During such periods large numbers of dairy cows are slaughtered, resulting in a decrease in the number of dairy cows. With the average productive life of 2.3 lactations for dairy cows, the numbers can not be easily increased if market forces change. This paper will indicate that the national dairy herd can only increase by 1% to 3% per life cycle through normal population growth. Short term financial issues that influence milk price or input costs can thus have a drastic and long term effect on the population dynamics of the national herd. The low surplus numbers implies that selection pressure from the female side is almost non existent, and selection can only be applied on the bull side. This raises the question as to whether South Africa really breed dairy cattle that are adapted to local conditions, since large quantities of semen are imported. World wide agricultural production is increasingly practiced in a systems relationship in order to optimize the entire production chain. The different stakeholders in the South African dairy industry should realize how interdependent they are, and start to think in terms of a systems approach, whereby the principles of breeding, nutrition, physiology, forage management, product technology and economics are integrated to ensure sustainable production.

Non-food-production functions of livestock and livestock systems

S.J. Oosting, B.K. Boogaard, S.R. De Bruin and A.K. Mekoya, Wageningen University, Animal Production Systems Group, P.O. Box 338, 6700 AH Wageningen, Netherlands

Livestock farming is a relationship between animals and humans. The food production function of livestock is obvious, but other functions of livestock for men and of livestock farming systems for societies are important and are increasingly becoming part of the livestock farming systems research agenda. Results of long term studies at our group at Wageningen University concerning the role of farms and livestock for care-needing children and elderly, the role of dairy farms in Dutch and Norwegian Society and the role of livestock functions and farm objectives for farmers' decision about adoption of multipurpose fodder trees in Ethiopia will be presented to picture the array of non-food-production functions of livestock. It will be discussed how solutions proposed to solve ecological problems associated with global livestock farming may be in conflict with non-food-production functions of livestock and livestock systems.

---

no. 8

Local knowledge on relevant traits in sheep and goats under pastoral management in Kenya: a contribution to characterization of animal genetic resources in their production system context

H. Warui, B. Kaufmann, C. Hülsebusch and A. Valle Zárate, University of Hohenheim, Garbenstrasse 17, 70593, Stuttgart, Germany

Livestock keepers, especially the pastoralists, base their decision making and management of livestock on their local knowledge. Therefore a study was carried out in the period December 2004–September 2005, to understand the traits of sheep and goats that Gabra and Rendille pastoralists in northern Kenya consider relevant, and to obtain information on the preferred trait levels. Data were collected using semi-structured interviews with 22 Gabra and 33 Rendille livestock keepers. Key informant interviews were used for triangulation. The traits that were provided by the interviewees when asked to describe preferred animals of different age and sex classes are considered as relevant traits in the production system context. For these traits they indicate the preferred trait levels. The relevant traits observed relate to the functions of the animals in the flock and the production conditions under which the animals are kept. Adaptive traits (e.g. strong body) and productive traits (e.g. has milk for the owner and the kid/lamb) were mentioned most often, and morphological traits were of lower importance. The livestock keepers had preference for the largest animal family in the flock. It is concluded that livestock keepers in the study area have preference for animals showing a suitable combination of productive and adaptive traits.

How to utilize the natural characteristics of the Greater Rhea (*Rhea americana*) in the farmed bird in order to maximize productivity

M. Sanchez[1] and C. Madeiros[2], [1]Secretariat of Agriculture, Livestock, Fisheries and Food (SAGPyA), Livestock, NaoTraditional Species, Paseo Colon, 982 2do Piso Of.207, Buenos Aires, Argentina, [2]West Bar Veterinary Hospital, 19, West Bar, OX16 9SA Banbury Oxfordshire, United Kingdom

The Greater Rhea (GR) is part of the Ratites Group, along with Ostrich, Emu, Cassowary and Kiwi. This specie has evolved in the Southern American continent and is represented by the Genus Rhea and Pterocnemia. Commercial Rhea breeding has increased since 1990 in different countries within South America as an alternative livestock enterprise. The objective of this exposition is to demonstrate the importance of giving consideration to the natural behavior of the GR to maximize farm production. This includes taking into account the biological and behavioral aspect, nutrition needs, and fundamental concepts for rhea breeding and chicks rearing. High mortality rates in farmed Rheas from day-olds to three months can be a serious limiting factor, when the natural requirements of the chick are ignored. The intensive breeding of Rheas is a dynamic system that seeks to obtain a continuous equilibrium between the innate characteristics of the bird and the requirements of the farm, such as docility and maintaining routine controls in time and management. Achieving minimal stress levels, especially in young chicks is essential. Seeking to respect the GR genetical make-up and its natural development and requirements, is the key to successful Rhea farming.

---

no. 10

Live animal and carcass characteristics of South African indigenous goats

L. Simela[1], E.C. Webb[2] and M.J.C. Bosman[2,3], [1]National Emergent Red Meat Producers Organisation, P O Box 36461 Menlo Park, 0102 Pretoria, South Africa, [2]University of Pretoria, Department of Animal and Wildlife Sciences, Hatfield, 0002 Pretoria, South Africa, [3]North-West University, School for Physiology, Nutrition and Consumer Sciences, Private Bag X6001, 2520 Potchefstroom, South Africa

Eighty-nine intact male, castrates and female South African indigenous goats in three age groups (0, 2– 6 and 8 permanent incisors) and two pre-slaughter conditioning groups (non-conditioned and conditioned) were slaughtered. The effect of sex, pre-slaughter conditioning and age on live animal and carcass dimensions and carcass composition were evaluated. The goats were large with live weight, carcass weight and carcass dimensions in the range of the large breeds of southern Africa. They had a high lean and low fat content. Intact males were suited for high chevon yield because they were heavy, had high lean and low fat content, and losses during dressing and chilling were reduced by improved nutrition. Goats between 2–6 teeth yielded heavy carcasses that were comparable to goats in the 8-teeth group, and had proportionately more lean. The hind limb was ideal for high lean and low fat high value cuts but the dorsal trunk was bony and yielded less lean. Pre-slaughter conditioning improved the overall size of the goats and reduced the losses during slaughter and chilling. It also improved the lean/bone and lean and fat/bone indices.

Comparative study on growth performance of crossbred and purebred mecheri sheep raised under dry land farming condition

A.K. Thiruvenkadan, K. Karunanithi, M. Murugan, K. Arunachalam and R. Narendra Babu, mecheri sheep research station, pottaneri, tamil nadu, 636 453, India

In order to improve the mutton production, Dorset x Mandya and Dorset x Nellore half-bred rams were mated with Mecheri ewes to produce Dorset-Mecheri-Mandya and Dorset-Mecheri-Nellore quarterbreds. Least-squares analysis of body weight of crossbred (n=541) and purebred (n=959) Mecheri sheep were made and the overall means for bodyweight of crossbred Mecheri sheep at birth, 3, 6, 9 and 12 months of age were 2.27±0.05, 7.97±0.24, 11.84±0.38, 14.73±0.48 and 17.55±0.56 kg respectively and for purebred Mecheri sheep the values were 2.27±0.02, 7.80±0.10, 11.48±0.15, 14.04±0.17 and 16.23±0.18 kg respectively. Crossbreds had higher body weight than purebred Mecheri sheep and they differed significantly ($P<0.05$) at 6, 9 and 12 months of age. In general, period of lambing and sex of the lambs had highly significant ($P< 0.01$) effect on body weight of both genetic groups at different ages. The estimates of heritability for body weight ranged between 0.177±0.129 and 0.638±0.376 among different age groups. Highest heritability estimates were obtained for body weight at 3 and 6months of age and this suggest that the selection at these ages would improve the body weights at subsequent ages both on phenotypic and genotypic scale.

---

no. 12

Comparison of the milk production between Holstein and Brown-Swiss cows grazed on the artificial temperate pasture

M. Hanada, M. Doi and M. Zunong, Obihiro University, Animal Science, Inada 2-11, Obihiro, 080-8555, Japan

Milk production of spring calved Brown-Swiss (BS) and Holstein-Friesian (HF) cows were compared from June to October in 2007 to investigate the adaptability of BS to Japanese pasture system which are based on artificial temperate pasture. Five cows for each species were grazed on the artificial temperate pasture in north Japan for 18 hours a day and offered supplement after milking. Average parity for BS and HF groups were 3.3 and 2.8 and average days in milk on 1st June for each group were 55.2 and 63.4, respectively. Milk yield and its composition were measured once a month and blood sample was obtained in June, August and October. Through the experiment, milk yield changed from 36.0 to 24.1 kg/d in BS group and from 40.5 to 23.9 kg/d in HF group. Milk fat was higher in BS group than in HF group ($P<0.05$), but milk protein and lactose did not differ between the species. Fatty acid composition in milk was not affected by the species and the proportion of conjugated linoleic acid ranged form 0.89 to 2.21 g/100gFA in BS group and from 1.19 to 1.49 g/100gFA in HF group. Urea nitrogen contents in milk and serum were higher in BS than in HF through the experiment ($P<0.05$). Glucose concentration in the serum did not differ between the species, but non- esterified fatty acid (NEFA) and γGTP in June were higher in HF group than in BS group ($P<0.05$). High urea nitrogen contents in milk and serum and low NEFA andγGTP values in BS group might be reflected by higher pasture intake in BS compared with HF.

Demographic analysis of Nguni cattle recorded in South Africa

E.L. Matjuda[1], V.R. Leesburg[2] and M.D. Macneil[2], [1]Agricultural Research Council, Livestock Business Division, Private Bag X2, Irene 0062, South Africa, [2]USDA Agricultural Research Service, Fort Keogh Livestock & Range Research Laboratory, 243 Fort Keogh Rd., Miles City, Montana 59301, USA

Pedigrees of Nguni cattle recorded from 1973 to 2006 (N=142,622) were used to estimate a sustainable age structure and effective population size. Average generation number of calves born 2004-2006 was 6.7.age and average ages of their parents were 6.1 yr for sires and 5.5 yr for dams. Change in inbreeding per generation was 0.00297±0.00014, implying an effective population size of 168. Equilibrium age distributions of females calving first at 2 or 3 and < 15 yr of age were estimated for herds of 1000 livestock units. Age-specific vectors for reproductive rate were: (0.0 0.0 1.0 0.46 0.74 0.76 0.75 0.68 0.77 0.76 0.68 0.75 0.66 0.51 0.11) and (0.0 0.0 0.0 1.0 0.54 0.68 0.69 0.74 0.71 0.71 0.70 0.68 0.61 0.52 0.26) for females first calving at 2 and 3 yr, respectively. Corresponding vectors for survival were: (0.57 0.90 0.74 0.91 0.86 0.84 0.84 0.86 0.84 0.81 0.83 0.74 0.51 0.11 0.00) and (0.80 0.90 0.93 0.66 0.81 0.81 0.84 0.82 0.83 0.82 0.77 0.69 0.57 0.27 0.00). Results indicate 21% more cows in reproduction with first calving at 2 yr of age rather than 3, with these cows having average ages of 5.5 and 6.7 yr, respectively. First calving at 2 years of age allows herd size to be maintained with a lower proportion of replacement heifers. This opens opportunities to market surplus females or to join some Nguni females with terminal sires and thereby increase the value of their calves.

---

no. 14

A comparison between broilers raised in a day range system and intensively

O.U. Sebitloane[1], A.T. Kanengoni[1], S.E. Coetzee[1], T. Botlhoko[1] and C. Manyanga[2], [1]1Agricultural Research Council, Animal Production Institute, Nutrition and Food Science, Pvt Bag X2, 0062, Irene, Pretoria, South Africa, [2]Gauteng Provincial Department of Agriculture, Conservation and Environment, Technology Development and Support Directorate, P.O. Box 8769, Johannesburg, South Africa

Poultry day range system, a pasture-based production, offers opportunities for producers diversifying operations, and providing niche products for consumers. A day-range system includes multiple camps allowing chickens access to fresh pasture. The study was conducted to evaluate the performance of broilers in a day range system or outdoor, compared to the intensive production system or indoor. Performance parameters measured were feed intake, Average Daily Gain (ADG) and Feed Conversion Ratio (FCR). There were no differences (P>0.05) between treatments, Cobb Outdoor (C-O), Cobb Indoor (C-I), Ross Outdoor (R-O) and Ross Indoor (R-I) in ADG, FCR and Performance Efficiency Factor (PEF) throughout the growth period (49 days) except on day 35 where the two outdoor treatments had a lower bodyweight (P<0.05) than indoor treatments and a similar pattern was found for feed intake. The R-O treatment tended to perform better than all treatments followed by C-O for the growth period. It was concluded that Ross chickens adjust very easily and perform better in outdoor conditions compared to the Cobb 500. About 6% savings in feed intake were experienced by keeping chickens in a day range system compared to intensive.

Reproduction and production potential of Nguni cattle on Dohne Sourveld

C.B. Nowers and J. Welgemoed, Dohne Agricultural Development Institute, Agriculture, Private bag X15, Stutterheim 4930, South Africa

Reproduction efficiency is the most economically important aspect of beef production but is generally perceived to be low on natural sourveld. It is therefore of vital importance that beef producers farm with functional efficient animals on sourveld. A Nguni stud herd was established in 1996 at the Dohne A.D.I., where selection is aimed at high fertility and resistance to tick-borne diseases under local sourveld conditions. The aim with this Nguni stud is to increase efficiency and more profitable beef production whilst quantifying reproduction and production norms on sourveld. Reproduction and production data is presented from 1996 until 2008. The use of Performance Testing to identify animals of superior genetic ability has also had a positive effect on increasing the reproduction parameters of animals in the Dohne Nguni Stud. The average calving percentage of the Dohne Nguni stud from 1997 until 2007 was 85.7% and weaning percentage 95.8%. The average mating mass for cows in the Dohne Nguni stud for the past six years was 337 kg. The 205 day corrected mass of the calves was the highest during 2006 (164.4 kg), averaging 153.5 kg since the start of the project in 1996. The cow efficiency for Nguni cows on sourveld seems to range between 42.5% and 46.5% and the average A.D.G. of calves from birth to weaning was 0.628 kg/day. The average data for the past nine years was 39.5 kg calf weaned per 100kg cow mated. This equation reflects on the reproduction efficiency of the herd, mothering ability, growth potential of the calf and calf survival.

---

no. 16

Indigenous chicken genetic resources in the Limpopo and Eastern Cape provinces of South Africa

B.J. Mtileni[1], F.C. Muchadeyi[2], A.N. Maiwashe[1], P.M. Phitsane[1], T.E. Halimani[3], M. Chimonyo[4] and K. Dzama[2], [1]ARC-Livestock Business Division, Animal Production, Private Bag X2, Irene, 0062 Pretoria, South Africa, [2]University of Stellenbosch, Animal Sciences, Private Bag X1, Matieland, 7602, South Africa, [3]University of Zimbabwe, Animal Science, P. O Box MP167, MP167 Mt Pleasant, Zimbabwe, [4]University of Fort Hare, Livestock and Pasture Sciences, P Bag X1314, 5700 Alice, South Africa

Indigenous chicken production systems were investigated in Limpopo and Eastern Cape provinces of South Africa. Semi-structured questionnaires were administered to 100 households in the two provinces. The study showed that livestock contributed less (10%) of the household incomes compared to social grants (51.1%) and wages (35.9%). About 50.3% of the households owned chickens in comparison to 49.7% that owned other livestock. The mean chicken flock size was 13.10±10.93/household and ranged from 1-56. There were no significant differences between provinces (P>0.05). Households with multi-species owned on average 6.1±10.60 goats, 3.4±13.30 sheep and 3.1±6.77 cattle. Results indicated that indigenous chickens perform significant functions in the livelihood of rural farmers, primarily for household meat and egg consumption, socio-cultural ceremonies, selling and manure. Sub-standard housing, poor disease control, and absence of organised vaccination and poultry extension services negatively affected chicken production in the provinces. The genetic structures of these chicken populations will be determined using microsatellite markers.

Prediction of body weight of indigenous chickens in Limpopo Province of South Africa

N.J. Mudau[1], A.E. Nesamvuni[1], K.A. Nephawe[1], D. Norris[2] and I. Groenewald[3], [1]Research & Training, Limpopo Department Agriculture, Private Bag X 9487, Polokwane, 0700, South Africa, [2]University of Limpopo, Department of Animal Science, Sovenga, 0700, South Africa, [3]University of Free State, Centre for Sustainable Agriculture, Bloemfontein, 0500, South Africa

Data on body weight (BW) and body measurements (BM) were collected from 254 indigenous fowls with the objective to investigate the possibilities of using BM to predict BW of birds. Measurements were obtained from 98 males and 156 females from March 2005 to March 2006. Using an electronic weighing scale and a measuring tape, body length (BL), chest circumference (CC), femur (F), crus (C), and tarsometatarsus (TM) were recorded for analysis. Data was subjected to step-wise regression & Correlation procedures of (SAS). Males showed higher BW and BM compared to females ($P<0.0001$). Males were 376 g bigger in BW, 4.27 cm longer in BL, 2.46 cm broader CC, 0.97 longer in F. Body measurements were correlated to BW ($P<0.0001$). Body weight was correlated to BL ($P<0.0001$) and CC ($P<0.0001$). The comparison of R^2 values for the regression equations indicated that, when step-wise regression techniques based on BL and CC were used there were improvements of about 25% to 50% in R^2 values for males, and about 37% to 50% for females. Thus, the CC and BL are the BM that is suitable for the prediction of the body weight.

---

no. 18

Goat farming in South Africa: findings of a national livestock survey

J. Bester[1], K.A. Ramsay[2] and M.M. Scholtz[1], [1]ARC: Animal Production Institute, Production Systems, Private Bag x2, Irene, 0062, South Africa, [2]Department of Agriculture, Directorate of Animal Production and Aquaculture, Private Bag x138, Pretoria, 0001, South Africa

During a national livestock survey held in 2003 under the FAO/UNDP/SADC project RAF/97/032, a total of 892 goat keepers completed questionnaires covering aspects of management and production. The majority of the keepers came from the communal farming areas (752), with 127 from the small scale or ermerging sector and 13 from the commercial sector. The comparison of production and product development between the different farm sectors highlighted the necessity of improving management in the communal and small scale sectors. Herds in the communal sector were generally small with 56.8% of the keepers owning less than ten goats as against 10% in the small scale sector. Presented in this poster is the comparison of production systems, health and herd management, reproduction and choice of breed between the different farm systems.

Smallholder goat production in the Eastern Cape Province of South Africa

F. Rumosa Gwaze[1], M. Chimonyo[1] and K. Dzama[2], [1] University of Fort Hare, Department of Livestock and Pasture Science, P Bag X1314, 5700, Alice, South Africa, [2] Stellenbosch University, Department of Animal Sciences, P Bag X1, 7602, Matieland, South Africa

The study was conducted to characterize goat production systems in a sourveld (Alfred Nzo) and a sweetveld (Amatole). Data were collected using participatory rural appraisal techniques, direct observations and structured questionnaires. Questionnaires were administered to 69 and 143 households in Alfred Nzo (ANZO) and Amatole districts, respectively. There were no differences (P>0.05) between mean goat flock sizes in Amatole (14.0±0.31) and ANZO (14.1±1.42). Fewer respondents (32% in ANZO and 27% in Amatole) owned bucks. Buck:doe ratios were 1:50 in ANZO and 1:90 in Amatole district. More (P<0.05) farmers in ANZO than in Amatole supplemented their goats. Higher twin births were reported in (78%) than in ANZO (51%). Ceremonies, followed by sales were the major functions of goats in the two districts. The most prevalent diseases were internal parasites (84% of respondents in ANZO, 77% in Amatole). Most farmers, regardless of village, did not vaccinate their goats. More (P<0.05) goat houses were roofed in ANZO compared to Amatole. Purchasing and exchanging of bucks, proper housing, supplementation and vaccination of goats against diseases, in both veldtypes, can significantly increase goat productivity at household level and hence reduce food insecurity.

---

no. 20

Effect of stocking rate on dairy farm efficiency in a seasonal calving system with supplemented Holstein x Jersey dairy cows

J. Baudracco[1], N. López-Villalobos[1], C.W. Holmes[1], L.A. Romero[2], E.A. Comeron[2], D. Scandolo[2] and M. Maciel[2], [1]Massey University, IVABS, Private Bag 11-222, Palmerston North, 5301, New Zealand, [2]Instituto Nacional Tecnologia Agropecuaria, AIPA, Ruta 34 Km 227, Rafaela, 2300, Argentina

Stocking rate (SR) has a great influence on the efficiency of pastoral dairy systems. This study reports one year results of a dairy systems trial established in INTA Rafaela Research Station, Argentina. A 43.8 hectares farm was divided into three homogenous farmlets and stocked at either 1.6 (LowSR), 2.1 (MediumSR) or 2.6 (HighSR) cows/hectare, with Holstein x Jersey cows that calved seasonally in spring, grazed on alfalfa pastures and were supplemented with sorghum silage produced on-farm plus imported concentrates (1,860 kg DM concentrate/cow/year). Harvesting efficiency of pasture increased (P<0.05) as SR increased: 55.8%, 65.3% and 73.1%, but intake of pasture per cow did not differ (P>0.05) between treatments: 9.2, 8.4 and 8.6 KgDM/cow/day for LowSR, MediumSR and HighSR, respectively. Per cow; Yields of milk and milksolids did not differ (P>0.05) and were 6,373, 6,065 and 6,202 kg milk/cow and 469, 453 and 457 kg milksolids/cow for LowSR, MediumSR and HighSR, respectively. No differences (P>0.05) were found for liveweight and body condition score changes. At the levels of feed offered in this study, as SR was increased cows apparently compensated for the lower pasture offered by increasing harvesting efficiency, with no significant effects on milk yield/cow, live weight or body condition score changes.

Effect of minimal supplemental feeding with lucerne during late gestation on doe weight gain, natal characteristics and pre-weaning performance of goats grazing on *Themeda triandra* and *Tarconanthus* veld

M.L. Mashiloane and O.M. Ntwaeagae, Vaalharts Agricultural Research Station, Northern Cape Department of Agric and Land Reform, Private Bag X9, Jan Kempdorp 8550, South Africa

The objectives of this study were to determine if minimal supplemental feeding with lucerne during late gestation has an effect on the weight gain of does, and the birth weight, natal body characteristics, weaning weight and pre-weaning growth rate of kids. Thirty six does in late pregnancy were randomly allocated to three levels of lucerne treatment on a minimal supplemental strategy in RCBD. The strategy required that animals be fed at least half of their daily dry matter requirement with a high nutrient feedstuff to supplement natural grazing. Effects of minimal supplementation with lucerne were analyzed using the GLM procedures of SAS 2003. There were no significant differences in doe weight gain between treatment classes or for kid performance traits. Minimal supplemental feeding with lucerne does not seem to have effect on doe weight gain, natal characteristics or pre-weaning performance in goats.

---

no. 22

Village chicken production and marketing systems in northwest Ethiopia

H. Hassen[1], F.W.C. Neser[2], A.D.E. Kock[2] and E.V. Marle-Koster[3], [1]ARARI, Andassa Livestock Research Center, P.O. Box 27, Bahir Dar, Ethiopia, [2]University of the Free State, Animal, Wildlife and Grassland Sciences, P.O. Box 339, Bloemfontein, 9300, South Africa, [3]University of Pretoria, Animal and Wildlife Sciences, P.O. Box 0002, Pretoria, South Africa

Surveys using both purposive and random sampling methods were carried out in northwest Ethiopia to describe village-based indigenous chicken production systems. The results revealed that almost all indigenous chickens are managed under extensive chicken management systems with an average flock size of seven chickens per household and producing 9 to19 eggs per clutch per hen with a maximum of 2 to 3 clutch per year per hen. The survey has also identified the poultry management profile of the region indicating that women and children were more involved in rural chicken management activities. There is an increasing shift towards the intensification of exotic chicken breeds by the ministry of agriculture, which will lead to genetic erosion of the indigenous chickens. Hence, characterization, conservation and utilization of the indigenous chicken genetic resources are very important.

no. 23

Study on growth performance of Murrah buffaloes raised under farm conditions

A.K. Thiruvenkadan, S. Panneerselvam and R. Rajendran,

Growth data from 590 Murrah buffalo calves (140 male and 450 female calves) maintained at the Central Cattle Breeding Farm, Alamadhi, Tamil Nadu, India, born in the period between 1990 and 2004 were used for this study. They were analysed using least-squares procedure. The adjusted birth weights of male and female calves were 33.0±0.49 and 31.9±0.27 kg respectively with an overall value of 32.43±0.30 kg. The mean body weight at 3, 6, 9 and 12 months of age pooled over periods, season and sex were 62.0±0.65, 87.9±0.95, 112.4±1.23 and 134.16±1.41 kg respectively. Year and season of calving influenced the weight significantly ($P<0.05$) at birth, three and six months of ages only. The influence of parity of the dam on body weight at different ages was highly significant ($P<0.01$). The calves borne in the dam's second parity were generally heavier than those born in other parities. Generally, males had a higher body weight than females at all age groups and they differed significantly. All the growth traits showed medium heritability estimates, which, ranged between 0.241 and 0.403. The genetic correlations were all medium and positive. Birth weight can thus be used as an indicator of early selection for growth considering its positive genetic correlations with succeeding growth traits.

---

no. 24

Species composition and seasonal occurrence of free-living ticks on Dohne Sourveld pastures grazed by Bonsmara or by Nguni in the Eastern Cape Province

N. Nyangiwe[1] and I.G. Horak[2], [1]Dohne Agricultural Development Institute, Agriculture, Private Bag x 15, 4930, Stutterheim, South Africa, [2]University of Pretoria, Veterinary Tropical Diseases, Faculty of veterinary Science, 0110, Onderstepoort, South Africa

The objective of this study was to establish the species composition, seasonal abundance and numbers of questing free-living tick larvae on Dohne Sourveld pastures grazed by Bonsmara or by Nguni cattle. Over a period of three years (2005, 2006 & 2007) ticks questing for hosts were collected monthly by dragging flannel strips attached to a wooden spar over the vegetation. At each occasion six replicate drags-samples were made in camps grazed by Bonsmara cattle, or by Nguni cattle. The most numerous of the questing larvae collected were those of *Rhipicephalus (Boophilus) microplus*, followed by *Rhipicephalus evertsi evertsi*, *Rhipicephalus (Boophilus) decoloratus* and *Rhipicephalus appendiculatus*. Other species were collected in the camps grazed by Bonsmara cattle than by Nguni cattle. The composite seasonal occurrences of larvae in the camps grazed by the two breeds of cattle were nearly identical, but seasonal peaks in larval numbers were always higher in camps grazed by Bonsmara cattle than in those grazed by Nguni cattle.

no. 25

Advances, constraints and potentials of the family poultry system

E.B. Sonaiya, Faculty of Agriculture, Department of Animal Science, Obafemi Awolowo University, 220005 Ile-Ife, Nigeria

Family poultry is not dependent on land ownership but has an effect on the economy and quality of life of especially rural families. What are the advances in family poultry development? There has been significant development in thermostable Newcastle disease vaccines: from a small scale trial in The Gambia; the coating of V4 strain on feed grains applied throughout South-East Asia; the production and direct ocular instillation of HRV4 in tests in Cameroon, Morocco, Nigeria, Malawi, Tanzania, South Africa and Zimbabwe; to the most extensive and sustained trials of HRV4 in Africa conducted in Mozambique. The evaluation of local poultry genetic resources in Africa, Asia and Latin America has led to the use of major genes traits, e.g. dwarf and naked neck, as an important genetic solution for alternative poultry production systems in the tropics. Ground breaking work on the assessment of the scavengeable feed available on the range has been done in Sri Lanka and tested in Nigeria and Tanzania along with the assessment of novel unconventional feed resources of family poultry in Africa. Economic analysis of the efficiency of the family poultry system in relation to labour and animal productivity is a priority area that is still outstanding for most countries. The paper evaluates the scientific advances and discusses the constraints and potentials of family poultry in the New World.

---

no. 26

Effect of different feeding regimes on growth rate and age of sexual maturity in Nili-Ravi buffalo heifers

M.A. Jabbar[1], T.N. Pasha[1], L. Jabbar[2] and S.A. Khanum[3], [1]University of Veterinary ad Animal Sciences, Food and Nutrition, Lahore, 54000, Pakistan, [2]Arificial Insemination Center, Livestock and Dairy Development, Okara, Okara, Pakistan, [3]Nuclear Institute for Agriculture & Biology, Faisalabad, Faisalabad, Pakistan

The effect of different feeding regimes on age at sexual maturity was studied using twenty eight one year old buffalo heifers. They were divided into four groups, A, B, C and D with seven animals in each group. Group A was kept on the conventional method of feeding i.e. *ad lib* green fodder according to seasonal availability. Groups B, C and D were offered concentrate mixture providing low, medium and high levels of metabolizable energy, respectively, along with 2-3 kg green forage per animal per day, for a period of 24 months. Growth rate data revealed that average daily weight gain was 211, 422, 534 and 573 g in groups A, B, C and D, respectively. The weight gain increased with increased energy levels in the ration. The time taken to attain maturity weight of 400 kg was 17, 15 and 15 months for groups B, C and D while it took 28.5 months for group A. Group A, reared on simple fodder, took an additional period of one year to attain sexual maturity weight. The cost of green fodder consumed by group A up to sexual maturity was US$ 287.9, whereas cost of the ration consumed by group s B, C and D was US$ 191.9, 184.9 and 238.7, respectively. There were significant differences among the groups for different feeding regimes on the onset of sexual maturity.

no. 27

Comparison of reproductive efficiency of beetal goats in different breeding systems

M. Younos, M. Aleem, A. Ijaz and M.A. Khan, University of Veterinary and Animal Sciences, Animal Reproduction, Lahore, 54000, Pakistan

In this study, reproductive efficiency of two adults beetal goat flocks was (Flock A=47, Flock B=25) compared under seasonal and year-round breeding systems. One year data of goats of comparable age were collected and analyzed during the same period of time. In flock A, a male was allowed to breed for 45 days during autumn when daylight length was decreasing. In flock B, a male was allowed to breed round the year. There was no significant difference between the conception rates and average age at puberty in both flocks. The average kidding interval was significantly shorter in flock B (119±15 days) compared to flock A (314±20). There was no effect on the weight of male and female kids. Goats in flock B produced 33% more kids compared to flock A. Total running cost of flock B was 9% higher than of flock A but flock B. The abortion and mortality rates were comparable in both flocks. It is concluded that beetal goats can reproduce efficiently round the year in Pakistan without any seasonal limitations.

---

no. 28

An overview of the management practices and constrains at the dairy camps in Khartoum State, Sudan

I.E.M. El Zubeir and A.G. Mahala, University of Khartoum, Faculty of Animal Production, Faculty of Animal Production, U. of K., Postal code 13314, Khartoum North, Sudan

This paper reports on the results of a survey that was designed to compare management, husbandry and constrains in dairy farms in the Khartoum State, Sudan. Information from 100 farms was collected by questionnaire, visits and direct interviews with the farm owners. The results indicated that dairy farm owners' in the studied sample exhibit high levels of illiteracy (36%) or only had informal education (22%) . Most of the dairy cows are primarily kept as the main milk producing animals. The herds predominant consist of dairy cross cows (60%) The crosses are mainly Friesian x local breed. The data illustrated that herd health, trained workers and availability of feeds are the major problems facing dairy herd owners. The lack of records and marketing of milk were also among the mangemental factors that need correction. Disease control was not satisfactory as most of the workers give the treatment without consultation with the veterinarian. Hence the present study recommended the training of animal keepers and workers (formal and vocational) to increase the awareness on house designing, rearing, management and feeding, Furthermore milk collection points supported by cooling facilities, veterinary services, and initiation of collaborative bodies (proceeing and marketing) are urgently needed.

no. 1

Metabolizable energy value of mixed-species meat and bone meal for pigs

O.A. Olukosi and O. Adeola, Purdue University, Animal Sciences, 915 West State Street, West Lafayette, IN 47907, USA

Apparent metabolizable energy (AME) and nitrogen-correct metabolizable energy (AMEn) of 7 samples of meat and bone meal (MBM) for pigs were determined. Treatments consisted of 1 standard maize-soybean meal (SBM) diet and 7 test diets in which each of the 7 MBM samples replaced maize and SBM in the standard diets such that the ratio of maize:SBM was the same in all the 8 diets. Each diet was fed to 9 barrows in a 5-d adjustment and 5-d feeding and total collection of feces and urine balance study. Correlation, regression, partial correlation and model development and selection analyses were used to determine relationship among proximate fractions and AME or AMEn and for choosing the optimum prediction equation. Apparent metabolizable energy and AMEn of the meat and bone meal ranged 2,320 to 3,267 kcal/kg and 2,212 to 3,170 kcal/kg, respectively. Nitrogen excretion and protein digestibility of the MBM explained > 50% of the variation in AME and AMEn of the MBM. The optimum prediction equation for AME is AME, kcal/kg=74,133 – (13.2×GE, kcal/kg) – (15.9×Protein, g/kg)+(68.6×Phosphorus, g/kg) – (144.2×Calcium, g/kg)+(81.2×Fat, g/kg), the model explained 99% of the variation in AME of the MBM. It can be concluded from the study that the individual chemical components explained less of the variations in MBM than the measures of nitrogen utilization of the MBM. However, the combination of the chemical components provides information that may be used to predict the energy value of MBM.

---

no. 2

Effectiveness of anthelmentic drugs on gastrointestinal nematodes species in sheep

M.A.A. Ahmed and I.V. Nsahlai, SASA, UKZN, Animal &Poultry Science, PMB, 3209, South Africa

The resistance of gastrointestinal nematodes (GIN) to anthelmentic drugs is affecting small ruminant production. This study determined the effectiveness of various drugs [Ivermectin 1% (D1), Closantel 7.5% (D2), and Abamectin 0.08%+Praziquantel 1.5%+Fenbendazole (D3)] currently being used in SA against GIN. The initial liveweight of 24 sheep (12 females & 12 males) and egg count (EPG_0) in rectal faeces were determined. Animals with *ad lib* access to infested pasture were blocked by gender, EPG_0 and initial liveweight before being randomly allocated to one of four drug administration treatments [untreated control (D0), D1, D2, and D3]. Rectal faeces was collected on days 0, 7, 14 and 21 to determine EPG. Part of the faeces was incubated to identify and determine the abundance of larval forms of *Haemonchus*, *Trichostrongylus*, *Strongyloides*, *Namatodirus* and *Cooperia* species. Differences among treatments changed ($P<0.05$) over time. On day 7, D1, D2, and D3 depressed EPG to 0.66, 0.37 and 0.80 of their respective starting values whilst EPG increased 1.39 times for D0. Thereafter, EPG increased consistently for all drugs; D2 recorded the lowest values. *Haemonchus*, *Trichostrongylus*, *Strongyloides*, *Namatodirus* and *Cooperia* species contributed respectively 60%, 30%, 6%, 3% and 1% of the larval forms on day 0; and 78%, 8%, 11%, 1%, 2% on day 21. Larval forms increased for *Haemonchus* spp. but decreased for *Trichostrongylus* species over time. Closantal was the most effective. *Haemonchus* spp. were least affected whilst *Trichostrongylus* spp were the most affect by all drugs.

no. 3

The removal efficiency of pollutants in the livestock manure with a revolving filamentous media and an advanced oxidation process

K.H. Jeong, D.Y. Choi, J.H. Kim, J.H. Kwag, C.H. Park and Y.H. Yoo, National Institute of Animal Science, Omokcheon-dong, Suwon Si Gwonseon-gu, Gyeonggi-Do, 441-706, Korea, South

The industry of livestock farming has sharply increased in South Korea over the past 20 years, and approximately 45 million tones of manure was generated in 2007. This study was carried out to develop a system to minimize the environmental effect of manure. The experimental treatment system use a biological reactor using filamentous media and an advanced oxidation process (AOP) consisting of ozone, ultra violet, ultrasonic, and hydrogen peroxide treatment. The sludge in the biological reactor equipped with filamentous media was decreased to 88% of it in the biological reactor with no media. On the other hand, the reduction efficiency of nitrogen and phosphorous were improved. In the AOP process, COD, SS, and T-N concentrations were changed from 210 to 200 from 820 to 760, from 309 to 262 mg/L, respectively. After 10 min. ultrasonic treatment, SS and T-N had tendency to increase but TP was not changed. Co-application of ozone an ultrasonic to wastewater for 30min, COD, SS, TN, and TP decreased from 238 to 165, 900 to 540, 400 to 263, and 5 to 4 mg/L, respectively. With co-application of ozone and ultra violet for 30 min., COD, SS, TN, and TP decreased from 321 to 151, 340 to 140, 204 to 111, and 15 to 7 mg/L, respectively, and color was also changed from 4,344 to 624.

---

no. 4

Evaluation of genetic structure of the ex-situ population of Japanese crested ibis in Japan by pedigree and principal component analyses

H. Iwaisaki[1], A. Arakawa[1], T. Nomura[2] and Y. Kaneko[3], [1]Kyoto University, Graduate School of Agriculture, Sakyo-ku, Kyoto-shi 606-8502, Japan, [2]Kyoto Sangyo University, Faculty of Engineering, Kita-ku, Kyoto-shi 603-8555, Japan, [3]Sado Japanese Crested Ibis Conservation Center, Niibo, Sado-shi 952-0101, Japan

The current ex-situ population of the internationally-conserved bird, Japanese crested ibis (Nipponia nippon), in Japan has originated from 3 foundation birds introduced from People's Republic of China in recent years. The purpose of this study was, using pedigree and principal component analyses, to assess the current genetic structure and the genetic diversity of this population. Population genetic parameters such as genetic contributions of founders to non-founders, founder genome equivalent, and total losses of genetic diversity relative to that in the infinite size of population were evaluated. Principal component analysis was performed to the relationship matrix for the current population. The first 2principal components extracted gave contributions of 40.4 and 6.6%, respectively, resulting in cumulative contribution of 47.0%. Consequently, all the birdsincluding a total of 3 founders were classified into 5groups, in which the actual pedigree structure of the population was well evaluated and reflected. It was found that with variously assumed relationships among the 3 founders, their ancestors were classified into the sixth group.

Cost and benefit of collecting, transporting and crushing of *Prosopis juliflora* pods by pastoral collective action groups

G. Gebru[1], D. Amosha[1], S. Desta[2] and D.L. Coppock[3], [1]GL-CRSP PARIMA C/O ILRI, P.O. Box 5689, Addis Ababa, Ethiopia, [2]GL-CRSP PARIMA C/O ILRI, P.O. Box 30709, Nairobi, Kenya, [3]Utah State University, Department of Environment and Society, Logan, UT 84322, USA

The more arid Afar region (northeast) is especially important rangelands for Ethiopia. *Prosopis juliflora* is one of the woody species gaining dominance in this area. Local communities engage themselves in off-farm activities such as *Prosopis* pods collection, and supply these to local pod crushers. No attempts were made to evaluate the economic value of *Prosopis* pods collection, pods transport, and pods crushing. Information generated will enable scaling up of the activity and also help pastoral collective action groups to diversify income and improve productivity of animals. Benefit-Cost Ratio (BCR) and Net Present Value (NPV) were used to calculate the feasibility of the intervention. Data was collected over a three year period. Direct benefits of *Prosopis* pods collection, transport and processing were also estimated. The undiscounted benefits of *Prosopis juliflora* pods crushing in the three years were 1982.75 ETB[1], 3987 ETB and 131,250 ETB. The discounted benefits in the respective years were 852.15, 1314.35 and 59163.22 ETB. The CBR is 1.53, and the NPV is 41,329.74 ETB. Opportunity and challenges in processing *Prosopis juliflora* pods are identified, and the roles of different stakeholders defined. [1] USD=9.54 ETB(Ethiopian Birr)

---

no. 6

Differences in physical traits such as coat score and hide thickness together with tick burdens and body condition score in four beef breeds in the Southern Free State

L.A. Foster[1], P.J. Fourie[1], F.W.C. Neser[2] and M.D. Fair[2], [1]Centrally University of Technology Free State, Agriculture, P Bag x20539, 9300 Bloemfontein, South Africa, [2]University of the Free State, Animal, Wildlife and Grassland Sciences, P.O. Box 339, 9300 Bloemfontein, South Africa

A study was conducted to determine differences between four breeds pertaining to coat score, hide thickness, tick burden and body condition score. Forty heifers from four breeds (10 from each breed) were included in the study. The participating breeds were Afrikaner, Braford, Charolais and Drakensberger. A system of subjective scoring of cattle coat scores, ranging from extremely short to very woolly was used. Body condition score was measured subjectively using a scale of one to nine, with one being emaciated and nine being obese. Hide thickness (in mm) and tick burdens (count) were also determined. Measurements were carried out on five occasions on the same ten animals of each breed from February to early March 2007. Highly significant ($P<0.0001$) differences in body condition score, hide thickness and tick counts were observed between breeds over all five measurement periods. There was also significant ($P<0.0001$) differences in body condition score within breeds. Hide thickness did not differ significantly within breeds. Coat scores differed significantly ($P<0.0001$) between breeds in the earlier and latter stages of the study becoming less significant midway in the study.

no. 7

Growth and biochemical responses of growing lambs injected with rbST and supplemented with calcium soaps of fatty acids

A.N.M. Nour El-Din[1], S.Z. El-Zarkouny[1], H. Ghobashy[2] and E.I. Abdel-Gawad[2], [1]Faculty of Agriculture, Alexandria University, Department of Animal Production, Alexandria, Egypt, 21545, Egypt, [2]Agricultural Research Center, Animal Production Research Institute, Cairo, Egypt, 21545, Egypt

Twenty male Rahmani lambs of similar initial body weight (27.9 kg) and age (161 d) were divided randomly into four equal treatment groups. Controls were fed the basal diet; second group (GH) was injected with 100 mg of rbST biweekly; third group (CSFA) was supplemented with 50 g/d of calcium soap and the fourth group (GH+CSFA) was injected with 100 mg of rbST biweekly plus 50 g/d of CSFA. Treatments with rbST, CSFA and both increased ($P<0.05$) average daily gain and final body weight. Insulin-like growth factor-I (IGF-I) was increased ($P<0.01$) in (GH) and (GH+CSFA) than CSFA and controls. Hemoglobin, RBC, and WBC did not change among groups. Animals injected with rbST had higher ($P<0.05$) total protein than other groups. Controls showed the least albumin concentration in comparison with other treated groups. Treatment with rbST and CSFA increased serum glucose but urea was not affected by injection of GH, while CSFA supplementation decreased ($P<0.05$) urea concentration compared to the control group. Serum triglycerides decreased ($P<0.01$) in rbST-injected lambs than other treatment groups. Supplementation with CSFA only or combined with rbST injection increased ($P<0.05$) serum cholesterol than control or rbST-injected lambs. The results suggest that rbST and CSFA may increase average daily gain and improve the physiological status of growing lambs.

---

no. 8

Chewing activity of Holstein dairy cows fed two type of dietary fat and different forage to concentrate ratio in alfalfa based diets

S. Kargar, G.R. Ghorbani and M. Alikhani, isfahan university of technology, animal sciences, College of agriculture, 84156 isfahan, Iran

This study was conducted to examined effects of source of fat and forage to concentrate (F:C) ratio on feeding behavior and chewing activity. Eight Holstein cows averaging 55d in milk were subjected to following treatments: 1) Control (without adding fat) and 34:66 F:C ratio, 2) diet with 2% of hydrogenated palm oil and 34:66 F:C ratio (HPO), 3) diet with 2% of yellow grease and 34:66 F:C ratio (YG), 4) diet with 2% of YG and 45:55 F:C ratio (YGHF). Dry matter intake was not affected by source of fat supplementation, but it was significantly ($P<0.001$) decreased in YGHF diet relative to other treatments. Forage NDF intake (fNDFI) did not affect by treatments, with exception that it was increased in YGHF diet compared with YG diet significantly ($P<0.0001$). Due to similar DMI and fNDFI, eating, ruminating and total chewing time were not affected by fat supplementation and source of fat but influenced by F:C ratio in diet. Rumination periods expressed as periods per day, duration and chews per bolus were not affected by treatments. These results indicated that fat supplementation and type of fat did not influence feeding behavior and chewing activity, but it was significantly affected by forage to concentrate ratio in diet.

Some effects on weaning performance of beef calves

F. Szabó, A. Fördős and S.Z. Bene, University of Pannonia Georgikon Faculty of Agriculture, Department of Animal Science and production, Deák F. str. 16, 8360 Keszthely, Hungary

The effect of herd, age of cows, year of birth, season of birth and sex of calves and sire on weaning performances, moreover variance, covariance components, heritability values and correlation coefficients, genotype-environment interaction on daily gain and 205-day weaning weight of beef type Simmental calves studied. Data of 7032 purebred Simmental calves (3650 male and 3382 female) born between 1981 and 2003 from 1452 cows mated with 113 sires were analyzed in two herds. Harvey's (1990) Least Square Maximum Likelihood Computer Program-mal, DFREML (Meyer, 1998) and MTDFREML (Boldman et al. 1993), SPSS. 9.0 (1996) were used for estimation. Sire and two animal models were used for estimation. The overall mean value and standard error of weaning weight, preweaning daily gain and 205-day weight were 214±3.01 kg, 980±17.31 g/day and 236±3.40 kg, respectively. According to the results significant effects of the influencing fact such as herd, age of cows, year of birth, season of birth and sex of calves and sire both on daily gain and on 205-day weaning weight were found. Direct heritability (h^2_d) of preweaning daily gain and 205-day weight was between 0.37 and 0.42. The maternal heritability (h^2_m) of these traits was 0.06 and 0.07. Contribution of the maternal genetic and maternal permanent environmental effect to the phenotype was smaller than that of direct heritability ($h^2_{m+}c^2 < h^2_d$). The results of the examination show important and significant sire x population interaction.

---

no. 10

Chemical composition, ruminal disappearance and *in vitro* gas production of wheat straw inoculated by pleurotus ostreatus mushrooms

R. Valizadeh, S. Sobhanirad and M. Mojtahedi, Ferdowsi University of Mashhad, Animal Science, P.O. Box 91775-1163, Mashhad, Iran

Chopped and pressure-pasteurized wheat straw in the form of compost was seeded with mushroom mycelium. Chemical composition, ruminal disappearance and *in vitro* gas production were carried-out on the samples taken on day, 0, 21, 42, 63 and 84 after seeding. NDF, ADF and Lignin content of the treated wheat straw reduced significantly with a rate, 0.20, 0.10 and 0.04 percent/day. Crude protein content increased significantly from 46.6 g/kg on day 0 to 50.90 g/kg on day 84. The a fraction for dry matter was highest ($P<0.05$) for the straw under the mushroom growing conditions e.g. 0.188 g/g for the day 0 vs. 0.342 for the day 84. Generally, the ruminal disappearance of different nutrients increased by growing the pleurotus ostreatus, although, the c rate was independent. In case of gas production mushroom growing led to a significant more gas production during the early hours (less than 24 hrs) of incubation. The b fraction of *in vitro* gas production as well as *in situ* results reduced significantly from day 0 to day 84. The c rates of gas production measurements were significantly increased between the noted days. It was concluded that straw seeding with p. ostreatus can be more effective for the longer period of growth.

no. 11

In vitro evaluation of wheat straw and alfalfa hay supplemented by different zinc sources

S. Sobhanirad, R. Valizadeh and A. Hossinkhani, Ferdowsi University of Mashhad, Animal Science, P.O. Box 91775-1163, Mashhad, Iran

Two experiments were performed to investigate the effect of zinc sulfate (ZnS), zinc oxide (ZnO) and zinc methionine (ZnM) on degradability of alfalfa hay and wheat straw. To the syringes one of the following treatments were applied:1) no added zinc (C),2) 500 ppm of Zn as ZnS; 3) 500 ppm of Zn as ZnO; 4)500 ppm of Zn as ZnM. The results indicated there were not significant differences between treatments in term of gas production measurements for wheat straw. In case of alfalfa hay, supplementation of rumen fluid with ZnM significantly ($P<0.05$) increased gas production after 48 h of incubation. At this stage ZnO had a little effect on gas production. Addition of ZnM led to a faster response of microbial fermentation compared to other Zn sources. However, supplementation with the rumen fluid zinc supplements had no effect on gas production during fermentation of wheat straw probably due to low nitrogen content of wheat straw which can not stimulate a normal microbial growth. Therefore, further works are required with other feedstuffs to identify the interaction between Zn sources and feed characteristics.

---

no. 12

Owners' constraints and community perception of urban draught horses in the south of Chile

T. Tadich[1], R.A. Pearson[1] and A. Escobar[2], [1]University of Edinburgh, Division of Veterinary Clinical Sciences, Royal (Dick) School of Veterinary Studies, Roslin, EH25 9RG, United Kingdom, [2]Universidad Austral de Chile, Fac. Cs. Veterinarias, Casilla 567, 509000 Valdivia, Chile

In the south of Chile urban draught horses are the main sources of household income for many families. These horses are often perceived as being mistreated and kept in poor conditions. The present study used surveys to obtain information about owners' constraints and community perceptions of these horses. This was to allow understanding of the needs of people that work directly with horses, and those who indirectly coexist with them. All owners (n=50) keeping horses were male and for half of them draught horses were their only source of income. The main advantages of having a cart were transporting their goods for free (38%) and flexibility of working hours (18%). Many owners did not see disadvantages (46%) in horse ownership and expenses involved in horse care were only mentioned by 14% of them. Owners' constraints included lack of working opportunities (67%) and lack of Municipal assistance (65%). The community (n=120) perceived horses as being mistreated (89%) and a hazard for people when left in green areas (82%), although they see carts as part of the Chilean folklore (78%) and a tourist's attraction (51%). Promotion of owners' organisations would facilitate communication with the Municipality improving their profile in the community. Awareness needs to be generated in the community about the advantages of animal power as a renewable source of energy.

no. 13

Bacterial consortium in a direct-fed microbial product based on 16S rRNA gene analysis

S. Ohene-Adjei, J. Baah, R.J. Forster and T.A. Mcallister, Agriculture and Agri-Food Canada, Research Centre, Lethbridge, AB, T1J 4B1, Canada

Phylogenetic analysis was conducted to evaluate the of bacteria in a direct-fed microbial (DFM) product that has been shown to improve feed efficiency and growth rate in livestock. The DFM (BE7; Basic Environmental Systems and Technology, Inc., Edmonton, AB, Canada) was developed from a mixed culture of bacteria isolated from the feces of dairy and feedlot cattle. Extracted DNA (20 pmol) from the DFM was amplified by PCR using universal 16S rRNA primers. The PCR product was cloned into a TOPO 10 cloning vector and used to transform 50 µL of *Escherichia coli* DH5a cells. Ninety six randomly picked clones were cultured overnight in Luria-Bertani medium amended with 75 µg/mL kanamycin, sequenced and analyzed. Results revealed two predominant genera in the bacterial community in the DFM; *Lactobacillus* (ca. 70%) and *Acetobacter* (ca.30%). Although all *Acetobacter*-related sequences were affiliated to *A. pasteurianus*, greater diversity was observed among the lactobacilli. The *Lactobacillus* clones were affiliated to *L. parafaraginis* (21%), *L. camelliae* (21%), *L. harbiensis* (12%), *L. parabuchneri* (12%), *L. buchneri* (10%), *L. acetotolerans* (10%), *L. ghizouensis* (7%), *L. coryniformis* (5%) and *L. kefir* (2%). *Lactobacillus* spp. produce lactic acid which can be used by *Acetobacter* spp. to generate acetic acid. Although the mode of action of the DFM may arise from specific compounds produced by at least one member of this consortium, the evidence suggests that the two major groups may be in a metabolic synergy.

---

no. 14

Day-time time budget of "Caballo Fino Chilote" (*Equus caballus*) stallions during summer

T. Tadich[1] and R. Pulido[2], [1] Universidad Austral de Chile, Programa Doctorado en Ciencias Veterinarias, Casilla 567 Valdivia, 5090000, Chile, [2] Universidad Austral de Chile, Instituto de Ciencia Animal y Tecnología de Carnes, Casilla 567 Valdivia, 5090000, Chile

The Caballo Fino Chilote is a unique equine resource in the world; morphological features and a genetic analysis confirm that its origin comes from equines of the Península Ibérica. This breed is considered the only pony with Spanish origins that has subsisted in South America. It has an average height to the withers of 120cm and an average weight of 218 kg±31 kg for stallions and 247 kg±32 kg for mares. The breed is known for its excellent temperament, but there are no studies in relation to its behaviour. There for we decided to evaluate the day-time time budget of the Caballo Fino Chilote horse in a group of 7 stallions kept on pasture in their natural habitat in Chiloé Island. Stallions were observed from 8:00hrs until 20:00hrs. Scan sampling every five minutes was used to collect data of 5 mutually exclusive behaviours (eating, resting, standing alert, locomotion and others). Stallions spent most of their time eating (56%), followed by resting (32%) and standing alert (7%). A small percent of time was used for locomotion (3%) and other behaviours (2%). Higher rates of eating ocurred in the early morning (8:00-9:00 hrs) and at dawn (19:00-20:00 hrs) The main specie grassed was *Blechnum pennamarina*, a common fern in this area, for browsing *Drimys winteri* was selected. The time budget of the Caballo Fino Chilote horse is similar to those described previously for other breeds.

Lactation performance and suckling lamb growth of Kermani fat-tailed ewe

R. Mangali Kahtuei, A. Zare Shahneh, M. Moradi Share Babak and M. Khodaei Motlagh, Tehran university, Departmant of Animal Science, Faculty of Agricultur, The University of Tehran, 0098, Iran

Milk composition is very important because it affected the quality and determines ratio processed products milk. The aim of this study were lactation performance of Kermani sheep during the lactation prior, to determine correlations of milk yield with udder measurements at two weeks postpartum, growth of suckling lambs and estimating production potential in Kermani sheep. Lactation performance of 3, 5 and 6 year–old ewes from 55 Kermani ewes giving birth to single lamb and their suckling lamb growth was studied. Milk production during the suckling period (14 weeks) was determined by a combined lamb-suckling and hand milking and daily milking from weaning to end of lactation. Average lactation length and milk yield were 15 weeks (103±1days). Total milk yield during the suckling and post-weaning periods averaged 62kg. The effects of ewe age, birth weight of lamb, sex, udder depth and width was significant on milk yield ($P<0.01$). Udder depth and udder circumference ($r=0.38$) at two weeks postpartum had the highest correlation ($P<0.01$) with lactation yield. Effect of ewe age and sex of lamb also was significant on daily gain and weaning weight. The lambs of 5 year-old ewes have the highest growth but the lambs of 3 year-old ewes had the lowest growth. Milk yield and lamb growth in this breed are similar to other Iranian breeds, that can be because of genetically and natural conditions.

---

no. 16

Application of cross sectional imaging techniques (CT, MRI) in cattle production

G. Hollo, I. Hollo and I. Repa, Kaposvár University, Guba Sándor street 40., 7400 Kaposvár, Hungary

The aim of this series of experiment was to examine the opportunity of application of X-ray Computer Tomography (CT) and Magnetic Resonance Imaging (MRI) technique in cattle production. Altogether 221 animals of different breeds and genders were used in the study. In Exp 1. the tissue composition of cuts between 11-13th ribs ($n=136$), was determined by CT and correlated with tissue composition of intact half carcasses prior to dissection and tissue separation. Results indicate that tissue composition of rib samples determined by CT closely correlates with tissue composition results by dissection of whole carcasses. In Exp. 2 tissue composition of rib samples by CT ($n=40$) were compared to the results of EUROP carcass classification. CT analysis has higher predictive value in estimation of actual tissue composition of cattle carcasses than EUROP carcass classification. The results of Exp. 3 ($n=15$) showed that calves with higher muscle tissue area at the age of 7 days determined by CT will also have more meat in carcass. On the other hand the cross sectional area of longissimus muscle can be used as a useful reference scan for the evaluation of meat of carcass. Significant longer T2 relaxation time was measured in rib samples ($n=30$) at post mortem 6. days (57.39 ms) in opposite to those measured at post mortem 24 hours (55.54 ms). T2 relaxation time measured pm 6.days showed postive relationship with lightness (L*) and crude fat content, whlist negative related to pH. It seems MRI can be used for the evaulation of beef quality traits.

no. 17

Effects of increasing prepartum dietary protein level using poultry by-product meal on productive performance and health of multiparous Holstein dairy cows

F. Kafilzadeh[1], M.H. Yazdi[1] and H. Amanlou[2], [1]Razi University, Animal Science, Faculty of Agric., Kermanshah, Iran, [2]Zanjan University, Animal Science, Faculty of Agric., Zanjan, Iran

The effects of two levels of crude protein using poultry by-product meal fed during late gestation on the performance, blood metabolites, some reproductive parameters and colostrum composition of Holstein cows was studied. Sixteen multiparous cows 26±6 d before calving were assigned to one of the two diets in a randomized block design to evaluate the effects of two isocaloric diets containing either 14 or 16% crude protein. postpartum diet was similar for both groups during the first 3 weeks of lactation. Yields of milk, protein, lactose, fat, and SNF were not affected by prepartum crude protein level. Colostrum composition, blood metabolites (Ca, Glucose, Total protein, Albumin, Globulin, Urea N and Cholesterol), some of the reproductive performance indexes (Pregnancy duration, Days open, First AI, service per conception, First service Conception rate%, and Pregnancy rate%), and disease incidence were not influenced by prepartum crude protein level.Colostrum fat percent was significantly affected. There was no significant difference in body weight and BCS in cows received the two different deits.Prepartum blood urea N concentrations were elevated in the cows fed 16% CP diets prepartum.Serum cholesterol during prepartum and postpartum periods was significantly different in the two groups of cows.

---

no. 18

Evaluating relationships among linear hip height measures of crossbred steers over time, using principal component analysis

H.R. Mirzaei[1], M.P.B. Deland[2] and W.S. Pitchford[2], [1]Zabol University, Zabol, Iran, [2]Adelaide Uni., Adelaide, SA 5371, Australia

A principal component analysis (PCA) was carried out on the matrix of correlations among hip height recorded on 452 crossbred steers to examine the changes in height with time; at birth, Pre weaning, weaning, feedlot in and feedlot out. Mature Hereford (766) were mated to seven sire breeds (Angus, Belgian Blue, Hereford, Jersey, Limousin, South Devon and Wagyu), resulting in steers born over 4 years and reared under various management practices. The first principal component, accounted for 51% of the total variability among steers, was defined as a measure of overall size. The second principal component accounted for 24% of the total variation in steers and contrasted pre-feedlot and feedlot growth. It seems that Wague, Jersey and Hereford tend to have a negative correlation with Limousine and South Devon.

no. 19

Characteristics of lactation function of Iranian buffalo ecotypes, using Wood's gamma function

H.R. Mirzaei[1], J. Rahmaninia[1] and H. Farhangfar[2], [1]Zabol University, Zabol, Iran, [2]Birjand University, Birjand, Iran

Characteristics of lactation function of Iranian buffalo ecotypes were estimated by Wood's Gamma function. The data consisted of 53222 test-day records in various stages of lactation and at different parities, collected from 743 herd in Iranian Animal Breeding Center (1993-2005). The characteristics of lactation function were milk production at the beginning of lactation (a), slope of the lactation curve up to the peak of milk yield (b), slope after the peak of milk yield (c), peak time, peak yield and milk persistency. The phenotypic correlation between milk yield with milk persistency and a were positively low and moderate, respectively. The phenotypic correlation between b and c with persistency were negative. The phonotypical trend of milk yield, milk persistency and peak time increased and were highly significant. It was concluded that the positive correlations between milk yield and persistency and peak yield were very important in phenotypical selection or culling animal at the beginning of lactation.

---

no. 20

Environmental factors affecting lactation function of Iranian buffalo ecotypes, using Wood's gamma function

H.R. Mirzaei[1], J. Rahmaninia[1] and H. Farhangfar[2], [1]Zabol University, Zabol, Iran, [2]Birjand University, Birjand, Iran

Environmental factors affectinglactation function of Iranian buffalo ecotypes were studied by Wood's Gamma function. The data included 53222 test-day records in various stages of lactation and at different parities, collected from 743 herd in Iranian Animal Breeding Center (1993-2005). The characteristics of lactation function were milk production at the beginning of lactation (a), slope of the lactation curve up to the peak of milk yield (b), slope after the peak of milk yield (c), peak time, peak yield and milk persistency. The effect of Herd and year were significant on all parameter and characteristics of lactation curve ($P<0.01$). Season had significant effect on milk yield, a, c, peak time and persistency ($P<0.01$). Lactation period had significant effect on milk yield, a, c, peak yield ($P<0.01$) and b, persistency ($P<0.05$) but had no significant effect on peak time ($P<0.05$). Calving age had only significant effect on c, persistency ($P<0.01$) and b ($P<0.05$).

no. 21

Describing variation in growth traits (body weights and fatness) of crossbred heifers, using principal component analysis

H.R. Mirzaei[1], M.P.B. Deland[2] and W.S. Pitchford[2], [1]Zabol University, Zabol, Iran, [2]Adelaide University, Adelaide, SA 5371, Australia

A principal component analysis (PCA) was conducted on the matrix of correlations among eight successive body weights and P8 fat measures of crossbred heifers to describe variation in size over time. The data collected from 1143 heifers from seven sire breeds: Angus, Belgian Blue, Hereford, Jersey, Limousin, South Devon and Wagyu, mated to Hereford (H) dams born over a 4-year period (1994 to 1997). The first principal component (PC1) described 75% of total variation and was interpreted as a measure of overall size. The second principal component (PC2) accounted for about 10% of variation in the data and described as a feedlot growth. A biplot (between PC1 and PC2) was conducted to characterise seven sire breeds. Considerable differences in terms of overall size and feedlot growth existed among Wagyu and Jersey sires on the one hand and Angus, Hereford, South Devon, Limousin and Belgian Blue sires on the other.

---

no. 22

Evaluating relationships among growth traits (body measures and fatness) of crossbred cattle

H.R. Mirzaei[1], M.P.B. Deland[2] and W.S. Pitchford[2], [1]Zabol University, Zabol, Iran, [2]Adelaide University, Adelaide, SA 5271, Australia

A principal component analysis (PCA) was conducted on the matrix of correlations among growth traits at different ages of steers. Data included body weight (Wt), hip height (Ht), length (Lt), heart girth (Gth) and P8 fat collected at pre weaning, weaning, feedlot in and feedlot out. Data collected from 1144 steers calves from seven sire breeds: Angus, Belgian Blue, Hereford, Jersey, Limousin, South Devon and Wagyu, born over a 4-years period. The first principal component explained 81%, 82%,62% and 0.81% of total variability at pre weaning, weaning, feedlot in and feedlot out, respectively and was interpreted as a measure of conformation. The second principal component accounted for 0.18%, 0.16%, 0.26% and 0.17% of total variability at pre weaning, weaning, feedlot in and feedlot out and described as a maturity type component. Wagyu and Jersey (with low frame score) had fatty light weight and short bodied characteristics, Limousin, south Devon, Belgian Blue stood out as being less fatty, heavy and tall-bodied at all ages. Hereford and Angus had fatty, heavy and short bodied at all ages.

no. 23

Describing variation in growth traits (body weights and fatness) of crossbred steers, using principal component analysis

H.R. Mirzaei[1], M.P.B. Deland[2] and W.S. Pitchford[2], [1]Zabol University, Zabol, Iran, [2]Adelaide University, Adelaide, Adelaide, Australia

A principal component analysis (PCA) was conducted on the matrix of correlations among thirteen successive body weights and P8 fat measures of crossbred steers to describe variation in size over time. The data collected from 1143 steers from seven sire breeds: Angus, Belgian Blue, Hereford, Jersey, Limousin, South Devon and Wagyu, mated to Hereford (H) dams born over a 4-year period (1994 to 1997). The first principal component described 61% of total variation and was interpreted as a measure of overall size. The second principal component accounted for about 23% of variation in the data and described as a feedlot growth. Considerable differences in terms of overall size and feedlot growth existed among Wagyu and Jersey sires on the one hand and Angus, Hereford, South Devon, Limousin and Belgian Blue sires on the other.

---

no. 24

Random regression analysis of cattle growth path: application of a piecewise linear model compare to a cubic model

H.R. Mirzaei[1], A.P. Verbyla[2], M.P.B. Deland[2] and W.S. Pitchford[2], [1]Zabol University, Zabol, Iran, [2]Adelaide University, Adelaide, SA 5371, Australia

A piecewise random regression analysis was conducted to model growth of crossbred cattle from birth to about two years of age, compare to a cubic model. Hereford cows (581) were mated to 97 sires from Angus, Belgian Blue, Hereford, Jersey, Limousin, South Devon and Wagyu, resulting in 1144 steers and heifers born over 4 years. The model for ln (wt) included fixed effects of sex, sire breed, age (linear and quadratic for pre- and post-weaning), as well as two-way interactions between the age parameters and sex or breed. Random effects were sire, dam, permanent environmental, linear age by animal, age by dam, management group, and management by age as linear and quadratic. The piecewise model is very similar to the cubic model in the fit to the data; with the piecewise model being marginally better. A piecewise model performs better than a cubic model at the end of the trajectory for higher values on these data.

no. 25

Seasonal variations in the crop contents of scavenging helmeted guinea fowls (*Numida meleagris*, L) in Parakou, Benin

M. Dahouda, University of Abomey-Calavi, Faculty of Agronomic Science, Department of Animal Production, BP 526 Cotonou, 00, Benin

An experiment was carried on with 120 helmeted guinea fowls during one year in Benin. Feed intake, ingredient and chemical composition, along with the nutritional adequacy of scavenging diets were measured during the rainy season and dry season in order to propose supplementation strategies. Ingredients found in crops were identified and allocated into six main categories (supplemental feed, seeds, green forages, animal materials, mineral matter and unidentified materials). Mean dry weight of crop contents were significantly higher in rainy than in dry season (P<0.001). Amounts and proportions of supplemental feed and seeds were not significantly different between seasons, whereas those of green forage, animal materials and mineral matter were higher (P<0.05) in rainy season. Supplemental feed, especially maize and sorghum, was the largest component of the crop content in both seasons. The most represented grass seeds were *Panicum maximum* (rainy season) and *Rottboellia cochinchinensis* (dry season). Dietary concentrations in organic matter, non-nitrogen extract and metabolizable energy were higher in dry season (P<0.05), while mineral concentrations were higher in rainy season (P<0.05). Scavenging provided insufficient nutrients and energy to allow guinea fowls to be productive. Therefore more nutritionally balanced supplementary feed would be required during both seasons.

---

no. 26

Successful culture of single bovine embryos using polydimethylsiloxane (PDMS) micro-well plates cured under low pressure

K. Saeki[1], N. Kato[2], D. Iwamoto[1] and S. Taniguchi[1], [1]Kinki University, Department of Genetic Engineering, Wakayama, Japan, 649-6493, Japan, [2]Kinki University, Department of Intelligent System, Wakayama, Japan, 649-6493, Japan

Mammalian embryos cultured individually or in small groups fail to develop to blastocysts. For the culture of single embryos, the well-of-well system was developed by Vajta et al. (2000). The problem with this method is that it is difficult to make wells with the same shape. We newly developed well-plates made with soft-lithography of PDMS under low pressure that induced nonporous PDMS. The wells are cylindrical (300 µm in diameter x 200 µm deep) with flat bottoms. We cultured IVF bovine embryos for 168 h after insemination using four different culture systems:, 1) well-plates cured under low pressure (nonporous, WP-LP), 2) well-plates cured under atmospheric pressure (porous, WP-AP) 3) group culture (25 embryos/50 µl drop, GC) and 4) well-of-well system (WOW). A total of 300 embryos were used for the experiments. The cleavage rates (65-75%) were not different among the culture systems. The blastocyst rate of WP-LP (24%) was the same as the rates of GC (25%) and WOW (15%, P>0.05). However, the rate of WP-AP (4%) was lower than the rates of the other culture systems (P<0.05). The results indicate that WP-LP plates can be used for the culture of single bovine embryos. Supported by Wakayama prefecture CREATE, JST.

Establishment of a method for collecting bovine fetal cells from amniotic fluid by transvaginal aspiration

S. Taniguchi[1,2], *J. Fukuhara*[3], *N. Hayashi*[4], *D. Iwamoto*[1], *M. Kishi*[1,2], *K. Matsumoto*[1,2], *Y. Hosoi*[1,2], *A. Iritani*[1,2] *and K. Saeki*[1,2], [1]*Department of Genetic Engineering, Kinki University, Wakayama, 649-6493, Japan,* [2]*Wakayama Industry Promotion Foundation, Wakayama, 649-6261, Japan,* [3]*Wakayama Prefectural Livestock Experiment Station, Wakayama, 649-3141, Japan,* [4]*Gifu Prefectural Livestock Research Institute, Gifu, 506-0101, Japan*

Bovine amniocentesis has been used for early diagnosis of the sex of fetuses. We suggested that fetal cells from transvaginally-collected amniotic fluid have the potential to produce clones. However, the collected fluid may include maternal cells as well as fetal cells. In this study, we investigated a method for collecting fetal cells from amniotic fluid using a cow pregnant with an embryo cloned from a bovine fibroblast transfected with a β–act/luc$^+$/IRES/EGFP gene. At each of 82, 90 and 99 days of gestation, we carried out four consecutive aspirations of 5 ml amniotic fluid. Then, we cultured the cells in each sample. At every passage, we observed EGFP fluorescence in the cultured cells and evaluated EGFP-positive/-negative cells as fetal/maternal cells, respectively. The cells in the first samples grew fast, and no EGFP-positive cells were observed even after 4 passages. The cells in the other samples grew slowly, and almost all the cells were EGFP-positive. These results indicate that fetal cells can be obtained from amniotic fluid by discarding the initial 5 ml in bovine transvaginal amniocentesis. Supported by Wakayama Prefecture CREATE, JST.

---

no. 28

Faecal moisture in parasitised Australian Merino sheep

A.R. Williams[1], *D.G. Palmer*[2], *L.J.E. Karlsson*[2], *P.E. Vercoe*[1], *I.H. Williams*[1] *and J.C. Greeff*[2], [1]*The University of Western Australian, School of Animal Biology, 35 Stirling Highway, Crawley 6009, Australia,* [2]*Department of Agriculture and Food Western Australia, Locked Bag 4, Bentley Delivery Centre 6983, Australia*

Nematode parasitism is a major constraint on profitable sheep production in temperate areas. Symptoms include weight loss and scouring. Scouring occurs in adult sheep that have high immunity to nematode parasites. In this experiment we measured the pattern of faecal moisture in immune Merino rams receiving an artificial infection of either *Trichostronglyus* sp., *Ostertagia* sp., both species or no infection (control). Faecal moisture was increased in infected sheep, but there were no differences within infected groups. All infected sheep were highly immune to the infection, indicated by mostly negative worm egg counts and very low numbers of total worm numbers at post-mortem indication. These results indicate that scouring can occur in adult sheep in the absence of a significant worm burden, and can be attributed to both the major parasitic nematodes of sheep in temperate areas.

no. 29

Improved dual purpose cowpea lines as ruminant feed: yield, nutrient changes in storage and pesticide residue in cowpea haulms on the activities of rumen microorganisms

C. Antwi[1], E.L.K. Osafo[1], H.A. Dapaah[2] and D.S. Fisher[3], [1]KNUST, Animal Science Department, Kumasi, Ghana, [2]CSRI, CRI, Kumasi, Ghana, [3]USDA, ARS, Watkinsville,30677-2373, Georgia

This research focused on the agronomic and nutritive characteristics of four improved cowpea lines IT93K2045, IT93K2309, SORONKO and IT86D716 to assess variation in nutritive value during storage and pesticide residues in cowpea haulms on rumen microbial functions. A complete randomized block with four replications was used to evaluate the yields of the cowpea lines while a strip-split-plot design with three replications, two storage methods – roof and shed (main plots), four varieties (subplots) sampled at 0, 4, 8 and 12 weeks (strip-plot in time) was adopted to assess storage method on dry matter loss, nutrient loss and retention of cowpea haulms as a function of time. No differences in grain yield were observed among the cowpea lines but the haulm yields of IT93K2045 and IT93K2309 were greater than SORONKO and IT86D716 ($P<0.05$). *In vitro* gas production, digestibility and nutrient composition differed among the cowpea lines. Analysis showed that storage methods had a significant effect ($P<0.001$) on DM loss, retention of CP ($P<0.05$) and ADIN, ADF and NDF in the haulms ($P<0.01$). Studies on pesticide residues revealed that, pesticide concentrations were not great enough to influence the rumen microbial functions.

---

no. 30

Egg loads of gastrointestinal strongyles in Nguni cattle on communal rangelands in semi-arid areas

M.C. Marufu, C. Mapiye and M. Chimonyo, University of Fort Hare, Livestock and Pasture Sciences, P. Bag X1314, 5700 Alice, South Africa

A one year monitoring study was carried out to determine the egg loads and seasonal prevalence of gastrointestinal (GI) strongyles of Nguni cattle on communal rangelands in the Eastern Cape Province, South Africa. Faecal samples were collected per rectum once a season and examined by the modified McMaster technique using a saturated solution of sucrose and sodium chloride. Results revealed that 58.7% of the cattle were infested by GI strongyles. The prevalence and mean egg counts of GI strongyles showed a seasonal sequence with the highest faecal egg counts occurring in the hot-dry season and the lowest in post-rainy season. The results suggest that GI strongyles have a moderate prevalence in Nguni cattle in the sweet and sour rangelands, and that strategic control of these parasites needs to be carried out during hot-dry season to prevent clinical nematodoses.

Tick prevalence in communal cattle raised on sweet and sour rangelands in semi-arid areas

M.C. Marufu[1], M. Chimonyo[1], K. Dzama[2] and C. Mapiye[1], [1]University of Fort Hare, Livestock and Pasture Science, P. Bag X1314, 5700 Alice, South Africa, [2]Stellenbosch University, Department of Animal Sciences, P. Bag X1, 7602 Matieland, South Africa

The objective of the current study was to compare tick prevalence in Nguni and non-descript cattle kept in communal areas in the sweet and sour rangelands of the Eastern Cape. Ixodid ticks were collected seasonally from 353 communal cattle raised on sweet and sour rangelands from August 2007 to April 2008. Three tick species were identified in the sweet rangeland namely *Rhipicephalus appendiculatus*, *Rhipicephalus (Boophilus) decoloratus* and *Rhipicephalus evertsi evertsi* with the following prevalences respectively 71.1%, 29.2% and 40.2%. Hyalomma species (19.0%) occurred only in the sour rangeland. Higher ($P<0.05$) prevalence and tick counts were recorded in the hot-wet season than in the cold-dry season. Cattle in the sweet rangeland had significantly lower tick counts than those in the sour rangeland in all the seasons except the hot-dry season in which they had similar ($P>0.05$) tick counts. The Nguni breed had lower ($P<0.05$) tick loads of *Rhipicephalus appendiculatus* in the hot-wet and post-rainy season and Hyalomma species in all seasons than the non-descript cattle. The use of the Nguni breed in the integrated control of ticks on cattle in communal areas of the Eastern Cape is recommended. Further studies to determine the associated serological prevalence of tick-borne diseases in the communal cattle breeds are required.

---

no. 32

The effect of season on aspects of in vitro embryo production (IVEP) in sub-fertile beef cows

J.M. Rust[1], D.S. Visser[2], J.E. Venter[2], M.P. Boshoff[2], S. Foss[2] and J.P.C. Greyling[3], [1]Döhne ADI, Stutterheim, 4930, South Africa, [2]Animal Improvement Institute, Irene, 0062, South Africa, [3]University of the Free State, Bloemfontein, 9300, South Africa

The effect of season on bovine IVEP has not been evaluated in South Africa. Most cattle presented for IVEP procedures have impaired fertility. Producing embryos from these animals is an intensive and costly exercise. It is important to establish whether the application of this technology should be restricted to specific seasons of the year. Various beef breed cows (n=40) were subjected to commercial IVEP procedures over a period of 4 years. All animals were of superior genetic merit and were unfit for conventional reproduction. Ultrasound guided oocyte retrieval (OPU) was performed on a weekly basis and all oocytes were entered into a standard IVEP protocol. Follicular population, oocyte recovery rate, oocyte quality, embryos produced per OPU session and embryo quality were recorded. Follicular populations were significantly higher during the colder months; no significant difference in either the recovery rate or oocyte quality was recorded and the number of embryos produced per OPU session showed a definite seasonal pattern with low production during the hot months, a steady production during cold months and a peak in spring/early summer. No significant differences were recorded in embryo quality. The results indicated that IVEP should be attempted in a systematic seasonal cycle to ensure maximum efficiency.

Survival of Holstein-Friesian heifers on commercial dairy farms in Kenya

D.K. Menjo[1], B.O. Bebe[1], A.M. Okeyo[2] and J.M.K. Ojango[1,2], [1]Egerton University, Box 536, Njoro, Kenya, [2]ILRI, Box 30709, Nairobi, Kenya

Herd health and adaptability are of concern in dairy herds in the tropics because of persistent exposure to multiple stresses of low quality and quantity feeding, heat stress, high disease and parasitic incidences, poor husbandry and breeding practices. Here the combined effect of mortality and culling is estimated to cause losses of between 40 and 60% of dairy heifers conceived or born. This study applies survival analysis techniques to evaluate important factors influencing survival to first calving in Holstein-Friesian cattle raised on large scale farms in Kenya. On average, 25% of all the heifers born were culled prior to calving for the first time, while 34% were culled prior to attaining four years of age. Though the highest proportion of losses was due to unspecified reasons, the relative risk of being culled was highest when an animal had a specific disease, and survival was most critical within the first 60 days of life. Daughters of sires from South-Africa and Israel tended to have better survival rates than those sired by bulls originating from other regions of the world. When selecting sires for breeding under tropical production systems, it is important for farmers to consider information related to survival and fertility under these systems.

---

no. 34

Effect of different weaning practices on post-weaning growth of Angora kids and reproduction of Angora ewes

J.H. Hoon, W.J. Olivier and P.J. Griessel, Grootfontein Agricultural Development Institute, P/Bag X529, Middelburg (EC) 5900, South Africa

The aim of the project was to find practical solutions for the problems encountered during the post-weaning period of Angora goats. The objectives were to determine the effect of different weaning practices on the post-weaning growth of kids, the body weight change of ewes and the conception rate of ewes in the next breeding season. The project was done over a four year period in the Middelburg, Eastern Cape, district under natural veld conditions. At weaning each year, the three treatments commenced: Weaned – kids and ewes separated; Swopped – kids separated from ewes, but kept with other mature animals; Not weaned – kids and ewes stayed together. Body weights of the ewes were recorded at weaning and mating, while body weights of the kids were recorded at weaning and monthly until 10-month age. The conception rate of the ewes was determined by ultrasound scanning. The body weight change and average daily gain of the kids from weaning until 10 month-age did not differ ($P>0.05$) between the Weaned and Not weaned kids, but both were higher than the Swopped kids ($P<0.05$). Also, no differences ($P>0.05$) in growth rate were observed between the ram and ewe kids. The weaning of kids had a positive effect ($P<0.05$) on body weight change of ewes from weaning until mating. The conception rate was in favour of the Not weaned and Swopped ewes, compared to the Weaned ewes. The results indicate that the non-weaning of kids did not have any negative effect on the reproduction of the ewes.

The effect of urea-treated barley straw on the reproductive performance and post-partum ovarian activity of Libyan Barbary sheep

F. Akraim[1], A. Majid[2], A. Rahal[2], A. Ahmed[2] and M. Aboshwarib[3], [1]Omar Al-Muktar UNIV., Animal production, 1118, AL-Biada, 1118, Libyan Arab Jamahiriya, [2]Al-Fateh Univ., Animal production, 13040 Tripoli, 13040, Libyan Arab Jamahiriya, [3]Al-Fateh Univ., Veterinary medecine, 13040 Tripoli, 13040, Libyan Arab Jamahiriya

Forty ewes (3-6 years of age, average weight 41 kg) were randomly chosen from the flock of Al-Fateh university sheep experiment station. Ewes were divided into two groups, control group(C) receive untreated barley straw and treatment group (T) receive barley straw treated with 10% urea solution applied as 40% (V/W). All barley straw sprayed by molasses when introduced to animals. Both groups receive commercial concentrate according to physiological state. Experiment started with introducing the rams in July. Average concentrations of progesterone didn't significantly differed and were 2.96 n/ml and 2.38 n/ml during 9 weeks for T and C respectively (P>0.05). In T group, 53.3% of ewes had been fertilized during the early period (two weeks after the introduction of rams) and maintained 3.5 ng/ml of progesterone. However, non of C group were fertilized in this period. Fertilization rate, fecundity and viability was 83.3%, 1,07 lamb/ewe, 93.75% and 78.94%, 1.13 lamb/ewe, 88.23% for T and C respectively, these differences were not significant (P>0.05). Progesterone concentrations were below 0.07 ng/ml during 9 weeks post-partum in both groups. Cereal straws can be treated with urea without adverse effects on reproductive performance.

---

no. 36

Seroprevalence of the neosporosis in the Dakar zoo (Senegal)

E. Dombou[1], A.R. Kamga Waladjo[1], O.B. Gbati[1], S.N. Bakou[1], G. Chatagnon[2], J.A. Akakpo[1], P.E.H. Diop[1], L.J. Pangui[1] and D. Tainturier[2], [1]Ecole Inter-Etats des Sciences et Médecine Vétérinaire de Dakar, P.O. Box 5077 Dakar Fann, SN, Senegal, [2]Ecole Nationale Vétérinaire de Nantes, Atlanpole – La Chantrerie BP 40706, 44307 Nantes cedex 3, France

The neosporosis is a recent discovery protozoose cause by *Neospora caninum*. It is manifested clinically by abortions and unrest among many neonatal domestic and wild animals. It would be responsible for about 42.5% of abortions in cattle in some countries. Neosporosis has been identified in virtually every country in which it was sought. Investigations have taken place in South Africa, Algeria and Zimbabwe. The objective of our study was to identify a contact between Dakar's zoo wildlife and *Neospora caninum*. Thus, 25 sera from 7 lions, 4 buffalos, 2 Oryx, 2 hyenas, 1 jackal, 8 Elan Cape, and 1 hypotrague were collected in Dakar's zoo and analysed by the ELISA technique at the Veterinary School of Nantes. Serologic prevalence of neosporosis in these different species is 100% for the lion, 50% for the Buffalo, and 0% for other species. *Neospora caninum* is present in Dakar zoo and circulates in the wildlife and maybe in other species. The observation of an association between the spatial abundance of wild and serologic prevalence *Neospora caninum* in farms in Texas reinforces the hypothesis that wild play a role in the cycle of *Neospora caninum*. The neosporosis could be partly responsible for the low yield of cattle on extensive farms in Senegal.

Proteomic analysis of bovine somatic cell nuclear transfer embryos at 14 days

T. Matsui[1], A. Ideta[2], K. Nagai[3], H. Ikegami[3], M. Urakawa[2], D. Iwamoto[1], S. Taniguchi[1,3], Y. Hosoi[1], K. Matsumoto[1], A. Iritani[1], Y. Aoyagi[2] and K. Saeki[1], [1]Depatment of Genetic Engineering, Kinki University, Wakayama, 649-6493, Japan, [2]Zen-noh ET Center, Hokkaido, 080-1407, Japan, [3]Wakayama Industry Promotion Foundation, Wakayama, 649-6261, Japan

Embryos cloned from bovine early G1 cells (eG1-NT embryos) achieved full-term development more frequently than embryos cloned from quiescent cells (G0-NT embryos, Kasinathan et al., 2001; Urakawa et al., 2004), although the reason is unclear. To understand why eG1-NT embryos develop more successfully, we compared the global protein expression profiles in embryos derived from eG1- and G0-NT blastocysts. In this study, we analyzed trophectoderms (TEs) of the NT embryos because many placental defects have been reported in the cloning of cattle. TEs were obtained from NT blastocysts that were transferred to recipiens and recovered on day 14. Proteins from the TEs were separated within a pI range of 4.0-7.0, quantified by 2-DE, and identified by peptide-mass fingerprinting. Parthenogenetic (Pg) and IVF embryos were used as controls. The statistical comparisons were done with Fisher's PLSD test following ANOVA. We identified three proteins that were differently expressed when comparing eG1-NT and IVF embryos with G0-NT and Pg embryos. Two proteins were up-regulated (1.97-fold and 1.36-fold) and one protein was down-regulated (0.7-fold) in eG1-NT and IVF embryos. We speculated that these proteins are associated with the developmental capacity of bovine NT embryos. Supported by Wakayama Prefecture CREATE, JST.

---

no. 38

Impact of using sodium or calcium salts of fatty acids as source of energy in late pregnant buffaloes dams

F.M. Abo-Donia, A.M. Aiad and A.M. Abd El-Aziz, Animal Production Research Institute, By-producte Utilization Section, 6 Nadi El-Said street, Dokki, Giza, 11433, Egypt

Thirty pregnant buffalo expected 60-75 day of parturition were divided to three balanced groups and lasted at parturition. Animal groups distributed randomly on three tested rations, the control ration contained concentrate mixture (CM) with corn grain plus berseem hay and rice straw. The second and third ration replaced CM with others CM containing either Na-SFA or Ca-SFA instead of corn grain. Content of AEE in Ca-SFA was lower than that Na-SFA, while TFA's in Ca-SFA was higher compared with Na-SFA. Chemical compositions of different concentrate mixture and tested rations were similar except AEE was higher in that contained either Na-SFA or Ca-SFA. Incubation teased rations in the rumen for 8, 16, 32, 48, 64 and 72 hrs shown decrease of DM, OM, CP, NDF and ADF disappearances. Feed intake didn't effect with feed rations containing Na-SFA or Ca-SFA. Changing in body weight during experimental period was significantly increase with feed ration containing fat than that control one, Overall mean of pH values, propionic acid and FFA's in the rumen higher ($P<0.05$) significantly when feeding ration containing Na-SFA compared that containing Ca-SFA or control one, while, significantly decreased TVFA's, acetic, Ac/Pr and NH_3-N. Add Na-SFA in the ration decreased total protein concentration in blood of late pregnant buffalo compare to Ca-SFA or control one.

The South African beef profit partnerships project: the estimated aggregate socio-economic results and impacts to date

T.P. Madzivhandila, Agricultural Research Council, P/Bag 2, 0062, Irene, South Africa

Too often agricultural research & development (R&D) investors and practitioners target research outputs rather than outcomes. They settle for developing new knowledge, information packages, publications, or a new or improved practice that, if used or 'adopted' could provide benefits. They fall short of setting targets of outcomes such as measurable improvements in enterprise, family and regional income, environmental sustainability, reduced poverty, improved livelihoods and better lives or beneficiary social infrastructure. The Beef Profit Partnerships (BPP) project aimed to achieve sustained improvement in profit in beef enterprises in two provinces in northern and north western South Africa. A number of beef market price and throughput key performance indicators were set and routinely assessed and recorded within each team. A subset of farmer teams also routinely calculated and recorded gross margins for their beef enterprises. Based on the recorded data, it is estimated that the BPP project increased revenue to the emerging farmers involved in the teams by more than 1.95 million Rand (R) over the period 2001-2006. This paper reports on the results achieved early in the project, as well as regularly and frequently throughout the life of the projects. The results are attributed to the design and management of this particular R&D intervention using the Sustainable Improvement and Innovation model.

---

no. 40

Higher survival and hatching rate of grade one bovine female embryos following vitrification

T.L. Nedambale[1], M.B. Raito[1], J. Xu[2] and F. Du[2], [1]Agricultural Research Council Animal Production, Germplasm & Reproduction Biotechnologies, Private Bag x2, 0062 Irene, South Africa, [2]Evergen Biotechnologies Inc., 392 Storrs Rd., 06269 Storrs Connecticut, USA

Selection of female blastocysts based on their graded stages can skew the survival and hatching rate post-vitrification *in vitro*. The current study assessed the effects of bovine female blastocysts at different developmental grading stages (C1 (excellent), C1- (good) and C2 (fair) on their ability to survive vitrification. A total of 203 Day 8 female blastocysts (FB) were collected and graded as C1 (69), C1- (69) and C2 (65). Thereafter, female blastocysts were vitrified in Evergen Biotechnologies Inc vitrification solution. Female blastocysts were then thawed per treatment group and evaluated for survival and hatching rate at 0, 4, 24, 48 and 72 h. Data was analysed by ANOVA. The result demonstrated no significant differences in all treatment groups at 0 and 4 h post-thawing. However, C1 group had highest survival at 24 h (95%), 48 h (87%), 72 h (83%) and hatching rate (71%) compared to C1- and C2 group. Vitrified C2 group resulted in the lowest survival and hatching rate of female blastocyst post-thawing. In conclusion, female blastocyst graded as C1 survive vitrification better and result in higher hatching rate. This study suggests that female blastocyst should be selected and graded before cryopreservation to increase the survival and hatching rate.

Development of bovine NT embryos using life-span extended donor cells transfected with foreign gene

S.S. Hwang[1], Y.G. Ko[1], B.C. Yang[1], K.S. Min[2], J.T. Yoon[2], M.J. Kim[1] and H.H. Seong[1], [1]National Institute of Animal Science, Rural Development Administration, Animal Biotechnology Division, Suwon, Gyeonggi-do, 441-706, Korea, South, [2]Hankyong National University, The Graduate School of Bio- & Information Tech., Anseong, Gyeonggi-do, 456-749, Korea, South

This study was performed to determine the developmental potentials of NT embryos using life-span extended cells transfected with foreign gene as donor cells. Life-span extended bovine embryonic fibroblast cell line transfected with an expression vector in which the human type α collagen (BOMAR) and ear cell were used as a donor cell. Cytogenetic analysis was performed to analyze the chromosomal abnormality of donor cells. The embryos lysed were significantly higher in 1.8 kV/cm for 20 μsec 1time compared to other groups ($P<0.05$). The blastocyst development in ear cell group was statistically significant compared to both BOMAR groups ($P<0.01$). Both BOMAR groups cultured more than 40 passages (>40 passages) had lower number of chromosomes, however, fresh GC and BOMAR group cultured less than 20 passages had normal numbers. The transfected foreign gene was expressed in all BOMAR groups, but not in GC group. Based on these results, the lower developmental potential of NT embryos using life-span extended donor cells transfected with foreign gene might be a cause of chromosomal abnormality in donor cells.

---

no. 42

A list of challenges that face beef cattle farmers in Eastern Cape Province of South Africa

S. Nini, Agriculrtural Research Council, LBD, Irene, South Africa

Using situation analysis step of Continuous Improvement and Innovation (CI & I) methodology, the Agricultural Research Council identified nine major elements that impact beef cattle marketing in the Sakhisizwe local municipality in Queenstown. These includes the following; Transport arrangements to and from the auction if not sold, Sales facilities, Price lower than the market price, Lack of loading facilities on the villages, Ownership clearance process (Branding); Weights and age of animals; Crime and security of animals & Payment methods. Using impact analysis step of CI&I, five farmer support partners that could influence these elements established partnerships with the seven farmer groups. A session which developed an action plan as in CI&I methodology devolved roles and responsibility to all eleven role players. This resulted in 214 animals becoming available for sale at an area where there is no history of an auction before and of these 198 animals was sold. This confirms that these challenges are really the major barriers for the emerging farmers to access the formal beef markets.

Mycotoxins and animal safety: Nigeria's stand in the new future

I.C. Okoli and A.A. Omede, Federal University of Technology,Owerri,Imo State,Nigeria., Department of Animal Science and Technology, PMB 1526, Owerri, 460001, Nigeria

Mycotoxin contamination of feeds and feedstuffs as well as the outbreak of mycotoxicoses are important livestock production problems in Nigeria. These problems are unrecognized or have received limited research attention, irrespective of evidences. This is unacceptable considering the global focus on combating mycotoxins and its related problems associated with animal health and safety. Regulatory measures have been established globally for this reason. It seems that in Nigeria, only Aflatoxins have been recognised still, the limit set for it's occurrence in animal feeds and feedstuffs can be deleterious in comparison to regulations obtainable in some parts Africa, while other mycotoxins are neglected. If this neglect continues, the safety of Nigerian livestock industry and its resultant contributions to poverty reduction, increased protein intake and sustaining livelihood will be lost. This neglect can generally be traced to lack of will-power of those concerned (quality control and assurance personnels/organizations, governmental and non- governmental agencies, educationists, researchers, and livestock producers) to accept mycotoxins and its related problems as dangers to animal health and safety. Nigeria must take a positive stand against mycotoxins and mycotoxicoses in the new future in order to protect and promote its animals' health and safety, sustain livestock productivity, and compete favourably in the global market through consciously advocated programs suited to its environment.

Assessment of hatching traits of four different commercial broiler breeder strains: a comparison and its trends

F.M. Khattak, T.N. Pasha, G. Saleem, M. Usman and M. Akram, University of Veterinary and Animal Sciences, Poultry Production, Shiekh Abdul Qadir Jelani (out Fall)Road, 54000, Pakistan

Studies were conducted to compare four commercial broiler breeder strains i.e. Arbor Acer (AA), Hubbard (HB), Hybro (HY) and Starbro (ST). Egg characteristics, hatchability, infertility and chick weights were evaluated. Nine eggs/strain were used for measurement of egg characteristics & 140 eggs/strain were incubated to compare the hatching traits. AA had ($P<0.05$) lowest and HB had the highest ($P<0.05$) egg weight and hatchability compared to all strains. After adjusting egg weight, HB had the highest ($P<0.05$) yolk weight, yolk diameter, yolk height, albumen height and ST had highest ($P<0.05$) albumen weight and height. Lowest chick weight ($P<0.05$) was in AA & HY. These results indicated significant strain variations. Negative linear regression was observed only for albumen weight, yolk & albumen height. Hatchability, infertility, dead in germ, dead in shell percentages showed significant ($P<0.05$) differences among these broiler strains. On the basis of this study it may be concluded that HB has better hatchability results as compared to AA, HY & STstrains.

no. 2

Egg Laying Performance of Local Ghanaian chicken and SASSO T44 chicken raised under improved Conditions

R. Osei-Amponsah[1], B.B. Kayang[1], A. Naazie[1], M. Tixier-Boichard[2] and R. Xavier[2], [1]University of Ghana, Animal Science, P.O. Box 25, Legon, Ghana, [2]INRA/Agro Paris Tech, Génétique et Diversité Animales, UMR 1236 Génétique et Diversité Animales, 78352, France

Egg production of 571 local Ghanaian chickens and 107 SASSO T44 chickens kept under improved management conditions were analysed. Percent egg production was significantly higher in the Savannah local chicken ecotype than the forest ecotype and SASSO T44 chicken. On the average, SASSO T44 had significantly ($P<0.05$) higher egg weights and feed intake compared to the local ecotypes. The local chicken ecotypes on the other hand had significantly ($P<0.05$) better feed conversion ratios than SASSO T44 chicken. Local Ghanaian chicken ecotypes have a more efficient feed conversion ratio per every gramme of egg produced compared to the control population. Compared to egg production studies under similar conditions, local chicken could be good genetic material for the development of an efficient egg production chicken breed adaptable to humid tropical climates.

---

no. 3

Variations in egg quality characteristics of local Ghanaian chicken and SASSO T44 chicken with storage time

R. Osei-Amponsah[1], B.B. Kayang[1], A. Naazie[1], M. Tixier-Boichard[2], R. Xavier[2] and H. Manu[1], [1]University Of Ghana, Animal Science, P.O. Box 25, Legon, Ghana, [2]INRA/Agro ParisTech, Génétique et Diversité Animales, UMR 1236 Génétique et Diversité Animales, 78352, France

Egg quality characteristics of local Ghanaian chicken eggs (140) were measured and compared to those of SASSO T44 chicken eggs (107) from day of lay to a period of 21 days in storage. On the average SASSO T44 chicken had significantly ($P<0.05$) higher egg weight, shell weight, albumen height, albumen weight, yolk weight and albumen ratio. Eggs of the local chicken ecotypes had significantly ($P<0.05$) higher yolk ratios whilst Haugh units were not significantly different from those obtained for the SASSO T44 chicken. Fluctuations in egg quality traits during the 3- week study periods were similar in the 3 ecotypes studied. Conversion of albumen height to Haugh units narrowed the gap in egg quality performance between local and SASSO T44 chicken. There was a negative effect of storage time on egg quality irrespective of the chicken ecotype. It was recommended that eggs should be kept at cooler temperatures in tropical climates or at best should be consumed within a week to minimise deterioration of its quality.

no. 4

Effects of dietary replacing soybean meal with increasing levels of canola meal supplemented with exogenous enzyme on laying hen performance

M. Torki and M. Davoodifar, Animal Science Department, Agricultural Faculty, Razi University, Imam Avenue, 67155-1158, Kermanshah, Iran

This experiment was conducted to evaluate effects of partially replacement of dietary soybean meal (SBM) with canola meal (CM) with and without a commercial exogenous enzyme preparation on performance of first-cycle laying hens. Two hundred and sixteen Hy-line leghorn hens after production peak were randomly divided in 36 cages (n=6). Six iso-energetic and iso-nitrogenous experimental diets (ME=2850 Kcal/kg and CP=15 g/kg) in a 2 × 3 factorial arrangement including three levels (0, 10, and 20%) of CM with and without a ß–mannanase-based enzyme were fed to hens with 6 replicates per diet during 10-week trial period. Collected data of feed intake (FI), egg production, egg mass and calculated feed conversion ratio as well as egg traits were analyzed based on completely randomized design using GLM procedure of SAS. Graded replacement of SBM with CM up to 20% and enzyme supplementation did not affect FI (g/day/hen), egg mass (g/day/hen), FCR (feed : egg) and production % (hen/day) ($P>0.05$). Egg weight of hens fed on diet included 20% CM decreased comparing with control (0% CM) diet ($P<0.05$). Egg traits including shape index, yolk index, specific gravity, shell weight and shell thickness were not affected by dietary treatment ($P>0.05$); however, yolk color after first month of trail decreased in dietary group with 10% of CM comparing to control diet ($P<0.05$).

---

no. 5

Influence of phytase inclusion to diets containing graded levels of canola meal on performance of first-cycle hens and egg quality traits

M. Torki and M. Davoodifar, Animal Science Department, Agricultural faculty, Razi University, Imam Avenue, 67155-1158, Kermanshah, Iran

To evaluate the effects of partially replacement of soybean meal (SBM) with canola meal (CM) supplementing by phytase on performance of laying hen and egg traits, 432 Hy-line leghorn hens after production peak were randomly divided in 72 cages (n=6). Twelve iso-energetic and iso-nitrogenous experimental diets (ME=2900 Kcal/kg and CP=15 g/kg) in a 3×2×2 factorial arrangement including three levels (0, 8, and 16%) of CM, two levels of available phosphorous (AP) with and without a phytase-based enzyme were fed to hens with 6 replicates per diet during 10-week trial period. Collected data of feed intake (FI), egg production, egg mass and calculated feed conversion ratio as well as egg traits were analyzed based on completely randomized design using GLM procedure of SAS. Graded replacement of SBM with CM up to 16%, decreasing dietary AP and phytase addition did not affect egg mass (g/day/hen), egg weight (g), FCR (feed : egg) and egg production % (hen/day) ($P>0.05$); however, FI (g/day/hen) of hens fed on CM-included diets increased comparing with control (0% CM) diet ($P<0.05$). Egg traits including shape index, yolk index, yolk color, specific gravity and shell weight were not affected by dietary treatment ($P>0.05$); however, after first month of trail shell thickness was improved by dietary phytase supplementation ($P<0.05$).

no. 6

Effect of an exogenous enzyme supplementation and diets containing graded levels of raunchy whole date palm *Phoenix dactylifera* L. on performance of laying hens

M. Torki, H.R. Zangiabadi and V. Kimiaee, Animal Science Department, Agricultural Faculty, Razi University, Imam Avenue, 67155-1158, Kermanshah, Iran

To study the effects of graded replacement of dietary corn by whole dates (WD) supplemented with enzyme on laying hen performance, 336 Hy-line leghorn hens after production peak were randomly divided in 56 cages (n=6). Eight iso-energetic and iso-nitrogenous experimental diets (ME=2800 Kcal/kg and CP=15 g/kg) in a 4 × 2 factorial arrangement including four levels (0, 10, 20 and 30%) of WD with and without a β–mannanase-based enzyme were fed to hens with 7 replicates per diet during 6-week trial period. Collected data of feed intake (FI), egg production, egg mass and calculated feed conversion ratio (FCR) as well as egg traits were analyzed based on completely randomized design using GLM procedure of SAS. Dietary inclusion of WD more than 10% (i.e. 20 and 30%) decreased egg production % (hen/day), egg mass (g/day/hen) and FI (g/day/hen) comparing to corn-based diets ($P<0.05$); however, increase in FCR (feed: egg) was only significant during weeks 1 and 2. Dietary inclusion of 10% WD increased egg weight (g) comparing to control diet ($P<0.05$). Enzyme supplementation had no significant effect on egg production, egg mass and FCR ($P>0.05$). Egg traits including yolk index, specific gravity, shell thickness and shell weight were not affected by enzyme inclusion ($P>0.05$). Dietary inclusion of 10% WD increased thickness of egg shell comparing to 20 and 30% WD-included diets ($P<0.05$).

---

no. 7

Apparent metabolizable energy content of whole dates and date pits supplemented by a β-mannanase-based enzyme in broilers

M. Torki and H.R. Zangiabadi, Animal Science Department, Agricultural Faculty, Razi University, Imam Avenue, 67155-1158, Kermanshah, Iran

The aim of the present study was to investigate the influence of β-mannanase supplementation on nitrogen-corrected apparent metabolisable energy (AMEn) of whole dates and date pits in broilers. The AMEn content of whole dates and date pits with or without a β-mannanase-based enzyme (0 or 0.4 gHemicell®/kg diet) were evaluated in 100 day-old chicks by a diet replacement method using a completely randomized experimental design for 21 d period. The metabolism trail was conducted at two substitution levels 40% and 20%. The experimental diets were fed for 7 days with the first 3 days serving as an adaptation period and the last 4 days for daily excreta collection. Dietary enzyme inclusion increased AMEn from 2.732 to 3.121 for whole dates and from 0.932 to 1.212 Kcal/kg DM for date pits ($P<0.05$). It can be concluded that dietary supplementation by β-mannanase-based enzyme could improve the AMEn value of whole dates and date pits.

no. 8

The effect of β-mannanase supplementation of whole dates on performance of broiler chicks

M. Torki and H.R. Zangiabadi, Animal Science Department, Agricultural Faculty, Razi University, Imam Avenue, 67155-1158, Iran

An experiment with 3 × 2 factorial arrangement was carried out to determine effects of dietary inclusion of whole dates with and without β-mannanase on broiler performance. The main factors were three dietary inclusion level of whole dates (0, 175 and 350 g/kg diet) and enzyme supplementation. Three hundred and sixty day old male Arbor Acre broiler chicks of both sexes were randomly allocated into six iso-nitrogenous and iso-caloric dietary treatment groups. Six pens of birds (n=10) were randomly assigned to each of twelve dietary treatment groups. Body weight (BW) and feed intake (FI) was measured on 21 and 42. Data were subjected to analysis of variance as a completely randomized design using the GLM procedure of SAS. Birds fed on diet with 17.5% dates had higher BW than birds in other dietary groups; however, the differences were not statistically significant. Feeding diet with the highest level of dates decreased FI of chicks through 4 to 6 weeks of age ($P<0.05$). Dietary inclusion of dates had no adverse effect on feed conversion ratio ($P< 0.05$). Dietary enzyme supplementation improved BW gain as well as FCR of birds ($P<0.05$). Mortality did not differ significantly among the dietary groups ($P> 0.05$). Gross examination of various organs in feeding trial revealed no abnormalities.

---

no. 9

Supplementation of corn–barley-soy-based diets with a commercial enzyme preparation: effects on laying hen performance

M. Torki, M. Davoodifar and M. Pourmostafa, Animal Science Department, Agricultural Faculty, Razi University, Imam Avenue, 67155-1158, Kermanshah, Iran

To evaluate effects of enzyme supplementation of corn–barley-soy-based diets in two energy levels on performance, 216 Hy-line leghorn hens after production peak were randomly divided in 36 cages (n=6). Four iso-nitrogenous experimental diets in a 2 × 2 factorial arrangement including two levels (2750 and 2850 Kcal/kg) of ME with and without a β–mannanase-based enzyme were fed to hens with 9 replicates per diet during 8-week trial period. Collected data of feed intake (FI), egg production, egg mass and calculated feed conversion ratio as well as egg traits were analyzed based on completely randomized design using GLM procedure of SAS. Hens fed on low energy diets had higher FI (g/day/hen) comparing to high energy diets ($P<0.05$); however, enzyme inclusion did not affect FI ($P>0.05$). Dietary treatment had no effect on egg mass (g/day/hen), egg weight (g), FCR (feed: egg) and egg production % (hen/day) ($P>0.05$). Egg traits including shape index, yolk index, specific gravity, shell weight and shell thickness were not affected by dietary treatment ($P>0.05$); however, yolk color after second month of trail decreased in low energy diets comparing to high energy diets ($P<0.05$). Interaction of dietary energy level and enzyme addition on weight and thickness of egg shell was significant, so that in low energy diet with enzyme was the highest and the difference with the corresponding diet with no enzyme was significant.

no. 10

Effect of extenders on cryopreservation of Venda cock semen

M.L. Mphaphathi[1], M.B. Raito[1], D. Luseba[2], F.K. Siebrits[2] and T.L. Nedambale[1], [1]ARC-API, Germplasm & Reproduction Biotechnologies, P/Bag X2 Irene, 0062, South Africa, [2]TUT, Department of Animal Sciences, Pretoria, 0001, South Africa

Cryopreservation of indigenous cock semen is necessary for the maintenance of gene pool diversity. The aim of this study was to find a suitable extender and improve survivability and motility rate of frozen-thawed (F-T) Venda breed semen. A total of six Venda cocks were used for semen collection and the best one was chosen for the experiment. The abdominal massaging technique was used to collect semen. Individual ejaculates were diluted with modified Kobidil$^+$ (mK$^+$) and modified Bracket & Oliphant (mBO) extender at a ratio of 1:2 (v/v). The spermatozoa survivability and motility rate was evaluated by contrast microscope (BHTU). Semen was then diluted with mK$^+$ or mBO extender plus 8% dimethyl-sulfoxide (DMSO) and equilibrated for a total of 4 h at 5 °C. Semen was loaded into 0.25 mL plastic straws and frozen from 5 °C to -20 °C at the rate of 1 °C/minute. Finally, exposed to liquid nitrogen vapour for 5 minutes and then plunged into liquid nitrogen for storage. After a week, Frozen-thawed semen were evaluated for spermatozoa motility and survivability rate at 5 °C for 0, 30, 60 and 90 min. Data was analysed by ANOVA. The results showed that there was a significantly higher survival and motility rate of semen diluted with mK$^+$ than in mBO group at 0 to 90 min. In conclusion, Venda cock semen can be cryopreserved successful with mK$^+$ plus 8% DMSO. More studies are still needed to improve cock semen survival and motility rate.

---

no. 11

Inducing moulting by using non food removal methods in laying hens

A.S. Chaudhry and M. Yousaf, Newcastle University, Agriculture, Food & Rural Development, Agriculture Building, NE1 7RU, United Kingdom

Induced moulting can help extend the productive lives of laying hens and improve the economics of poultry production. Therefore, non-food removal methods were compared to induce moulting in hens: ad-libitum feed intake (FI) with either 16h (A) or 8h (B) photoperiods or 8h photoperiod with restricted feed (C) or C plus Aluminium (D). Ninety six hens were allocated to 32 groups of 3 hens each and housed in 32 2-tier cages under pre-planned photoperiods. Each method (Treatment) was tested with 24 hens in 8 cages per method. The hens were compared for weekly FI, LW, egg production and moult scores at $P<0.05$. Treatment B hens ate more and gained more LW than Treatment A hens for each week. Although the egg production reduced with weeks, Treatment B had more eggs than Treatment A in most weeks ($P>0.05$). Treatment C and D hens received restricted FI and lost more LW in the first week but less afterwards. Treatment A & B hens did not moult whereas Treatment C and D hens did moult with greater scores for D v C where D took longer than C to cease egg production. Conversely, the Treatment D hens started egg production earlier than Treatment C. It is possible to use feed restriction without or with aluminium at 8h photoperiod to induce moulting and extend the productive lives of laying hens.

no. 12

Metabolizable energy content of maize distillers' dried grains with solubles in practical or semi-purified diet for broiler chickens determined using the regression method

O. Adeola, Purdue University, Animal Sciences, 915 West State Street, West Lafayette, IN 47907, USA

This study determined the metabolizable energy content of maize distillers' dried grains with solubles (DDGS) in practical or semi-purified diets for broiler chickens by the regression method. Ross 308 broiler chickens were assigned to 6 diets consisting of diet type at two levels (practical or semi-purified nitrogen-free diet) and DDGS at three levels (0, 300, or 600 g/kg) arranged in a 2 x 3 factorial. Six birds per cage and 8 replicate cages per diet were fed for 7 d. Dry matter (g/kg), gross energy (kcal/kg), crude protein (g/kg), crude fat (g/kg), crude fiber (g/kg), and ash (g/kg) in the DDGS were 895, 4.811, 265.7, 107.6, 61.3, and 41.8, respectively. There were linear reductions ($P<0.0001$) in metabolizable and nitrogen-corrected metabolizable energy (kcal/g) contents of the diets from 3.615 and 3.414 to 2.753 and 2.642, respectively as DDGS increased from 0 to 600 g/kg in the practical diets. Corresponding linear reduction ($P<0.0001$) values for semi-purified diets were 3.21 and 3.227 to 2.732 and 2.697, respectively. Regression analyses of DDGS-associated metabolizable or nitrogen-corrected metabolizable energy intake in kilocalories against grams of DDGS intake suggest that the respective metabolizable and nitrogen-corrected metabolizable energy values (kcal/g) of the DDGS sample evaluated were 2.904 and 2.787 in practical diet, and 3.013 and 2.963 in semi-purified diet.

---

no. 13

The prevalence of gastro-intestinal parasites in village chickens in South Africa

M. Mwale and P.J. Masika, University of Fort Hare, Livestock and Pasture Sciences and Agricultural and Rural Development Research Institute (ARDRI), P. Bag X1314, Alice, 5700, South Africa

Village chickens are the predominant poultry species in African communal areas. They significantly improve the nutritional status and income of rural communities. However, chickens are mainly infested with gastro-intestinal parasites. A study was, thus, conducted in Centane district of South Africa to determine prevalence of these parasites in village chickens. Faecal samples (70) were randomly collected from chickens in two villages; Qolora by-sea and Nontshinga. Ninety-nine percent of the chickens were infested with gastro-intestinal parasites. Coccidia were the most prevalent in both villages; 41.43 and 25.71% in Qolora by-sea and Nontshinga respectively. Nematodes (1.43% prevalence), *Prosthogonimus* species and *Syngamus trachea* were found in Qolora by-seawhile *Subulura brumpti* and *Gongylonema ingluvicola* were in Nontshinga. The least prevalent were cestodes (1.43%); *Amaebotaenia sphenoides* in Qolora by-sea and, *Choanotaenia infundibulum* and *Davainea proglottina* in Nontshinga. The prevalence of parasites varied between the two villages ($P<0.05$). There was a high prevalence of gastro-intestinal parasites in village chickens in Centane district. Designing and implementing sustainable methods for controlling gastro-intestinal parasites is important for improved village chicken productivity and hence rural livelihoods. Future work will examine the seasonal prevalence of gastro-intestinal parasites in Centane district.

no. 14

The prevalence of gastro-intestinal parasites in village chickens in South Africa

M. Mwale and P.J. Masika, University of Fort Hare, Livestock and Pasture Sciences and, Agricultural and Rural Development Research Institute (ARDRI), P. Bag X1314, Alice, 5700, South Africa

Village chickens are the predominant poultry species in African communal areas. They significantly improve the nutritional status and income of rural communities. However, chickens are mainly infested with gastro-intestinal parasites. A study was, thus, conducted in Centane district of South Africa to determine the prevalence of these parasites in village chickens. Faecal samples (70) were randomly collected from chickens in two villages; Qolora by-sea and Nontshinga. Ninety-nine percent of the chickens were infested with gastro-intestinal parasites. Coccidia were the most prevalent in both villages; 41.43 and 25.71% in Qolora by-sea and Nontshinga respectively. Nematodes (1.43% prevalence), *Prosthogonimus* species and *Syngamus trachea* were in Qolora by-seawhile *Subulura brumpti* and *Gongylonema ingluvicola* were in Nontshinga. The least prevalent were cestodes (1.43%); *Amaebotaenia sphenoides* in Qolora by-sea and, *Choanotaenia infundibulum* and *Davainea proglottina* in Nontshinga. The prevalence of parasites varied between the two villages (P<0.05). There was a high prevalence of gastro-intestinal parasites in village chickens in Centane district. Designing and implementing sustainable methods for controlling gastro-intestinal parasites is important for improved village chicken productivity and hence rural livelihoods. Future work will examine the seasonal prevalence of gastro-intestinal parasites in Centane district.

---

no. 15

Antinutritional and nutritional values of differently processed *Alchornea cordifolia* seeds fed to broilers

O.O. Emenalom, A.B.I. Udedibie, B.O. Esonu, E.B. Etuk, B.A. Obiora and L.C. Nwaiwu, Federal University of Technology Owerri, Animal Science and Technology, PMB 1526 Owerri Imo State, Nigeria, Nigeria

Some antinutritional and nutritional characteristics of *Alchornea* (*Alchornea cordifolia*) seeds were investigated. Dehulled, Dehulled and fermented and undehulled and fermented, *Alchornea* seeds were screened for phytochemical compounds and incorporated into broiler diets fed 0-35days of age to replace 10% of maize. The matured seeds contained moderately high levels of phytic acid (973.8mg/100g) but this was reduced by dehulling and fermentation. The other antinutritional factors (tannins and anthraquinone) were eliminated by fermentation and dehulling respectively while saponins, alkaloids, steroids, flavonoids and cardiac glycosides but not cyanides were present in processed seed samples. Broilers fed raw *Alchornea* seed grew significantly (P<0.05) slower and gained 72.2% of the control. Feed intake declined significantly (P<0.05) only with the raw seeds but feed conversion ratios were unchanged. Inclusion of dehulled and fermented seeds promoted much better growth and feed intake than undehulled and fermented seed meal diet but the values were similar to that of control. Blood plasma levels of alanine amino transferase (ALT), alkaline phosphatase (ALP) and aspartate aminotransferase (AST) increased with raw seeds while serum calcium decreased. Raw *Alchornea* seeds hard the most detrimental effect as broilers fed this diet gained less weight during the expirement.

no. 16

Effect of transportation stress on blood parameters and weight loss in broilers

S.H. Raza, S. Mahmood, M.R. Virk and M. Rasheed, University of Agriculture, Faisalabad, Dept. Livestock Management, Dept. Livestock Management, University of Agriculture, Faisalabad, 38400, Pakistan

The study was conducted to determine the effect of transportation stress on blood parameters and weight loss in broilers. Twelve broiler farms were selected for the study depending upon the distance from the wholesale market. These farms were allotted to three treatment groups A, B and C (4 farms/group). Each group had overcrowded and normal spaced birds during transportation. The farms in group A were located within about 10 km radius from the poultry market of the city, whereas the farms in group B and C were located within 10-20 and >20 km radius, respectively. The data regarding the blood glucose, cholesterol, hemoglobin, heterophil to lymphocyte ratio, differential leukocyte count, Meat pH, cooking losses of breast meat were subjected to the analysis of variance (ANOVA) in Completely Randomized Design and means were compared by Duncan multiply range test. The transportation stress significantly decreased the weight of birds. There was linear decrease in the weight of broilers with increase in transportations distances. The transportation stress significantly effect on the hemoglobin, lymphoctes, monocytes, eosinophils, heterophils percentage and H/L ratio. The hemoglobin, monocytes, lymphocytes decreased while heterophils eosinophils perecentage increased during the stress conditions. However, glucose, cholesterol level in the blood and on the basophils percentage, meat pH and cooking losses remained unaffected.

---

no. 17

Avian influenza and vertically integrated contract poultry farming system in Bangladesh

M.J. Alam[1,2] and I.A. Begum[2], [1]Ghent University, 653 Coupure Links, A-098, 9000 Ghent, Bangladesh, [2]Bangladesh Agricultural University, Mymensingh, 2202, Bangladesh

Vertically integrated contract poultry farming system has already been recognized a system for small farm's income generation in Bangladesh. But outbreaks of Avian influenza (bird flu) in 2007 result in tremendous economic losses in the sector. Now the crucial question is whether small poultry farm can continue within the system or not. The paper attempts to assess the impact of avian influenza on small contract farm in Bangladesh. Vertically integrated contract farming system is well known in Bangladesh to expand commercial poultry farms, because the system help small farmers to get significant benefit in terms of income, employment and capital access. Moreover, the system offers an internal insurance scheme to cover the risk of loss in case of immature death of chicks by disease and other cogent reason. But in the period of bird flu incidence both integrator and contract farmers faced losses, many of farmers has quit from the system. The objective of the study is to find out how the contract farming system works after bird flu incidence especially contractual agreement, risk reduction and net return. Also to find out what can be incentive for small farmers to continue poultry farming under the contracting system? The primary data were collected from Aftab Bahumukhi (multipurpose) Farm Ltd., the pioneer vertically integrated farm in Bangladesh to figure out the answers of the objectives.

Carcass and digestive effects of Xtract and avilamycin on broiler in Mediterranean conditions

L. Mazuranok and D. Bravo, Pancosma Research, Voie-des-Traz 6, 1218 Geneva, Switzerland

This study investigates the effects of a natural plant extract on broiler carcass paramters in Mediterranean (Tunesia) conditions. The animals were allocated to 3 treatments fed on a maize and soya based diet: a negative control (C), avilamycin (AV) and a combination of 5% carvacrol, 3% cinnamaldehyde and 2% Capsicum oleoresin (XT) 100 g/t. The trial was set up over a 42-day period. XT statistically increased the hot and cold carcass yields in comparison to negative control (74.36% vs. 72.92%, P<0.009; 74.13% vs. 72.72%, P<0.006 respectively) and improved only numerically these 2 parameters compared to AV (74.36% vs. 73.35%, P<0.06; 74.13% vs. 72.72%, P<0.07 respectively). AV increased these 2 parameters versus C (73.35% vs. 72.92%; 73.21% vs. 72.72% respectively). Concerning the fat yield, there was no influence of the treatment (C 1.54%, AV 1.68%, XT 1.81%). For the liver percentage, there was also no statistical differences (C 1.88%, AV 1.84%, XT 1.96%). The digestive tract percentage was statistically improved by XT vs. C (6.51% vs. 7.69%, P<0.0004) and vs. AV (6.51% vs. 7.34%, P<0.009). AV decreased numerically this parameter in comparison to C (7.34% vs. 7.69%). To conclude, in Tunisian conditions, in maize and soya based diet, hot and cold carcass were improved by AV and XT vs. C. Natural plant extracts seem to be a reasonable alternative to AGP to promote performance in broiler.

Parentage verification of South African Angora goats using microsatellite markers

H. Friedrich, C. Visser and E. Van Marle-Koster, University of Pretoria, Department of Animal and Wildlife Sciences, 0002, Pretoria, South Africa

Incorrect and inaccurate pedigree information used by industry compromise the value of EBVs and decrease selection efficiency. This study aimed to evaluate a panel of microsatellite loci for its suitability in parentage verification in South African Angora goats. A total of 192 individuals representing different family structures were genotyped with 18 microsatellite markers. The power of resolution of the panel was assessed firstly when the genotypes of both parents and offspring were available, secondly when only the paternal and offspring genotypes were available and lastly when only the maternal and offspring genotypes were available. The statistical parameters were calculated with CERVUS 3.0 and Microsatellite Toolkit. The number of alleles per locus varied from two to eleven (with an average of 7.89), while the PIC values ranged between 0.332 and 0.784. The observed (H_O) and expected (H_E) heterozygosity ranged between 0.34 and 0.78 and 0.34 and 0.81 respectively. A final panel of 10 microsatellites was compiled based on their respective allele frequencies, number of alleles per locus, PIC and H_O. The average number of alleles per locus increased from 7.9 to 9.1 when the number of markers was decreased from 18 to 10. The total combined exclusionary power of this panel was 0.999 when both parents' genotypes were available and 0.989 when only one parent was genotyped. Our results indicate that this panel of markers is highly suitable for accurate parentage verification in the Angora goat.

Quantifying the relationship between birth coat score and wool traits in Merino sheep

W.J. Olivier[1], J.J. Olivier[2] and A.C. Greyling[3], [1]Grootfontein ADI, P/Bag X529, Middelburg (EC), 5900, South Africa, [2]ARC:API, P\Bag X5013, Stellenbosch, 7590, South Africa, [3]Cradock ES, P.O. Box 284, Cradock, 5880, South Africa

The aim of this study was to quantify the relationship between birth coat score and wool traits in a fine wool Merino stud. The data collected on lambs born within this stud from 1988 to 2003 were used for this analysis. The traits included in the analysis were birth coat score (BC), clean fleece weight (CW), mean fibre diameter (MF), staple length (SL), clean yield (CY), number of crimps (Crim), Duerden (Duer), coefficient of variation (CV), comfort factor (CF), wool quality (WQ) and variation over the fleece (WV). Birth coat score is done on a scale from 1 to 4 with 1 being woolly and 4 being hairy. Variance components and genetic parameters were estimated with the ASREML software. The estimated direct additive and maternal heritabilities and the permanent environmental effect of the dam for BC were 0.45±0.04, 0.03±0.02 and 0.06±0.02 respectively. The genetic correlations between BC and CW, MF, SL, CY, Crim, Duer, CV, CF, WQ and WV were 0.12±0.06, 0.09±0.06, 0.06±0.07, -0.13±0.06, -0.15±0.07, -0.06±0.07, 0.27±0.06, -0.34±0.06, -0.19±0.07 and -0.26±0.08 respectively. It is evident from the results of this study that the woollier lambs tend to have lower CV's and less fibres over 30μm (CF), as well as better wool quality and less variation over the fleece compared to the hairier lambs. BC can therefore possibly be used as early selection criteria to improve these traits.

---

no. 3

Can repeated superovulation and embryo recovery in Boer goats limit a donor participation in MOET programme?

K.C. Lehloenya[1], J.P.C. Greyling[2] and S. Grobler[2], [1]Tshwane University of Technology, Department of Animal Science, P/Bag X680, 0001, Pretoria, South Africa, [2]University of the Free state, P.O. Box 339, 9300, Bloemfontein, South Africa

This study evaluated the effect of repeated superovulation and embryo recovery rates in 15 Boer goat does. Does were synchronised for oestrus with controlled internal drug release dispensers (CIDR) for 17 days and superovulated with pFSH during the breeding season (autumn). Cervical inseminations (fresh undiluted semen) were performed 36h and 48h following CIDR removal and the embryos surgically flushed 6 days following the second AI. Does superovulated for the first time showed a shorter (P<0.05) mean duration of oestrus (20.8±10.1h), when compared to those repeatedly superovulated (30.4±6.7 h). The mean number of structures and embryos recovered were significantly (P<0.05) lower in the repeat-treated does (6.0±8.7 and 3.8±8.4) than does superovulated for the first time (12.9±5.0 and 11.7±5.0), respectively. The mean number of unfertilised ova per donor was significantly (P<0.05) higher in repeatedly-treated does (5.5±7.8), compared to does superovulated for the first time (0.1±0.3). The fertilisation rate and the number of transferable embryos were significantly (P<0.05) lower in the repeatedly-treated does compared to does superovulated for the first time. These results indicate that the number of times that a Boer goat doe can be utilised as donor would seem to be limited to three times.

Reproduction parameters for Holstein cows towards creating a fertility index

C.J.C. Muller[1], J.P. Potgieter[2], E.F. Dzomba[2] and K. Dzama[2], [1]Institute for Animal Production, Private bag x1, 7607 Elsenburg, South Africa, [2]University of Stellenbosch, Department of Animal Sciences, Private bag x1, 7602 Matieland, South Africa

Profitable milk production and genetic improvement in dairy herds depend largely on cows becoming pregnant again after calving. The reproductive performance of Holsteins in South Africa has declined as milk recording data indicate an increase in calving interval from 386 days in 1986 to 412 days in 2004. For herd management purposes, reproduction data on individual cows are collected on a routine basis. In this paper, reproduction parameters as possible indicators of fertility in dairy cows are presented. Insemination records (n=39200) of 14284 lactations of 5455 Holstein cows in eight herds were collected and a number of reproduction parameters determined. The mean (±sd) number of days from calving to first insemination was 87±47 days with 53% of cows inseminated within 80 days post calving. The number of inseminations, interval from first insemination to conception, percentage of cows pregnant after first insemination were 2.43±1.86, 59±93 days and 42%, respectively. The interval from calving to conception was 145±100 days and pregnancy rate 83%. From records obtained, it would be possible to determine a fertility index for dairy cows based on days open.

---

no. 5

Modelling the long-term consequences of undernutrition of cows grazing semi-arid range for the growth of their progeny

F.D. Richardson, University of Cape Town, Department of Mathematics and Applied Mathematics, Private Bag X 3, 7701 Rondebosch, South Africa

Under-nutrition of cows grazing semi-arid range as a result of drought or heavy stocking rates reduces prenatal and pre-weaning growth of their calves. A mechanistic model has been developed that simulates the productivity of vegetation and cattle on semi-arid savanna rangeland in Southern Africa. A soil moisture module partitions daily rainfall between runoff, infiltration and drainage and also simulates the loss of soil moisture by evaporation and transpiration. Forage production is modelled in relation to soil moisture and the present potential for growth. The model combines three mechanisms of food intake regulation: the rate at which the animal can eat forage, physical capacity of the digestive system, and, in young animals, their growth potential. Metabolisable energy intake is partitioned between maintenance, accretion/depletion of body protein and fat, conceptus growth and milk production. When cows are stocked at 0.123 cows/ha instead of 0.278 cows/ha the model predicted that empty body weights of their progeny at birth and weaning would be reduced by 3.6 and 55 kg respectively. If all young animals are stocked at 0.222 animals/ha after weaning, differences in empty body weight at weaning are predicted to persist almost unchanged for a further two years. Comparison of simulated results with data from an experiment in south-western Zimbabwe indicates that the model accurately simulates the long-term effects of rainfall and stocking rate on cattle growth.

Body weight growth trends in Döhne Merino lambs from birth to 16 month age

T. Rust[1], S.W.P. Cloete[2] and J.J. Olivier[3], [1]Dohne ADI, Animal Science, Stutterheim, 4930, South Africa, [2]Elsenburg ADC, Animal Science, Elsenburg, 7607, South Africa, [3]Irere AII, Animal Science, Private Bag X1, Matieland, 7602, South Africa

Repeated weight records of 997 Döhne Merino lambs born during 2001 up to 2007 were used to study live weight growth trends in Döhne Merino lambs from birth to 16 month age. ASREML was used for the analysis of the growth trend in a repeated record analysis. Fixed effects included in the operational model included the linear trends in a year of birth sex interaction, birth status and age of dam means. Simple orthogonal polynomials were fitted to model the overall curve and deviations from the curve at both genetic and permanent environmental levels. A cubic spline was used to model changes in growth data with age. The interaction of year of birth and sex of animal with the linear and curvilinear components of the spline was fitted. It was observed that the female animals grow at a reduced pace and that the tempo of growth slows down earlier in female animals over the 20 month period when compared to their male counterparts. By multiplying the covariance/variance/correlation animal matrix with the estimated coefficients for age, the animal variance component at each stage of growth, up to 16 months of age, is estimated.

---

no. 7

A farmer centred approach to the development small holder livestock systems in South Africa

M.D. Motiang and L.E. Matjuda,

Eemerging beef farmers own 40% of South Africa's beef herd but contribute only 5% to the beef market. This is mainly because of their production system that is characterized by low throughput and profit. Current stakeholder support systems tend to be discipline centred and seldom converge towards addressing business objectives of farmers. The Agricultural Research Council, Department of Agriculture and emerging farmers implemented the Beef Profit Partnerships Project, in Limpopo and North West, from 2001 till 2006, to study the effect of participative approaches on the emerging beef enterprises. Using a profit thinking framework, a multidisciplinary farmer support team (FST) was selected to implement the program in the two provinces. Four hundred farmers were organized into teams, which met once in 30 days to identify needs, set objectives, take action and report every 90 days to measure progress. Results show that farmers recognized that profit should be the main focus of a beef enterprise with all other activities oriented towards this goal. Farmers were eager to take action that had evidence that it improves profit. Prices for beef were improved by accessing new markets and price information while costs were reduced by collective transportation and eliminating less economic practices. FST also obtained in-depth understanding of technical challenges in farmers operations and assisted farmers to implement appropriate solutions. Results show that a farmer centred approach assist researchers and extension workers improve the performance of emerging beef farmers

The effect of tannin-rich feeds on gas production and microbial enzyme activity

F.N. Fon, I.V. Nsahlai and N.A.D. Basha, SASA, UKZN, Animal and Poultry Science, PMB, 3209, South Africa

Leaves and pods of *Acacia sieberiana* and *A. nilotica* were subjected to chemical analysis and *in vitro* fermentation with and without polyethylene glycol (PEG; 35 mg/g). Gas production (GP), true degradability (TD); and the proteolytic and fibrolytic enzyme activities in digesta were determined. Data analyses accounted for the effects of feed type, feed fraction, PEG and their interactions. Leaves and pods had similar NDF and ADF contents in *A. sieberiana* but differed in both attributes in *A. nilotica*. Leaves had higher protein but lower condensed tannin (CT) contents than pods. *A. siberiana* had more CT than *A. nilotica* but both had similar maximum GP. PEG increased ($P<0.01$) the maximum GP. The effect of PEG and feed type x PEG interaction tended to affect the global rate of GP ($P<0.1$); thus PEG increased the rate of GP only in *A. sieberiana*. PEG stimulated ($P<0.05$) GP from the soluble fraction. PEG increased the TD for leaves and pods of *A. siberiana* but suppressed TD in fractions of *A. nilotica* than in the untreated. PEG radically increased ($P<0.001$) the activity of protease enzymes, carboxymethyl cellulase and xylanases, indicating that tannin-rich feeds have suppressed activity of these enzymes. *A. nilotica* elicited lower protease activity but higher CMCase and xylanase activities than *A. sieberiana*. Pods elicited higher activities of the three enzymes than leaves.

---

no. 9

Influence of dietary energy level and production of breeding ostriches

T.R. Olivier[1,2], T.S. Brand[1,2], Z. Brand[3] and B. Pfister[3], [1]University of Stellenbosch, Department of Animal Science, Private Bag X1, 7602, Matieland, South Africa, [2]Institute for Animal Production: Elsenburg, Department of Agriculture, Private Bag X1, 7607, Elsenburg, South Africa, [3]Institute for Animal Production: Oudtshoorn, Department of Agriculture, P.O. Box 351, 6620, Oudtshoorn, South Africa

Six different diets varying in ME content (7.5, 8.0, 8.5, 9.0, 9.5 and 10 MJ ME/kg) were provided to breeding ostriches at an average rate of 3.4kg/bird/day. Dietary protein and lysine levels were held constant respectively at 12% and 0.58%. The trial ran over one breeding season and production records taken included egg production, chick production, number of infertile eggs, number of dead-in-shell eggs, number of eggs set and live weight change of breeders. No significant differences ($P>0.05$) were observed for total eggs produced per female per season (44.81 ± 7.8), number of chicks hatched (15.43 ± 4.1), number of infertile eggs (12.1 ± 4.0) and number of dead-in-shell eggs (12.1 ± 3.2). Average live weight change of female breeders was 10.65 ± 3.6kg, which did not differ significantly ($P>0.05$) between treatments. Significant differences ($P<0.05$) in the live weight changes of male breeders were observed which ranged between 6.27kg and 18.43kg over the breeding season. It was concluded that the production of breeding female ostriches was not influenced by dietary ME levels in this trial. The highest dietary ME content resulted in the biggest live weight change of male breeders, probably due to higher fat accretion in the body.

In vivo validation of ethno-veterinary practices used by rural farmers to control cattle ticks in the Eastern Cape Province of South Africa

B. Moyo and P.J. Masika, University of Fort Hare, Department of Livestock and Pasture Sciences, ARDRI, Faculty of Science and Agriculture, P. Bag X 1314, Alice 5700, South Africa

Ticks are one of the major health problems of livestock; they are vectors of TBDs and cause tick worry to livestock. Ticks are commonly controlled using conventional acaricides, which are expensive to the rural farmers, resulting in them resorting to alternative control methods. The objective of this study was to validate the acaricidal properties of some materials (*Ptaeroxylon obliquum*, *Aloe ferox*, *Lantana camara*, *Tagetes minuta*, used engine oil and Jeyes fluid (carbolic acid 13%)) used by rural farmers in the ethno-veterinary control of cattle ticks in the Eastern Cape Province of South Africa. A total of 52 cattle at Mdeni diptank in Amathole District Municipality were divided into 13 experimental groups with 4 cattle in each group. The Jeyes fluid at 76.8% and used engine oil had an efficacy greater than 70% that compared well with the positive control Ektoban® (Cymiazol 17.5 and cypermethrin 2.5%) ($P<0.05$). *Lantana camara* at 40% had an efficacy of 57% while *A. ferox*, *P. obliquum* and *T. minuta* were not effective. This study has revealed that the materials rural farmers use as acaricides vary in efficacy to control ticks, some are as effective as the conventional acaricides whereas others are not effective at all. Despite being effective some of the materials have potential toxic effects in animals and are also environmental contaminants.

---

no. 11

Opportunities for improving Nguni cattle production in the smallholder farming systems of South Africa

C. Mapiye[1], M. Chimonyo[1], K. Dzama[2] and M. Mapekula[1], [1]University of Fort Hare, Livestock and Pasture Sciences, P. Bag X1314, 0027, South Africa, [2]Stellenbosch University, Department of Animal Sciences, P. Bag X1, Matieland 7602, 0027, South Africa

A total of 218 structured questionnaires were administered to identify the constraints and opportunities of smallholder cattle farmers in the Eastern Cape Province of South Africa. Cattle were ranked as the most important livestock species and were mainly used for cash, milk and ceremonies. Cattle herd sizes were higher ($P<0.05$) in the small-scale (23±5.2) compared to the communal (9±3.1) areas. Over half of the farmers reported the shortage of grazing in winter. There was an association between rangeland type and proportion of the farmers who provided supplementary feeding ($P<0.05$). Non-descript and Nguni were the common cattle breeds in the smallholder areas. African tradition worshippers had higher ($P<0.05$) Nguni herd sizes (6±3.2) than Christians and non-worshippers (1±0.5). Tick-borne diseases were the popular causes of cattle mortality, especially in summer. Although the importance of constraints varied with production systems and rangeland types; shortage of feed and prevalence of tick-borne diseases were ranked as the most important constraints. Farmers suggested use of locally available indigenous feed and medicinal resources as possible options for improving cattle production in the smallholder areas. It was concluded that farmers' socioeconomic and pedo-climatic situations should be considered when planning strategies for smallholder livestock development.

Genetic parameter estimates for certain production traits of ostriches

M.D. Fair[1], J.B. Van Wyk[1] and S.W.P. Cloete[2], [1]University of the Free State, Animal, Wildlife and Grassland Sciences, P.O. Box 339, 9301, South Africa, [2]Elsenburg Agricultural Development Institute, Private Bag X1, Elsenburg 7607, South Africa

Data involving monthly records of egg production (EP), chick production (CP), hatchability (H), mean egg weight (MEW) and mean day old chick weights (MCW) were analysed using REML procedures. All traits were treated as hen traits. The random effects of animal, temporary environment (TE – unique hen within a year) and service sire (SS) were estimated from the data. Heritability estimates (±s.e.) were 0.12±0.02 for EP, 0.14±0.02 CP, 0.55±0.03 for MEW, 0.61±0.02 for MCW and 0.14 ±0.02 for H. Corresponding estimates for TE were 0.16±0.01, 0.15±0.01, 0.11±0.01, 0.08±0.01, and 0.14±0.01. The effect of SS was significant but relatively low for all traits, ranging from 0.04±0.01 for MCW to 0.07±0.01 for CP. Monthly EP and CP were highly correlated at all levels, ranging from 0.75±0.06 for the SS correlation to 0.88±0.03 for the genetic correlation. EP was favourably correlated with H on a genetic level (0.55±0.19). The genetic correlations of EP and CP with MEW and MCW were variable and low but not antagonistic (negative) as is often found in poultry. The genetic correlations of H with MEW and MCW were positive (0.18±0.08 and 0.20±0.08 respectively). As expected, the genetic correlation of MEW and MCW was very high at 0.90±0.01. Selection for reproduction is unlikely to be complicated by unfavourable correlations with H, MEW and MCW.

---

no. 13

Evaluation of the genetic performance of goats in an on-field research environment

C.S.O. Otoikhian, Ambrose Alli University Ekpoma, Animal Science, p/m.b. 14, 00234 Ekpoma, Nigeria

Birth weight records of kids produce by breeding dams covering a period of twelve years (1987- 1999) at Ambrose Ali university teaching and research farm Ekpoma, Edo State were analyzed. The genetic factors which included type of birth, birth weight and sex of kids were determined. The result revealed a significant difference (P< 0.05). the average birth weight within the experimental period was 1.18kg. it was also observed in this study that there were variation on birth weight between years with male s and females born single recording the highest birth weight of 1.89kg and 1.59kg, respectively, while average litter size for single, twins and triplet, were 1.45, 1.03 and 1.21 respectively. Twinning had the highest occurrence with 38.46%, sex ratio of 1.1 male to female was observed with male kids showing heavier birth weight than female kinds within and between Dam/sive analysis. Based in these estimate and on the fact that there is a positive correlation between birth weight, type of birth and growth rate, substantial improvement can be achieved when our indigerious breed (WAD) goats with good and promising genetic make-up, higher birth weight and improved type of birth (Triplet) are selected and reared in a controlled environment as a nucleus breed towards the production of improved breeding stock that could alleviate the present demand for small ruminant animal products as human and animal protein source.

Getting through the chasm: adoption and institutionalization mechansisms for improving and maximizing the benefits from agricultural research and development interventions

T.P. Madzivhandila, N.B. Nengovhela and R.A. Clark, New South Wales Government, Primary Industries, The UNE, 2351, Armidale, Australia

The real benefits from agricultural research and development (R&D) interventions depend on: (1) the size of the improvements flowing from the adoption of the outputs of the R&D per enterprise; (2) the rate of adoption; and (3) the extent or scale of adoption. If R&D projects result in outputs but no outcomes then, however efficient the projects may have been, they are not effective, and it is therefore difficult to justify public investment in them. Separately, the simple fact is that a new innovative practice has no actual economic value to an industry until it is adopted by individual practitioners and end-users. Any sustainable improvement and innovation effort is fundamentally a change effort, and change is difficult. Change in human activity systems is achieved by people changing their decisions and practices, and a purposeful change requires a process specifically designed to achieve the required outcomes. In addition, getting a new idea adopted is difficult. Many innovations require a lengthy period to be widely diffused and adopted even when they have obvious advantages. There should be a way to speed-up the adoption of a new practice. This paper identifies and discusses a range of mechanisms for improving and maximizing the benefits from agricultural R&D interventions in animal production industries.

---

no. 15

Welfare assessment of urban draught horses in the south of Chile

T. Tadich[1], R.A. Pearson[1] and A. Escobar[2], [1]University of Edinburgh, Division of Veterinary Clinical Sciences, Royal (Dick) School of Veterinary Studies, Roslin, EH25 9RG, United Kingdom, [2]Universidad Austral de Chile, Fac. Cs. Veterinarias, casilla 567 Valdivia, 5090000, Chile

Animal power continues to be an important resource in agriculture and urban transport in developing countries. This source of power needs to be used without risking the animal's integrity. This study considered aspects of health, behaviour and husbandry in order to assess the welfare status of 61 draught horses. Horses were of a local sturdy type. They were generally in good body condition (59%) and alert (92%), with only a few showing avoidance/aggressive responses (31%) when behaviour was assessed. The majority of horses did not present lesions (57%) and were in good health condition. The presence of molar hooks (34%) on the teeth was one abnormality found, which could influence food processing. Husbandry problems were related to hoof management and feeding practices. No association was found between use of a farrier and good hoof conformation or with owners' experience of keeping horses. The use of two-wheeled shafted carts is the common practice. Harnessing was not a major problem, but design could be improved by the use of swingletrees. The low incidence of health problems and negative behaviour responses found are positive outcomes. They show that owners are aware of the need of keeping their horses in good conditions to achieve efficiency at work. Improvements could be achieved by facilitating veterinary services and by training owners in aspects of harnessing and farriery.

no. 16

Validating indigenous veterinary practice in Cameroon

H. Taboh[1], H. Njakoi[1], J. Ajuh[1] and N. Tsabang[2], [1]Heifer Internation Cameroon, BP 467, Bamenda, Cameroon, [2]Institute of Medical Research and Studies of Medicinal Plants, BP 6163, Yaounde, Cameroon

Heifer Cameroon-HC works with resource limited families enhancing their socio-economic status through integrated animal agriculture in environmentally sustainable ways. Animal productivity constraints, animal welfare problems, expensive/erratic orthodox veterinary services, unsustainable exploitation of medicinal plants and knowledge transfer gap are challenges amidst the rich plant diversity and indigenous animal healthcare practices. HC uses these potentials to overcome the constraints through ethnoveterinarian's (EV) indigenous knowledge while conserving biodiversity and promoting environmentally friendly practices. Ethnobotanic study conducted in twelve groups with goats, sheep and pigs in Centre Province of Cameroon. Twenty nine EV selected and interviewed by ethnobotanist and veterinarians. Collected plant species identified by scientific and vernacular names, dried and mounted, in herbarium. Veterinarians identified animal diseases from symptoms and treatment described. Identified seventy five plants, twenty seven diseases and one hundred and eight recipes. *Euphorbia hirta, Ocimum gratissimum* and *Bidens pilosa* have anti-amoebic and anti-diarrhoea extract; cough suppressant extract (thymol) and antibiotic extract; *Alchornea cordifolia* effective against anaemia in piglets. Potential first line treatment of disease conditions in goats, sheep and pigs. Intervention of veterinarian still has its role.

---

no. 17

Effect of different extenders and preservation period on the survival rate of South African indigenous ram semen stored at 5 °C

P.H. Munyai[1], M.L. Mphaphathi[1], M.B. Raito[1], M.B. Makhafola[1], C.W. Mantiziba[1], L.M. Schwalbach[2], J.P.C. Greyling[2] and T.L. Nedambale[1], [1]Arc-Irene, Germplasm & Reproduction Biotechnologies, P/Bag X2, 0062 Irene, South Africa, [2]University of Free State, P.O. Box 339, 9300 Bloemfontein, South Africa

A long-term storage effort to develop a simple and suitable extender for cryopreservation of ram semen is crucial. The aim of the study was to determine the effect of breed, extenders (EYC (egg yolk Citrate), YET (egg yolk Tris) and mK (modified Kobidil$^+$) and preservation periods following storage of ram semen at 5 °C. Four ejaculates were collected by use of electro-ejaculator technique from each Damara, Pedi, and Zulu ram breed. Collected semen was transported to the laboratory for evaluation within an 1 h. Semen was diluted equally with EYC, YET and mK at a ratio of 1:2, v/v) and stored at 5 °C for 0, 3, 6, 9, and 24 h. Sperm parameters were recorded. Data was analysed by ANOVA. Preliminary results of this study demonstrated no significance differences between breeds (Damara, Pedi, and Zulu) and extenders (EYC, YET and Kobidil$^+$) at 0 to 9 h. Extended ram semen with mK resulted in a lowest survival and motility rate at 24 h compared to EYC and YET groups. In conclusion, this study suggest that breed and extenders do not affect survival and motility rate of ram semen stored at 5 °C from 0 to 9 h. Modified Kobidil$^+$ is not a suitable extender for ram preservation at 5 °C. Further studies are needed to freeze ram semen from EYC group.

Piglets born from frozen-thawed epidiymal sperm obtained *post mortem*: a case study

S. Foss[1], N.L. Nedambale[1], J.P.C. Greyling[2] and L.M.J. Schwalbach[2], [1]ARC-Animal Production Institute, Indigenous Germ Plasm and Biotechnology Development, Private Bag X2, 0062 Irene, South Africa, [2]University of the Free State, Department of Animal, Wildlife & Grassland Sciences, P.O. Box 339, 9300 Bloemfontein, South Africa

The aim of this study was to investigate the possibility of collecting and freezing epididymal sperm from a boar post-mortem, for AI in sows. Minutes after a valuable Duroc boar had been humanely euthanized due to injury, the testes were removed and transported to the laboratory in a saline solution at 37 °C. In the laboratory, the sperm in the epididymis were aspirated, evaluated and extended using Acromax® diluent-solution containing 33% egg yolk and 13.8% glycerol in the ratio of 2:3 and cryopreserved and stored in liquid nitrogen. Eighteen months later, two semen doses were thawed at 50 °C for 13 seconds and inseminated into two sows, 22 hours after standing oestrous. Pregnancy diagnoses were performed 35 days later by ultrasonography and one of the sows confirmed pregnant. This sow subsequently farrowed 3 live piglets, from which two piglets were successfully weaned. The results indicate that post-mortem epididymal porcine sperm can be successfully harvested, cryopreserved and used for A.I to produce viable piglets. The trial has demonstrated that it is still possible to salvage gametes post-morten and induce pregnancy in sows by AI using this frozen-thawed sperm. Further research is warranted to improve the efficiency of these techniques in boars.

---

no. 19

Brilliant cresyl blue staining improve the quality of selected indigenous cattle oocytes and their developmental capacity to blastocyst *in vitro*

M.B. Raito[1], M.L. Mphaphathi[1], L.M. Schwalbach[2], J.P. Greyling[2] and T.L. Nedambale[1], [1]ARC-API, Germplasm & Reproduction Biotechnologies, P/Bag X2 Irene, 0062, South Africa, [2]UOFS, Department of Animal Science, Bloemfontein, 9300, South Africa

Subjective morphological assessment of oocytes based on the thickness and cumulus cells compactness has been a popular way of evaluating oocyte quality. The aim of this study was to test the effect of Brilliant Cresyl Blue (BCB) staining on selecting competent cattle oocytes for further development *in vitro*. Following aspiration, 623 oocytes were exposed to 26 μM BCB for 90 min at 39 °C. Then, 278 oocytes were matured immediately and served as a control. Exposed oocytes were classified according to their cytoplasm colouration; 223 oocytes classified as BCB^+ (blue cytoplasm, grown oocytes) and 400 oocytes were classified as BCB^- (without a blue cytoplasm, growing oocytes). After *in vitro* maturation, fertilization and culture of all oocytes; cleavage, morula and blastocyst rate were recorded. Blastocysts were stained with Hoechst 33342 for cell number counting. Data were analysed by ANOVA. Cleaved and 8-cells embryo rate were higher for BCB^+ (83 and 36%) compared to BCB^- (60 and 18%) group. Significantly higher morula and blastocyst rate was recorded in BCB^+ (30 and 18%) than in BCB^- (10 and 6%) group. The average cell numbers of Day 7 blastocysts from control and BCB^+ group were higher than in BCB^- group. In summary, this study demonstrated that BCB staining is a useful tool to select better quality oocytes, resulting in better quality blastocysts.

no. 20

Effect of addition of bsa to extenders on survival time and motility rate of South African landrace boar semen

C.W. Mantiziba[1], M.B. Raito[1], L.M. Schwalbach[2], J.P.C. Greyling[2] and T.L. Nedambale[1], [1]ARC Animal Production, Germplasm & Reproduction Biotechnologies, P/Bag x2, 0062 Irene, South Africa, [2]University of Free State, Animal, Wildlife and Grassland Sciences, P.O. Box 339, 9300 Bloemfontein, South Africa

Long-term storage efforts to develop a simple, contamination free and efficient extender for cryopreservation of boar semen remain limited.The aim of the study was to determine the effect of addition of bovine serum albumin (BSA) or egg yolk (EY) on semen survival time and motility rate at 5 °C and 25 °C. Four ejaculates were collected from landrace boar with a gloved hand technique. Within 1 h of collection, semen was transferred to the laboratory for evaluation. Semen was diluted equally with Citrate + BSA, Citrate + EY, Kobidil + BSA, and Kobidil + EY (1:1, v/v) and stored at 5 °C or 25 °C for 0, 3, 6, 9, and 24 h. Sperm survival and motility rate was examined every 3 h. Data was analysed by ANOVA. The effect of addition of BSA on Citrate extender significantly resulted in a lowest sperm survival and motility rate at 5 °C. The survival time and motility rate of the sperm diluted in Citrate or Kobidil with BSA or EY was higher regardless of time storage at 25 °C. In conclusion, addition of BSA to Citrate or Kobidil had similar sperm survival and motility rate compared to egg yolk group at 25 °C. The results of this study suggest that landrace spermatozoa are sensitive to storage at 5 °C regardless of extender and source of protein supplement.

---

no. 21

Effect of different levels of rapeseed meal with enzyme in diets formulated based on digestible amino acid on performance and carcass characteristics of broiler chicks

M. Toghyani[1], A. Mohammadsalehi[1], A.A. Gheisari[2] and S.A. Tabeidian[1], [1]Islamic Azad University, Khorasgan Branch, Department of Animal Science, Esfahan, 8159659897, Iran, [2]Esfahan Agricultural Research center, Department of Animal Science, Esfahan, 81598, Iran

One hundred and ninety two day old chicks (Ross 308) were used in a completely randomized design with four treatments and four replicates for each treatment. The experimental treatments included the levels of 0 (control), 5, 10 and 15% of rapeseed meal in corn-soybean diets fed to chickens at starter period (0-11 d). Except for the control group, these levels for grower period (11-28 d) were increased to 10, 15 and 20% and for finisher period (28-42 d) increased to 15, 20 and 25% respectively. A commercial multi enzyme (contains: amylase, protease, lipase, cellulase, phytase, xylanase and β-glucanase) was added to all the diets (except of control) at a rate of 100 ppm. The diets were formulated on the basis of digestible amino acid according to Ross manual catalogue. At 42 d, two birds from each replicates were selected, weighed, slaughtered and carcass, abdominal fat, liver and pancreas removed. The results from this study indicated we can use rapeseed meal with enzyme at levels of 15, 20 and 25%, respectively, in starter, grower and finisher in diets formulated based on digestible amino acids, without having any adverse effects on performance and carcass characteristics of broiler chicks.

Ruminal and post ruminal digestibility of two different dietary fats by dairy bulls

S. Kargar, G.R. Ghorbani and M. Alikhani, isfahan university of technology, Animal sciences, College of Agriculture, 84156 isfahan, Iran

Two Holstein dairy bulls fitted with ruminally and duodenally cannula were used to examine the effects of supplementing hydrogenated palm oil (HPO) and yellow grease (YG) and different ratio of forage to concentrate (F:C) on *in situ* digestibility of nutrients. Treatments consisted of 1) Control, without supplemental fat and 34:66 F:C ratio, 2) 2% of HPO and 34:66 F:C, 3) 2% of YG and 34:66 F:C ratio, and 4) 2% of YG and 45:55 F:C ratio (YGHF). Feeding HPO decreased ruminal digestibility of DM compared with control diet ($P<0.05$). Also, digestibility of NDF, ADF and Cellulose were not affected by both type of fat and different ratio of F:C, with exception of cellulose digestibility that increased significantly in YGHF diet. Post ruminal digestibility of DM was decreased in YGHF diet ($P<0.01$). Total tract digestibility of DM, ADF did not influence by treatments, but due to lower digestibility of NDF in HPO treatment, feeding fat significantly decreased NDF digestibility in total tract. Furthermore, level of F: C significantly affected total tract digestibility of NDF, ADF and cellulose. Results of this study indicated that compared with HPO, feeding yellow grease in diets based on alfalfa hay as sole source of forage did not have detrimental effects on ruminal, post ruminal and total tract digestibility of fiber and its constitute.

---

no. 23

A genetic persistency index for the South-African dairy industry

B.E. Mostert, R.R. Van Der Westhuizen and H.E. Theron, ARC Animal Production Institute, Private Bag X2, 0062 Irene, South Africa

Persistency of milk production refers to the flatness of the lactation curve. The economic value of persistency in dairy cattle from the expected impact of improved persistency on feed cost, milk returns, health and reproductive costs, are estimated to be around 7% of the economic value of milk. Procedures have been developed for calculation of a genetic persistency index, based on the Canadian index, for South African dairy breeds. This index indicates the decline in production from 60 until 280 days of lactation. Data used in this study is test-day records of the first three lactations, as included in the National Dairy Genetic Evaluations of South Africa, of the Ayrshire, Guernsey, Holstein and Jersey breeds. Interpolation, using the Wilmink curve, is done on these test-day records to calculate 60-day and 280-day yields for each cow and lactation. Variance components have been estimated for these yields to be used in breeding value estimations, using bivariate evaluations and a repeatability model. To obtain independence of persistency from total milk yield, 305-day milk yield breeding values were included as covariates in the model for estimation of these breeding values. Persistency indices were then calculated, using these breeding values for all sires having at least one daughter with a completed lactation.

no. 24

The effect of supplementary light on the productive performance of Dorper lambs fed intensively

P.J. Fourie, P.J.A. Vos and S.S. Abiola, Central University of Technology, Free State, Agriculture, 20 Pres Brand Street, 9300 Bloemfontein, South Africa

The objective of this research study was to quantify the differences in average daily gain (ADG), back fat thickness (BFT), eye muscle area (EMA), fat thickness (FT) on different body parts, feed conversion ratio (FCR) and body dimensions (by means of body measurements) of Dorper lambs exposed to supplemented light. For this study hundred and twenty Dorper lambs (115±10 days old) weighing (29.76±5.01kg) were used. The lambs were randomly divided into three homogeneous groups (20 castrates and 20 intact males). The three groups were then exposed to different levels of supplemented light at 145 lux (16h, 24h and normal photoperiod). The animals were fed *ad libitum* with pellets containing 9.5 MJ ME/kg DM and 12% CP in open pens. The animals were weighed every 7 days while ultrasound scanning of the EMA and the BFT was done at the beginning and the end of the 35 day trial. The ADG, FCR and feed intake (FI) were calculated at the end of the trial. Linear body measurements including shoulder height, body length and heart girth and were taken at day 1 and day 35 respectively. All the animals were slaughtered at the end of the trail. The carcasses were then weighed, graded and the FT was measured with a caliper. The final results of the study concluded that no significant differences exist between the three treatment groups in terms of body measurements, ultrasound scanning ADG and FCR.

---

no. 25

Effect of heat stress on six beef breeds in the Zastron district: the significance of breed, coat colour and coat type

L.A. Foster[1], P.J. Fourie[1], F.W.C. Neser[2] and M.D. Fair[2], [1]Central University of Technology, Free State, Agriculture, Private Bag X20539, 9300 Bloemfontein, South Africa, [2]University of the Free State, Department of Animal, Wildlife and Grassland Sciences, P.O. Box 339, 9300 Bloemfontein, South Africa

This study was done to ascertain whether breed, coat colour and or coat type has the greatest influence on an animals susceptibility to heat stress. A system of subjective scoring of cattle coats, ranging from extremely short to very woolly and coat colours in the range from white to black has been described. Ten heifers of each breed participated in the study: Afrikaner, Bonsmara, Braford, Charolais, Drakensberger, and Simmentaler. Sampling for coat score and coat colour was carried out twice – in August (winter) and December (summer) – the same 10 animals of each breed being sampled in each season. Temperature Humidity Index (THI) combines effects of temperature and humidity into one value and is used to determine environmental warmth. Rectal temperature (RT) was used as a parameter to determine heat stress. Rectal temperatures of the same 60 heifers which did not experience heat stress during August 2007 (THI < 70, 10-day period) and which experienced heat stress during December 2007, (THI > 70, 10-day period) were determined. During both winter and summer significant differences (P<0.0001) in RT were observed between breeds. However, no explanation for differences in RT between breeds could be determined as none of the parameters measured had a significant influence in RT.

no. 26

Prediction of carcass traits using live animal ultrasound

I.C.W. Maribe[1], W.J. Olivier[2] and N.H. Casey[3], [1]Northern Cape Department of Agriculture and Land Reform, P/Bag X5018, Kimberley, 8300, South Africa, [2]Grootfontein ADI, P/Bag X529, Middelburg (EC), 5900, South Africa, [3]Department of Animal and Wildlife Science, University of Pretoria, Pretoria, 0002, South Africa

The aim of the study was to evaluate the use of live animal ultrasound measurement of the *Longissimus dorsi* area as a possible predictor of carcass quality and to determine the accuracy of these measurements. Data recorded on the animals of the Carnarvon Afrino flock, Carnarvon Dorper flock and Carnarvon Namaqua Afrikaner flock. The phenotypic correlations and repeatabilities were estimated with SAS statistical software. The traits included in the analysis were body weight (BW), ultrasound muscle area (UMA) and ultrasound fat thickness (UFT) measured at the 13th rib with a Pie Medical 100 Falco Ultra Sonic Scanner. Slaughter weight (SW), carcass weight (CW), dressing percentage (DP) and the actual *L. dorsi* area (MA) were also recorded on the Dorper lambs. The repeatabilities estimated for UMA were 0.18±0.02, 0.17±0.02 and 0.50±0.02 for the Afrino, Dorper and Namaqua Afrikaner lambs respectively. The respective values for UFT were 0.00±0.01, 0.00±0.01 and 0.10±0.02 and for BW 0.73±0.01, 0.76±0.01 and 0.78±0.01. The phenotypic correlations between SW and UMA, SW and MA and UMA and MA were 0.22±0.00, 0.14±0.04 and 0.16±0.01 respectively. The low correlation between the ultrasound muscle area (UMA) and the measured muscle area (MA) needs some further investigation.

---

no. 27

Evaluation of cashmere production of the meat producing Boer goat

J.A. Roux[1], W.J. Olivier[2], C. Trethewey[3] and A.F. Botha[4], [1]Cradock ES, P.O. Box 284, Cradock, 5880, South Africa, [2]Grootfontein ADI, P/Bag X529, Middelburg, 5900, South Africa, [3]Adelaide ES, P.O. Box 142, Adelaide, 5760, South Africa, [4]CSIR, P.O. Box 1124, Port Elizabeth, 6000, South Africa

South Africa does not produce cashmere commercially and presently there is no local cashmere industry. The Boer goat is well known for its meat production and is an established farming enterprise in South Africa. The aim of this study was to evaluate the cashmere production of the experimental Boer goat flock at the Adelaide Experimental Station. Data collected on the Adelaide Boer goat flock from 1996 to 2000 were used for the study and analysed with SAS statistical software. The traits included in the analysis were cashmere length on the shoulder (SL), rib (RL) and hind leg (HL), cashmere production (CP), fibre diameter of cashmere and guard hair, cashmere yield and percentage fibre diameter classes. The CP per Boer goat was 15.25±0.48 g, which is very low and not economically worthwhile to harvest. The fibre diameter of the cashmere was 16.93±0.06 µm and falls within the definition of cashmere (< 18.5 µm). The SL, RL and HL were 20.25±0.39 mm, 21.34±0.38 mm and 24.18±0.33 mm respectively, which is shorter than the minimum length of 40 mm required by the industry. Initiating a cashmere industry with pure Boer goats does not seem to be a viable option. However, the production per goat could possibly be increased through crossbreeding with a known cashmere producing breed.

no. 28

Production and breeding performance of South African dairy cattle

H.E. Theron and B.E. Mostert, ARC-Animal Improvement Institute, P/Bag X2, Irene, 0062, South Africa

Technological advances in dairy cattle farming have significantly changed production processes on dairy farms worldwide. There is a distinct trend toward more cows per farm and mechanized confinement feeding on the one hand and a lower cost and production option that rely on grazing systems on the other hand. Comparisons between production and breeding potential on different feeding systems (Total mixed ration (TMR), Mixed and Pastures) in South Africa were made. Data of active cows participating in performance testing as in November 2007 were used. Holstein cows numbered 68280 in 254 herds and Jersey cows 51275 in 248 herds. Average milk production, number of lactations and inter-calving period was 8147±2260, 2.9±1.8 and 318±198 for Holstein and 5347±1156, 3.1±2.0 and 306±184 for Jersey, respectively. Holstein herds are equally spread among the regions, while Jerseys are concentrated in the Western and Eastern Cape. Most Holstein and Jersey herds (53%) are on mixed rations, followed by TMR (28%) and pastures (19%). Milk production was 9967±2022; 6996±1623 and 7143±1549 kg for Holstein and 6385±1233; 5155±955 and 4753±1022 kg for Jersey cows respectively, for TMR, Mixed and Pasture systems. Most sires used are local. Imported Holstein sires are mostly from USA and Netherlands, and foreign Jersey sires are mostly from the USA. Sires do not seem to have been selected on production breeding values. Farmers seem to select the same sires on the different feeding regimes. Differences between feeding regimes were significant.

---

no. 29

Variation in *Longissimus dorsi* muscle area in South African Merino and Dorper sheep

I.C.W. Maribe[1], W.J. Olivier[2] and N.H. Casey[3], [1]Northern Cape Department of Agriculture and Land Reform, P/Bag X5018, Kimberley, 8300, South Africa, [2]Grootfontein ADI, P/Bag X529, Middelburg, 5900, South Africa, [3]Department of Animal and Wildlife Science, University of Pretoria, Pretoria, 0002, South Africa

The aim of this study was to determine the variation in *M. Longissimus dorsi* area in South African Merino and Dorper sheep at selection age. The *L. dorsi* muscle area and fat depth of 702 Merino lambs aged 14-16 months and 304 Dorper lambs aged 10-12 months was measured with ultrasound at the 13th rib with a Pie Medical 100 Falco Ultra Sonic Scanner. This was done on one Merino farm in the Cradock district and two Merino and three Dorper farms in the Middelburg district. The data recorded on these animals were ultrasound muscle area (UMA) and ultrasound fat thickness (UFT). The data were analyzed with SAS statistical software. The average UMA for the Merino lambs were 8.99±0.05 mm, ranging from 4.25 mm to 11.96 mm, and the average UFT was 0.26±0.00 mm, ranging from 0.05 mm to 0.49 mm. The corresponding UMA value for the Dorper lambs was 9.66±0.15 mm, ranging from 5.12 mm to 14.92 mm, and the UFT value was 0.26±0.00 mm, ranging from 0.15 mm to 0.33 mm. The coefficient of variation for UMA and UFT of the Merino lambs were 10.60 and 9.06 respectively and the corresponding values for the Dorper lambs were 10.02 and 13.45. It is evident from the results of this study that there is a large variation in the UMA and UFT of both Merino and Dorper lambs in commercial flocks.

no. 30

Effect of two different levels of supplementary feed and two different stocking rates on the production of grazing ostriches

M. Strydom[1,2], T.S. Brand[1,2], B.B. Aucamp[1] and J.M. Van Heerden[3], [1]Elsenburg Animal Production Institute, Private Bag X1, Elsenburg, 7607, South Africa, [2]Department of Animal Science, University of Stellenbosch, Private Bag X1, Matieland, 7602, South Africa, [3]Agricultural Research Council, Private Bag X5026, Stellenbosch, 7599, South Africa

Two hundred and five ostriches were allocated to 5 groups. Four groups rotationally grazed lucerne pasture with a stocking rate of 10 or 15 birds/ha and received either 0g or 800g supplementary feed/day. A fifth feedlot group received a complete finisher ration. No interaction was found between supplementary feed and stocking rate regarding end weight, ADG or weight change of the birds (P=0.70). Levels of supplementary feed had a significant effect (P=0.00) on ostrich performance. Average weight of the group receiving 0g supplementation was 84kg at slaughter while it was 100kg for the 800g group. ADG differed between the 0g supplementary groups (119g/day) and the 800g group (197g/day). Average weight change of birds receiving 0g supplementary feed was 24kg compared to 41kg for the 800g group. Parameters measured for the two stocking rates did not differ significantly from each other (P=0.99). It had no significant effect on end weight, ADG or weight change. Groups receiving supplementary feed reached slaughter weight (± 95kg) at 209 days, while groups receiving 0g supplementary feed did not reach slaughter weight on pasture alone. This indicates that ostriches being kept on grazing with a supplement will grow better than grazing ostriches receiving no supplement.

---

no. 31

The effect of supplementary feeding of grazing ostriches (*Struthio camelus*) on the yield of irrigated lucerne pastures

M. Strydom[1,2], J.M. Van Heerden[3], T.S. Brand[1,2] and B.B. Aucamp[1], [1]Elsenburg Animal Production Institute, Private Bag X1, Elsenburg, 7607, Private Bag X1, Elsenburg, 7607, South Africa, [2]Department of Animal Science, University of Stellenbosch, Private Bag X1, Matieland, 7602, South Africa, [3]Agricultural Research Council, Private Bag X5026, Stellenbosch, 7599, South Africa

Two hundred and fifty ostriches were randomly allocated to five groups. Four of these groups grazed irrigated lucerne pastures with supplementary feed supplied at 1500 g, 1000 g, 500g and 0 g/bird/day respectively. The fifth group was maintained in a feedlot and received a complete finisher diet. The lucerne pasture was grazed in a rotational system at a stocking rate of 15 birds/ha. Pasture samples were collected after each grazing period and manually divided into lucerne, grass, broad leaf weed, clover and dry/dead material fractions. Plant material was composed mainly of lucerne. Average lucerne production and lucerne intake declined with increase in supplementary feed provided. With less supplementary feed, ostriches consumed more lucerne and stimulated the yield of lucerne. However, the group receiving no supplementary feed over-grazed the lucerne leading to very low lucerne yield. The residual lucerne after grazing was significantly higher in the 1500 g supplementary group than in the zero supplementary group. This indicate a high level of supplementary feed substitution in the high supplementation group or over-grazing in the zero supplementary group. Over-grazing can be due to either a too long grazing period or too high stocking rate.

Growth of the reproduction organs of breeding female osriches during a breeding season

T.R. Olivier[1,2], T.S. Brand[1,2], B. Phister[3] and Z. Brand[3], [1]Institute for Animal Production: Elsenburg, Department of Agriculture, Private Bag X1, Elsenburg 7607, South Africa, [2]University of Stellenbosch, Department of Animal Science, Private Bag X1, Matieland 7602, South Africa, [3]Institute for Animal Production: Oudtshoorn, Department of Agriculture, P.O. Box 351, Oudtshoorn 6620, South Africa

A study was conducted to determine the nutrient content of the reproductive organs of breeding ostriches which can be used in a prediction model for estimating nutrient requirements. Forty breeding female ostriches were slaughtered over an eight week period (five per week) starting at the onset of the breeding season. The ovary and oviduct were collected and weighed at slaughter. Analysis of variance on the data collected showed no significant difference ($P>0.05$) in either parameter. Mean weights (g) and standard error (±) of the ovary and oviducts were respectively 1557±534g and 749±189g. A linear regression ($y=1734+29x$; $R^2=0.59$; SE estimation=480g; $P<0.05$) fitted on the mean values for the eight week intervals showed that the organs grew at a rate of 29g/day during the first 49 days of the breeding season. This data will be used in an optimising model (Brand & Gous, 2006) to predict the nutrient requirements of female breeding ostriches.

---

no. 33

Effect of breed on cryopreservation of cock semen

M.B. Makhafola[1,2], K.C. Lehloenya[2] and T.L. Nedambale[1], [1]Agricultural Research Council, Germplasm & Reproductive Biotechnol, Private Bag X2, 0062, Irene, South Africa, [2]Tshwane University of Technology, Department of Animal Science, Private Bag X680, 0001, Pretoria, South Africa

This study evaluated the effect of breed on cryopreservation of cock semen. Semen from three breeds (White Leghorn (WL), Ovambo (OV) and Potchefsroom Koek Koek (PK)) was collected through the abdominal massage technique. The semen was diluted (1:2 v/v) with egg yolk citrate (EYC) and then EYC + 5% DMSO. The equilibration after each dilution was 2 h each at 5 °C. Then diluted semen was evaluated for sperm concentration, motility, survivability and pH. Semen was then, loaded into straws and cooled in a programmable freezer from 5 °C to -20 °C at the rate of 1 °C/minute. Thereafter, the semen straws were plunged directly into liquid nitrogen (-196 °C) and stored for 24 h. Frozen straws were thawed at 5 °C and evaluated at 0, 30, and 90 min post-thaw. Preliminary results from this study demonstrated no significant difference in survival rate and motility between fresh diluted and frozen-thawed semen at 30 and 90 min post-thaw in all breeds. Survival of PK sperm was similar to that of OV sperm but, significantly ($P<0.05$) higher than that of WL sperm immediately after thawing. There was no significant difference based on sperm survival between the OV and WL immediately post-thaw. Prolonging storage period of frozen-thawed semen decreases spermatozoa survival and motility rate of all cock breeds.

Effects of cryoprotactant on cryopreservation of Kolbroek semen

M.H. Mapeka[1,2], K.C. Lehloenya[1], B. Sutherland[1] and T.L. Nedambale[2], [1]Tshwane University of Technology, Department of Animal Sciences, Private Bag X680, 0001, Pretoria, South Africa, [2]Agricultural Research Council, Germplasm & Reproductive Biotechnologies, Private Bag X2, 0062, Irene, South Africa

This study evaluated different cryoprotectants on post-thaw survival and motility rates of Kolbroek spermatozoa. Semen from Kolbroek boars were collected with the gloved hand technique. Ejaculates were diluted with Beltsville thawing solution (BTS) at a ratio of 1:1 prior to freezing. Semen was then diluted with egg yolk tris plus three different cryoprotectants i.e (14% glycerol, 14% DMSO and 7% glycerol + 7% DMSO), loaded into 0.5 mL straws and cooled with a programmable freezer. Thereafter, the semen straws were plunged directly into liquid nitrogen (-196 °C) and stored for 48 h. Frozen straws were thawed at 39 °C for a minute and evaluated at 0, 30, 60 and 90 min post-thaw. The post thaw survival rate of semen frozen using glycerol was significantly ($P<0.05$) higher immediately after thawing compared to DMSO however, similar to the combination of glycerol and DMSO. The post-thaw survival rate of spermatozoa at 30, 60, 90 min, was significantly ($P<0.05$) higher in semen cryopreserved with glycerol than with other cryoprotectants. There was no statistical difference on motility rate immediately (0 min) post-thaw between the three cryoprotectants. Spermatozoa frozen with glycerol had higher ($P<0.05$) post-thaw motility rate at 30, 60 and 90 min than with other cryoprotectants. In conclusion, glycerol seems to be a better cryoprotectant for cryopreservation of Kolbroek boar spermatozoa.

---

no. 35

Effects of breed and diet on growth and killing out characteristics of beef cattle

A.J. Mwilawa[1], A.E. Kimambo[1], L.A. Mtenga[1], J. Madsen[2], T. Hvelplund[3], M. Weisbjerg[3], G.H. Laswai[1], D.M. Mgheni[1], M. Christensen[2] and S.W. Chenyambuga[1], [1]Sokoine Univ. of Agric., Animal Science and Production, P.O. Box 3004, Chuo Kikuu, Morogoro, Tanzania, [2]Univ. of Copenhagen, Fac. of Life Sciences, 1958 Frederiksberg C, Denmark, [3]Univ. of Aarhus, Fac. of Agric. Sc., Foulum Research Centre, 8830 Tjele, Denmark

Sixty Tanzania Short Horn Zebu (TSHZ) and 60 Boran steers (1.5 to 2.5 years old) were randomly allocated to five dietary treatments during the wet season (Diet 1 (grazing alone as control), diet 2 (control + 50% adlib concentrate intake), diets 3, 4 and 5 were adlib hay + 60, 80 and 100% of the adlib concentrate intake respectively) for a 100 days fattening trial. The concentrate contained 124.9 g CP and 11.8 MJ ME per kg DM, and was formulated from maize meal, cotton seed cake, molasses, urea and minerals. Growth rates (ADG), feed conversion ratio in kg feed/kg gain (FCR) and dressing percentage (DP %) were assessed. Although breed did not show significant different Boran steers had slightly higher ADG, (735 vs. 623 g/d) FCR (9.5 vs. 9.0) and DP (52.4 vs. 49.2%) than TSHZ. Steers (regardless of breed) on diet 5 had higher ($P<0.05$) ADG (836 g/d), FCR (8.1) and DP (56.3%) than those fed diet 4 (668 g/d, 9.4, 49.0%) and diet 3 (549 g/d, 10.4, 50.1%), respectively. Steers in diet 2 ranked the second in ADG (717.5 g/d) but not in other parameters, and steers in diet 1 had higher ADG (625 g/d) than diet 3. It is concluded that both Boran and TSHZ cattle respond similarly to the fattening diet and minimal effect can be obtained with supplementation at 60% adlib concentrate for animals fed on hay.

Promoting gender equality and empowering women

A. Waters-Bayer[1] and B. Letty[2], [1]ETC EcoCulture, POB 64, 3830 AB Leusden, Netherlands, [2]Institute of Natural Resources, UKZN, Pietermaritzburg, South Africa

Despite years of gender sensitisation in many research and extension organisations, the role of women in livestock production and in marketing of animals and their products continues to be underestimated. Some encouraging projects do focus on women livestock-keepers, but most projects still tend to assume that the major actors in livestock (especially ruminant) systems are men. By making this assumption, such projects often strengthen the position of men versus women in households and communities, and may even deprive women of traditional realms of responsibility, social recognition and income. We examine key issues that impact on gender equality when livestock-related interventions are made in the name of development – issues related to roles and responsibilities (both perceived and real), rights of ownership over livestock, access to livestock services and markets, and decision-making powers regarding inputs and outputs of livestock production and resulting income. We highlight initiatives that lead to greater equality between the genders. These entail various forms of empowering women, such as through recognising and stimulating their innovativeness in livestock-keeping and enhancing their ability to organise themselves so as to strengthen their negotiating position and access to benefits. We pay particular attention to the multifunctional and changing roles of livestock at household level, especially in improving the lives of women in marginalised groups, such as pastoralists, the rural and urban poor and families affected by HIV/AIDS.

---

no. 2

Sustainable intensification: wishful thinking or realistic prognosis?

A.J. Van Der Zijpp and H.M.J. Udo, Wageningen University, Animal Production Systems group, P.O. Box 338, 6700 AH Wageningen, Netherlands

Sustainable intensification: Wishful thinking or realistic prognosis? The demand for livestock products is increasing as predicted by the Livestock Revolution. Sustainable intensification has been the solution to decrease poverty of smallholder farmers and provide more food at acceptable prices for a growing population. Why does increased demand not easily translate into incentives for small farmers? The response of poor farmers depends on the household resources and the context of their farming system. The family situation, labour, land (fertility and water), feed and cash availability determine development. The context presents a wide range of issues: infrastructure and markets, services like extension, animal health, breeding, microcredit, feed access and farmers associations. They effect the management options of the smallholder. These options may be further effected by competition of smallholders, medium and large scale farmers and cheap imports. Poor farmers appreciate livestock as assets and for financial insurance and food. Market participation requires investment and changes the asset and insurance function of livestock. Livestock species vary in their contribution to income, investment needs and gender association. To assist poor farmers to offset the risks of market led production appropriate policies are needed linking resources with local/regional and national contexts.

Pathways from poverty through value chains and innovation

H.M. Burrow, CRC for Beef Genetic Technologies, UNE, Armidale 2351, Australia

Developing supply chains that deliver to specifications of new or existing markets offers a way to ensure poor livestock farmers benefit from growth opportunities from the "Livestock Revolution". To develop such supply chains, several research and industry development issues must be addressed: 1) Proof of concept that livestock raised by poor farmers can meet the requirements of commercial domestic or international markets (to provide the confidence required by commercial investors); 2) Definition of production and marketing systems that can be targeted by the rural poor, ideally based on low input production systems or where commercial enterprises accept partial risk and/or funding of higher input production systems; 3) Capacity building of poor farmers focused on supply chain requirements, including profitability and productivity and continuous improvement and innovation; 4) Capacity building of commercial supply chain sectors so they can best utilise livestock delivered by poor farmers; 5) New win-win partnerships established across two or more segments of the supply chain (e.g. poor farmers and commercial feedlotters, processors and/or retailers); 6) Implementation of a value-based marketing system across supply chain segments, with value at each stage reflecting the product's ultimate value at consumer level; 7) Possible market segmentation based on consumer demand; and 8) Measurable and agreed performance indicators across all sectors of the supply chain. Each of these issues will be discussed by example.

---

no. 4

Vulnerability and risk: multiple roles of livestock in agrarian economies

M. D'Haese, S. Speelman and L. D'Haese, Department of Agricultural Economics, Ghent University, Coupure links 653, 9000 Gent, Belgium

Livestock is a key factor in African economies. Two case studies illustrate the main roles of livestock keeping in Africa: (a) a study in Burundi shows that households keep livestock as a management strategy to reduce vulnerability to risk of failure in future income streams; and (b) a study in South Africa explains how collective action in a farmers' association enables its members to keep sheep to generate additional income. Burundi is one of the poorest countries in Africa. Data was collected in villages in two provinces. Results confirm that households depend on subsistence agriculture to cover their daily food needs. Investment in livestock and its management are limited. Keeping livestock is identified to be an important way for people to diversify their livelihoods. It is a way of ex-ante risk management, which is a deliberate household strategy to anticipate failures in crop yields or other income streams. The case of wool farmers' associations in South Africa illustrates how livestock keeping contributes to village development. Data was collected in three villages in the former Transkei area among members and non-member of the association. For its members, wool production has become a sustainable income generating activity. The farmers' association is found to successfully create market access and enhance production. The collective action in the association reduces the vulnerability of individual farmers in market arrangements.

A case study on the existing and potential market opportunities for quality beef in Tanzania

F.M. Mapunda[1], L.A. Mtenga[2], G.C. Ashimogo[1], A.E. Kimambo[2], G.H. Laswai[2], D.M. Mgheni[2] and A.J. Mwilawa[2], [1]Sokoine University of Agriculture, Department of Agricultural Economics and Agribusiness, P.O. Box 3007, Morogoro, Tanzania, [2]Sokoine University of Agriculture, Department of Animal science and Production, P.O. Box 3004, Morogoro, Tanzania

A study on identifying the existing and potential market opportunities of quality beef was carried out in two major cities (Dar es Salaam and Arusha) in Tanzania. Data was collected from 198 respondents using a structured questionnaire and was handled using descriptive as well as multiple response analysis techniques. Beef attributes desired by consumers in Dar es Salaam were freshness (31.4%), tenderness (27.5%), and amount of fat (27.5%) while in Arusha; consumers were attracted by tenderness (61.5%), freshness (28.1%) and safety of the product (4.2%). Special meat cuts packed and sold in supermarkets, were preferred by 24% of the respondents and were purchased by tourist hotels, mine workers and middle to high class people. This niche market for quality beef is growing fast in Tanzania. The level of marketing profitability of quality beef varied with types of beef cuts and type of market the products sold. The growing niche market for quality beef observed in Tanzania showed high potential suggesting that there is reasonable market for selling quality beef in specific areas in Tanzania.

---

no. 6

Evaluation of agricultural research and development interventions for socio-economic service delivery: what are the considerations?

T.P. Madzivhandila, Agricultural Research Council, Animal Production, P/Bag 2, 0062, Irene, South Africa

There is an increasing demand by government, funders, and donors asking managers and practitioners for more performance and impact measurement (here evaluation) of the public services they provide. The demand is to: 1) account for resources consumed and outcomes delivered (to prove) and 2) learn for the betterment of future interventions and more effective socio-economic development (to improve). Evaluations are not an end in themselves; they are valid ways for an organization to increase the quantity and quality of its service delivery. It is important to assess the effectiveness, efficiency and relevance of the public service, and related programmes and projects, in undertaking evaluations for social betterment. However, it is difficult to ensure that evidence is integrated into policy and is used in programme design and implementation. Managers and practitioners know very little about the ways to respond to the demand. This paper focuses on (1) knowledge relevant to understand evaluation, (2) conceptual framework(s) that enable understanding of evaluation implementation process, (3) model(s) of the process of organizational evaluation, and (4) the main ways of intervening through the study to increase influence. The context is the socio-economic intervention in agricultural research and development in animal production industries and programmes.

Cooperation between universities and farmers to accelerate development

J. Madsen[1], M. Christensen[1], I.K. Hindrichsen[1], C.E.S. Larsen[1], T. Hvelplund[2], M.R. Weisbjerg[2], A.E. Kimambo[3], L.A. Mtenga[3], G.H. Laswai[3], D.M. Mgheni[3], A.J. Mwilawa[3], D.B. Mutetikka[4] and D. Mpairwe[4], [1]University of Copenhagen, Faculty of Life Sciences, Groennegaardsvej 2, DK-1958 Frederiksberg C, Denmark, [2]University of Aarhus, Faculty of Agricultural Sciences, Postbox 50, DK 8830 Tjele, Denmark, [3]Sokoine University of Agriculture, Animal Science and Production, Postbox 3004, Morogoro, Tanzania, [4]Makerere University, Faculty of Agriculture, Postbox 7062, Kampala, Uganda

To promote research based development, cooperation between East African and Danish Universities have for many years (for over one decade) been financed by the Danish Foreign Ministry. Recently, research has to a large extent been based on cooperation with African farmers. The cornerstone of the collaboration has been based on capacity building within universities aiming at increasing the amount and relevance of research, teaching and extension both in the South and in the North. Cooperation with farmers is time consuming, trials are difficult to control and results can be harder to get into mainstream periodicals. Despite these disadvantages it is experienced that farmers have a lot to add to theoretical knowledge. Hence collaboration with farmers enriches the academic environment and farmers are much more positive towards research trials when they are partners and not only spectators. On farm experiments in Denmark have, through many years, had tremendous impact on the farmers' uptake of new opportunities and technologies and such development is emerging in Africa too.

---

no. 8

Are current arrangements for beef industry improvement able to help the low income sector?

N.B. Nengovhela[1,2], R.S.J. Beeton[1] and R.A. Clark[1], [1]University of Queensland, School of Natural & Rural Systems Management, Gatton, 4545, Australia, [2]ARC, Animal Production Institute, Irene, 0062, South Africa

Beef sectors world wide have low and high income participants. The gap in income reflects many causes. A frequent assertion is that the uptake of new knowledge from interventions is the major factor influencing the industry wide performance. This South African study provides an ideal case study to understanding the issues associated with the gap and the role of uptake. It focused on barriers and promoters to interventions targeting the low income sector of the South African industry in Limpopo province. This paper reports on findings from the supports staff both at the Agricultural Research Council (ARC) and Limpopo Provincial Government Agriculture Department (LDA). The theoretical underpinning for this work is grounded theory which looks for emerging themes from qualitative data. Focus groups and interviews were conducted with the research and province managers, researchers, extension officers and technicians within the ARC and the LDA livestock divisions. The data analysis revealed 151 themes and sub-themes that impact on the efficacy of the support system for the low income beef sector in South Africa. The most important among these were historic legacy of the ARC and LDA, technology dependent culture, cultural stereotyping and disconnection with the intervention target.

Designing livestock poverty eradication projects to achieve outcomes and sustainability: Kgalakgadi Dipudi Enterprise (KDE) project review

N.B. Nengovhela[1,2], M.D. Motiang[2], R.A. Clark[1], J. Timms[1] and T.P. Madzivhandila[2,3], [1]University of Queensland, School of Natural & Rural Systems Management, Gatton, 4545, Australia, [2]ARC, Sustainable Rural livelihood, Irene, 0062, South Africa, [3]University of New England, School of Business, Economics and Public Policy, Armidale, 2351, Australia

Poverty eradication initiatives in South Africa are being called into question for lesser than desired achievements of outcomes and ongoing sustainability after the project's lifespan. There is a need to identify key components for the design, implementation and management of projects to achieve sustainable poverty eradication interventions. Sustainability in all situations is based on individuals and teams being self-making and self starting. Focusing techniques and projects designs can stimulate these attributes in the targeted individuals and teams. This paper presents an evaluation of the design and management of the KDE project towards achieving defined poverty eradication outcomes. It looked at the inclusion or exclusion of the needed key components of the project design and management to ensure that impact is achieved during the life of the project and afterwards. The project was evaluated on elements of Technology Integration, Focusing Methodology, Shared Mental Models, Capacity Building, Infrastructure, Policy Management, Design& Management of Key Systems & Processes and Performance Management were found to be highly unbalanced in the KDE project.

---

no. 10

The recent improvement in financing animal production in Brazil: a quantitative approach comparing regions

A. Minniti Igreja[1], M. Brasil Rocha[2], F.M. Mello Bliska[3], E. Pinatti[4] and S. Santana Martins[5], [1]Instituto de Zootecnia, Laboratório de Metodologias Quantitativas, Rua Heitor Penteado, 56 Nova Odessa, 13.460-000, Brazil, [2]Instituto de Economia Agrícola, Av. Miguel Stefano, 3900, 04301-903, Brazil, [3]Instituto Agronômico, R. Barão de Itapura, 1481, 13.001-970, Brazil, [4]Instituto de Economia Agrícola, Av. Miguel Stefano, 3900, 04301-903, Brazil, [5]Instituto de Economia Agrícola, Av. Miguel Stefano, 3900, 04301-903, Brazil

Rural financing is basically the remaining instrument of economic policy for supporting agribusiness agents and corporations in a market-oriented policy style, based on efficiency and on agribusiness governance, although actions implemented for countervailing rural poverty were implemented as part of the financing system. Statistics from Central Bank of Brazil show that credit has grown its amount significantly in Brazil, from an index of 100 in 1999 to 376,14, in 2006. Non-uniform standards were observed regarding regional aspects. One important objective of this paper is to focus the faster expansion of the credit for animal production. Regions that can be highlighted are: North, Southeast and Northeast. The paper brings full report of statistics relating credit to land use, and also rating credit as a leverage factor in the recent agricultural and animal production and exports growth.

no. 11

Continuous improvement and innovation as an approach to effective research and dvelopment : a case study in Mokopane, Makhuvha and Bronkhorstspuit

M.S. Thaela-Chimuka, B.N. Nengovhela, O.U. Sebitloane, B.J. Mtileni and J. Grobblelaar, Agricultural Research Council, Animal Production, Private Bag X 02 Irene, 0062, South Africa

Small-scale broiler poultry production has the potential of overcoming rural poverty and increase food security and household incomes in South Africa. Despite significant progress that has been made in recent years to improve living standards and bring vital services to communities the official statistics still present a profile of underdevelopment. In the case study presented below, the methodology called Continuous Improvement & Innovation technique (CI & I), which is a six cyclic process to support the best practice process was adopted to improve the ability to identify and concentrate on benefit oriented focus areas that should lead to economically sustainable ventures. The specific target outcome of the project was to achieve sustained improvement in profit per broiler enterprise, per year, in a growing number of enterprises Limpopo and Gauteng province. After CI&I guided intervention, the Makhuvha village indigenous chicken project at Venda, Limpopo increased production from 4 to a maximum of 30 chickens per household. Mokopane poultry projects (broilers) in Limpopo reduced the mortality rates from 10% to 5% while the Bronkhorstpruit project (broilers) in Gauteng reduced mortalities rates from between 15 – 20% and in some cases up to 100% to 6%. The CI&I methodology in these projects increased the relevance, effectiveness, efficiency and sustainability with which target outcomes are achieved..

---

no. 12

Investment potentials in the livestock sector in sub Saharan Africa: a case study of Rwanda eastern province

I. Mpofu, J. Mupangwa, P. Chatikobo and S.M. Makuza, Umutara Polytechnic University, Veterinary, P.O. Box 57, None, Nyagatare, Rwanda

The objectives of the study were to use SWOT and Gap analysis to identify investment livestock potentials in Gatsibo District, Rwanda and to determine implementation strategies. The study was done using participatory data gathering techniques like key informant interviews, informal (participatory rural appraisal) and formal (questionnaire surveys). Data from the formal questionnaire survey was subjected to statistical analysis using SPSS. The main animal products identified were milk, meat and eggs from dairy, beef and poultry, respectively. Hides for tanneries, hooves and bones for the feed and ornamental industries were ranked insignificant by respondents despite their potential. Dehydration, pelleting and fermentation of the most abundant resources like bananas and cassava were found to be potential investment areas in livestock feed processing. Commercial soybean production, a wonder legume was found attractive for processing into stock-feeds. Investment was also identified in the areas of animal breeding to increase to high genetic material by farmers; construction of milk collection centers, and milk processing. In conclusion, the study unraveled investment potentials in the livestock sector in the district studied and in Sub Saharan Africa in general. These investment potentials have a huge potential of helping governments in Africa in attaining Millennium Development Goals on reducing poverty and hunger.

no. 1

Milk production from Kikuyu over-sown with Italian, westerwold or perennial ryegrass

L. Erasmus[1,2], R. Meeske[1], P.R. Botha[1] and L.J. Erasmus[2], [1]Dept of Agriculture Western Cape, Outeniqua Research Farm, 6530 George, South Africa, [2]Dept of Animal and Wildlife Sciences, University of Pretoria, 0001 Pretoria, South Africa

Compared to temperate pasture species, the nutritive value of kikuyu (*Pennisetum clandestinum*) is poor and hence has a low milk production potential. The strategic incorporation of ryegrass species into kikuyu pastures can improve pasture quality. The aim of this study was to determine the productivity of Jersey cows grazing on kikuyu over-sown with three different ryegrass species. Nine hectares of irrigated kikuyu was divided into 24 paddocks and randomly allocated to the three treatments: italian ryegrass (*L. multiflorum* Lam var. *italicum*), westerwold ryegrass (*Lolium multiflorum* Lam. var. *westerwoldicum*) and perennial ryegrass (*L. perenne*). Forty-five Jersey cows were randomly allocated to the pasture treatments; cows were milked twice/day and received 2 kg concentrate (13.7% CP, 12.9 MJ ME/kg DM) during each milking. Milk production was measured daily and milk composition was determined once/month. Milk production/cow did not differ between treatments and was 4829, 4944 and 4944 kg for the italian, westerwold and perennial treatments, respectively. The perennial ryegrass yielded more ($P<0.10$) milk/hectare/year (32288 kg) compared to the westerwold (29761 kg) and italian ryegrass (30446 kg) and had a higher ($P<0.001$) stocking rate than the westerwold and italian ryegrass treatments (6.93 vs. 6.49 and 6.44 cows/hectare/year, respectively). Kikuyu over-sown with perennial ryegrass proved to be the most economical option.

---

no. 2

Dairy cattle breeding in Tanzania: Mpwapwa cattle breed

S.M. Das[1] and H.A. Mruttu[2], [1]Ministry of Livestock Development and Fisheries, Central Veterinary Laboratory, P.O. Box 9254, Dar es Salaam, None, Tanzania, [2]Ministry of Livestock Development and Fisheries, Animal Identification, Registration and Traceability, P.O. Box 9152, Dar es Salaam, None, Tanzania

Among other aspects, breeding objectives in Tanzania were to increase milk and meat production performance. In 1930 crossbreeding between *Bos taurus* dairy cattle and indigenous Zebu cattle resulted in high mortality of crossbreds and low dairy potential. This was mainly due to tick borne diseases and pneumonia in calves. It was therefore decided to include *Bos indicus* breeds from India and Pakistan, such as Red Sindhi and Sahiwal. The resultant breed over years of crossbreeding was known as the "Mpwapwa cattle breed", which was dual purpose (dairy and beef) for semi-arid areas of Tanzania. The composition of the Mpwapwa breed is reported to be 32% Red Sindhi, 30% Sahiwal, 19% TSZ, 9% Boran, 8% *Bos taurus* and traces (2%) of Ankole. The Mpwapwa cattle breed is hardy and performs better in terms of milk yield and weight gain than local zebu cattle in semi-arid conditions. Recording scheme and methods of selection based on physical characteristics were the tools used to develop this breed. This poster describes the initiatives and innovations of Tanzanian Animal Scientists that ended up with dual purpose cattle known as the Mpwapwa breed. It also highlights the strategies and specific objectives of breeding *Bos taurus* and *Bos indicus* cattle in the country.

no. 3

Mathematical modeling of individual lactation curves of Boran cows

S.D. Mulugeta, North-West University, Animal Science, Private Bag x2046, 2735 Mmabatho, South Africa

Wood, Wilmink, Quadratic and Simple linear functions were fitted to model individual cow lactation curves of 108 Boran cows with 179 lactation records. Milk yield was sampled daily starting from day 3 post calving to the end of lactation at 305 days. The models were compared through logistic probability regression on the frequency of adjusted R^2 values, which was grouped into four categories (1=less than 0.4; 2=0.4< to <0.6; 3=0.6< to <0.8; and 4=>0.8). The adjusted R^2 values for Wood's model were 12, 13, 22 and 53% for categories one to four, respectively. The frequency of curves with adjusted R^2 values greater than 80%, were significantly higher for Wood's model than for Wilmink and simple linear functions. Using the Wood's incomplete gamma function, 66% of the lactation curves could be characterized as standard curves, 25% as continuously decreasing or atypical, 8% as reversed standard and 1% as continuously increasing curves. Parity and year of calving were not significant factors influencing the frequency of the different types of curves. This study showed that three parameter functions, such as Wood's incomplete gamma function, could be used to model individual lactation curves of Boran cows. However, the presence of 25% atypical curves could not be explained with factors considered in this study.

---

no. 4

Performance of imported Jersey cattle in the hot and humid climate of Tamil Nadu, India

A.K. Thiruvenkadan and N. Kandasamy, Veterinary College and Research Institute, Department of Animal Genetics and Breeding, Namakkal, Tamil Nadu, 637002, India

The objective of the study was to evaluate the performance of imported Jersey cattle in the hot and humid climatic conditions of Tamil Nadu, since they have been recommended for cross breeding in the plains to augment milk production. The records of 150 purebred Jersey heifers (imported from Australia in 1978 and 1979) and 120 heifers (imported from Denmark in 1995) constituted the material for the study. Least-squares analyses of the performance of these animals were done separately, since they were reared in different periods. In imported Australian Jersey cattle, the observed total lactation milk yield was 2431.8±50.7 kg in a lactation of 371.7±5.9 days. The mean service period, calving interval and dry period were 200.6±11.8, 490.6±12.7 and 142.4±11.8 days, respectively. The overall means for total lactation milk yield, lactation length, service period, calving interval and dry period of imported Danish Jersey cattle were 1544.6±59.3 kg, 315.9±11.7 days, 212.9±54.4 days, 489.5±55.0 days and 218.2±32.8 days, respectively. In general, the production and reproduction performance of the imported Danish Jersey cows were low and might be due to the fact that they have not adapted well to the harsh climatic conditions and perhaps also due to less than optimum management.

Heat stress in Tunisia: effects on dairy cows and management strategies for its alleviation

M. Ben Salem[1] and R. Bouraoui[2], [1]INRA Tunisia, Laboratory of Forage and Animal Productions, rue Hédi Karray, 2049 Ariana, Tunisia, [2]ESA Mateur, Mateur, 7030, Tunisia

Tunisia has a Mediterranean climate characterized by high ambient temperatures for a long period. The objectives of this work were to characterize the environmental conditions to which Holstein cows are exposed in Tunisia, to examine heat stress effects on lactating cows using the Temperature Humidity Index (THI) and to discuss management strategies available to producers to minimize its effects. To define stress intensity, THI values were calculated using monthly temperature and relative humidity data over a 10-year period from different weather centers throughout the country. Milk per cow and reproductive indices were then examined for the same period using data from four selected herds. Results showed summer heat stress in Tunisia for four to five months each year going from May trough September with THI values greater than 72. On average, milk production per cow dropped by about 10% between March and September. First conception rate (CRI_1) and overall conception rate (CR) were lowest in the summer and highest in the winter. Regression equations between THI and CRI_1 and THI and CR suggest a strong relationship between heat stress and reproduction. These observed negative effects are often aggravated by current management practices. Maintaining cow performance under hot weather conditions requires environmental control techniques, appropriate feeding strategies, and genetic improvement programs which may include cross breeding to enhance heat tolerance.

---

no. 6

Analysis on the productive and reproductive traits in murrah buffaloes maintained in the coastal region of india

A.K. Thiruvenkadan, S. Panneerselvam, R. Rajendran and N. Murali, veterinary college and research institute, department of animal genetics and breeding, namakkal, tamil nadu, 637002, India

Data of 698 Murrah buffaloes sired by 43 bulls maintained at the Central Cattle Breeding farm, Alamadhi (Tamil Nadu, India), and calved between 1979 and 2006 were analysed to study the performance of Murrah buffaloes. Least-squares means and genetic parameters were estimated for different production and reproduction traits of the first lactation. The average age at first calving was 1578.7±20.3 days. The averages for 305-day and total lactation milk yield were 1616.3±39.6 and 1686.2±44.4 kg, respectively,with a heritability value of 0.255±0.124 and 0.204±0.115 respectively. The means for lactation length, service period, dry period and calving interval were 312.8±5.7, 253.7±17.3, 250.5±15.9 and 559.6±17.3 days, respectively. The heritability estimates for these traits were 0.101±0.096, 0.141±0.121, 0.187±0.130 and 0.131±0.119, respectively. The mean milk yield was considered higher than earlier estimates. The low heritability for service period, dry period and calving interval suggested improvements in these traits through better management. Performance traits, such as age at first calving, service period, dry period and calving interval would need further improvement in the herd under study. It is therefore imperative to emphasise improvements in the husbandry and introduction of genetic evaluation programmes at the same time.

no. 7

Performance analysis of farmbred Jersey cattle under tropical climatic conditions of Tamil Nadu, India

A.K. Thiruvenkadan and N. Kandasamy, Veterinary College and Research Institute, department of animal genetics and breeding, laddivadi, namakkal, 637002, India

Data on production and reproduction performance of 373 farmbred Jersey cattle maintained at the Exotic Cattle Breeding Farm, Eachenkottai, Tamil Nadu, India, for a period of 29 years (1978 to 2006) were collected. They were analysed to study the effects of various non-genetic factors on these traits and to estimate the genetic parameters. The least-squares means for 305-day and lactation milk yields were 1491.6±25.9 kg and 1560.9±29.9 kg, respectively, and the corresponding heritability estimates were 0.176±0.143 and 0.0403±0.109, respectively. The averages for lactation length, service period, dry period and calving interval were 303.1±4.0, 177.0±10.0, 160.2±9.7 and 461.0±9.7 days, respectively, and heritability estimates for these traits ranged between 0.077±0.147 and 0.191±0.147. Years grouped into five periods had a significant ($P<0.01$) influence on all the traits. Parity was found to influence 305-day milk yield, service period, dry period and calving interval significantly ($P<0.01$) and also contributed ($P<0.05$) to source of variation for lactation milk yield. The study revealed that the performance of the animals was much lower than those maintained under high altitude conditions of Tamil Nadu and hence better management and feeding practises are suggested for improved performance.

---

no. 8

Heat stress in dairy cattle

J.H. Du Preez, Institute for Dairy Technology, Milk Producers Organization of South Africa, P.O. Box 1284, Pretoria, South Africa, 0001, South Africa

Heat stress hampers the performance of dairy cattle in South Africa. Losses in the primary dairy industry caused by heat stress amount to more than R500 million annually. Monthly temperature-humidity indices (THI) of all weather bureau stations (563) have been calculated for South Africa and Namibia and mapped to indicate the variation in the monthly mean THI and conception rate (CR). A multiple regression model ($THI [predicted]=33.22 - 0.001409A+2.11MT - 0.02157 [MT]^2$, where A=altitude and MT=maximum temperature) was calculated to obtain THI values. The proposed regression model for conception rate was related to mean monthly THI by: $CR\%=-890.2+31.15THI - 0.25THI^2$. The THI threshold for reproduction, as measured by CR, is 65, which is lower than the 72 for milk production. Body temperature of dairy cows suffering from heat stress increased and milk butterfat and protein percentages, milk yield and plasma cortisol concentrations decreased. Body temperature, THI and respiratory rate are the most practical parameters for determining heat stress in dairy cows. Careful planning of facilities, and sound husbandry and management are required to protect dairy cattle from heat stress. Precautions against heat stress are the following: provision of shade, *ad libitum* availability of water, direct evaporation cooling, adapting rations, farming with a well-adapted dairy breed, selecting the most suitable individuals within a breed and avoiding stressful handling during the hottest parts of the day.

no. 9

Milk production potential of Kohi camels in Balochistan, Pakistan

A. Raziq and M. Younas, University of Agriculture, Livestock management, Livestock management department, 38040, Pakistan

Camels are potential milch animals and produce more and higher quality milk than any indigenous cattle breed in Pakistan. This study was therefore conducted on 40 lactating camels of a pastoral herd in northeastern Balochistan and milk was recorded at the end of every second week for a complete lactation period in 2006. The daily milk yield ranged from 6.05-11.73 kg/day with a mean daily yield of 10.2±0.4 kg/day (mean±SE). The lactation length ranged from 231-275 days with a mean of 259±7 days. The lactation yield ranged from 1566-3168 kg with an average of 2590.5 kg. Factors affecting daily and total lactation yield included age and parity, stage of lactation, season of production, type/breed of camel and calving season. The fifth parity animals had the highest lactation yield (3168 kg) while the lowest yield (1566) was observed in first parity animals ($P<0.05$). The mean daily yield was lower in the first month of lactation (9.6±0.87 kg/day) and reached a peak (11.2±0.31 kg/day) in the fourth month of lactation ($P<0.05$). The white Kohi camel produced significantly ($P<0.05$) more milk (11.4 kg/day) than the Spole Kohi (8.9 kg/day). Camels that calved in April had the highest milk yield (11.2 kg/day) and those that calved in May the lowest milk (8.9 kg/day) ($P<0.05$). This study proved the camel to be an efficient milch animal but further studies under different management and environmental conditions are recommended.

---

no. 10

Effect of recombinant bovine somatotropin on some physiological parameters and milk production

T. Khaliq, Z. Rahman, I. Javed, I. Ahmad, A. Malik and R. Ullah, Department of Physiology and Pharmacology, University of Agriculture, Faisalabad, 38040, Pakistan

The project was conducted at the Livestock Experiment Station, University of Agriculture, Faisalabad, to observe the effect of rbST (500 mg) on milk production, composition and some physiological parameters of lactating Nili-Ravi buffaloes. Two groups, with eight buffaloes in each group were formed. The buffaloes in Group I were administered 500 mg of rbST subcutaneously, twice during the experimental period at an interval of 16 days. The animals in Group II acted as a control and a similar amount of sterilized saline was injected into them. The experiment lasted for 32 days. The following observations were recorded from each buffalo: rectal temperature, respiration rate and pulse rate. Respiration rate differed significantly between days, irrespective of group. Milk was analysed for protein, fat, total solids, solids-not-fat, lactose and plasmin. Milk production increased ($P<0.05$) with 31% and fat with 1.03% in bST treated buffaloes. Milk lactose increased ($P<0.05$) after 32 days of treatment. The bST injected buffaloes showed an increase ($P<0.05$) in plasminogen as compared to control animals. Plasmin increased after the first injection then decreased significantly after the second injection, irrespective of group.

no. 11

Milk composition and enzymes of camels (*Camelus dromedarius*) injected with oxytocin

Z. Rahman, A. Malik, M.,A. Jalvi, H. Anwar, I. Ahmad and R. Ullah, Department of Physiology and Phar Macology, University of Agriculture, Faisalabad., 38040, Pakistan

The experimental camels (Marrecha) were selected from the Breeding & Research Station, Rakh Mahni District, Bhakkar, Pakistan. One group was early in lactation (1-2 months), while the second group was at the end of lactation (12-14 months). Both groups were treated with oxytocin (5 i.u.) intramuscularly for fifteen days. The determination of milk composition i.e. fat, density, protein, lactose and solids was estimated with the aid of a Milky Lab Analyzer. The enzymatic profile of milk included lactoperoxidase activity, thiocyanite, and alkaline phosphatase by using a spectrophotometer and lipase activity was determined with a titration method. Oxytocin affected ($P<0.01$) the composition of milk along with level of milk enzymes. Milk fat, protein, lactose and solids increased in oxytocin injected camels. Milk enzymes, lactoperoxidase, lipase and alkaline phosphatase activity was higher ($P<0.01$) in oxytocin injected animals. Stage of lactation also affected milk composition and enzyme activity. There was a significant decrease in milk fat, solids and protein as stage of lactation increased. Milk enzymes, lactoperoxidase, thiocyanate and lipase activity increased while alkaline phosphatase activity decreased with stage of lactation. These results showed that repeated doses of oxytocin affect milk composition and quality of camels.

---

no. 12

Constraints and possibilities for milk production in Burkina Faso

V. Millogo[1], G.A. Ouedraogo[2], S. Agenäs[1] and K. Svennersten-Sjaunja[1], [1]Swedish University of Agricultural Sciences (SLU), Department of Animal Nutrition and Management, P.O. Box 7024, Ultuna S-750 07, Uppsala, Sweden, [2]Polytechnic University of Bobo-Dioulasso (UPB), Department of animal production, 01 P.O. Box 1091, Bobo-Dioulasso, Burkina Faso

The Zebu cow (*Bos indicus*) is well adapted to the environment but their low milk production is a problem, both for farmers and consumers. The first study showed that daily milk yield was 1-2 litres per cow in sedentary traditional farms and 2-4 litres per cow in semi intensive farms. Milk temperature at dairy farm level (32.5±4.6 °C) was a factor that reduced milk quality before it reached the collection centre. The use of crossbred cows was also related to higher daily milk yield per cow (Chi-square=31.80; P=0.001). In the second study, the estimated average milk yield was 2.02±0.23 litres per day, with 18% day-to-day variation. Milk composition was 4.88, 3.46 and 4.84% in the morning and 5.52, 3.35 and 4.73% in the evening milking for fat, protein and lactose, respectively. The day-to-day variation was 23-24% for fat, 12-14% for protein, 6% for lactose and 7.8% for somatic cell count, higher for hand milked compared to machine milked cattle. It was concluded that more extensive supplementation of diets and cross-breeding would improve milk production in Burkina Faso and that the effect of different hand milking techniques and hygienic aspects of raw milk should be investigated in future work.

Authors index

C

G

H

I

J

K

L

M

O

P

T

Z

Printed in the United States
by Baker & Taylor Publisher Services